Foreword

CIRIA's research project on small embankment reservoirs for water supply and amenity use is intended to provide guidance for landowners and for the construction industry.

This project is a continuation of CIRIA's work on Reservoir Storage and Safety which is supported by funding from the Department of the Environment, Water Directorate.

Current additional projects deal with the protection of the upstream face of dams, with new developments in the design of reservoir spillway channels and with the safety of concrete and masonry dams.

This report was prepared by Mr M F Kennard, Mr C G Hoskins, and Mr M Fletcher of Rofe, Kennard and Lapworth, Consulting Engineers under contract to CIRIA.

Following CIRIA's usual practice the research study was guided by a Steering Group which comprised:

Mr J R Claydon (Chairman)	Yorkshire Water
Mr A Birtles	National Rivers Authority
Dr J A Charles	Building Research Establishment
Mr J G Durward	Forestry Commission (to December 1991)
Mr I D M Hampson	Forestry Commission (from December 1991)
Mr A A George	North West Water Ltd
Mr F M Law	Institute of Hydrology
Mr A MacDonald	Babtie Shaw and Morton
Mr C B Stansfield	ADAS
Mr A Pym	County Landowners Association
Mr M Tutton	Robert West and Partners
Mr C E Wright	Department of the Environment

CIRIA's research manager for this project was Mr R Freer.

The project was funded by:

Department of the Environment, Water Directorate
Anglian Water Services Ltd
North West Water Ltd
Severn Trent Water Ltd

Southern Water Services Ltd
Thames Water PLC
Yorkshire Water Services Ltd
Forestry Commission

CIRIA and Rofe, Kennard and Lapworth are grateful for the help given to this project by the funders, by the members of the Steering Group and by the many individuals and organisations who were consulted.

Thanks are also expressed to the following who commented on various aspects of the guide and for providing technical material:

Mr J Anderson	Health and Safety Executive
Mr J Armstrong	Sovereign Woodlands Ltd
Mr M Cooper	Southampton University
Mr J Dodd	NRA, Southern Region
Mr G Harris	ADAS
Mr S Hawes	Independent Consultant
Mr M Hough	Forestry Commission
Ms H Hougton-Carr	Institute of Hydrology
Mr J House	O & J House Ltd
Mr P Nicolle	British Association for Shooting and Conservation
Mr R Orford	Miles Group
Ms P Rice	University of Derby
Ms D Ward	Royal Society for the Protection of Birds
Mr H Pocock	Shephard Hill Civil Engineering Ltd

Contents

Figures 14
Boxes 17
Tables 18
Glossary 20
Notation 30
Abbreviations 32
Conversion factors 33

PART 1 FEASIBILITY STAGE 35

1   INTRODUCTION 35
    1.1   Background 35
    1.2   Scope 37
    1.3   Definitions 38

2   RESERVOIR FEASIBILITY 42
    2.1   Initial assessment 42
    2.2   Reservoir purpose 43
    2.3   Water requirements and size of reservoir 46
    2.4   Water sources 48
    2.5   Available land 49
    2.6   Reservoir situation 49
    2.7   Topography and land use 50
    2.8   Ground conditions 51
    2.9   Floods and waves 51
    2.10 Available construction materials 52
    2.11 Planning, licences and consents 53
    2.12 Environmental constraints and considerations 54
    2.13 Costs   55
    2.14 Inspection, monitoring and maintenance considerations 55
    2.15 Alterations/extensions to existing reservoirs 55
    2.16 Specialist advice 56

3   LICENCES AND STATUTORY REQUIREMENTS 58
    3.1   Introduction 58
    3.2   Planning considerations 59
    3.3   Licences and consents 64
    3.4   Fisheries 71
    3.5   Environmental aspects 72
          3.5.1   Species protection 72
          3.5.2   Felling consents and tree preservation orders 73
          3.5.3   Pollution 74

	3.6	Responsibilities	77
		3.6.1 Owner's liability	77
		3.6.2 Employer's liability	78
		3.6.3 Liability during construction	78
		3.6.4 Liability of others	79

4 ENVIRONMENTAL CONSIDERATIONS 80
    4.1 General 80
    4.2 The site and its environment 80
    4.3 Environmental effects of constructing the reservoir 83
    4.4 Enhancement and mitigating measures 83
        4.4.1 General 83
        4.4.2 Preservation of existing habitats 84
        4.4.3 Habitat creation 84
        4.4.4 Aquatic and shoreline planting 85
        4.4.5 Terrestrial vegetation 85
        4.4.6 Landscape enhancement 87

5 COST CONSIDERATIONS 89
    5.1 Introduction 89
    5.2 Site investigation 89
    5.3 Pre-construction stage 90
    5.4 Construction stage 91
    5.5 Visual observation and monitoring 92
    5.6 Maintenance 93
    5.7 Remedial works 93
    5.8 Other costs 93

PART 2 DESIGN 95

6 YIELD AND RESERVOIR OPERATION 95
    6.1 Sources of water 95
        6.1.1 General 95
        6.1.2 Surface water 95
        6.1.3 Groundwater 96
        6.1.4 Public supply 98
    6.2 Inflow and yield 99
        6.2.1 Infiltration, runoff and losses 99
        6.2.2 Estimation of available water 103
    6.3 Normal operation 105
        6.3.1 Reservoir losses, dead storage and compensation flow ... 105
        6.3.2 Water quality 107
        6.3.3 Operating levels and first filling 108
        6.3.4 Sedimentation 109
    6.4 Groundwater implications 110
        6.4.1 Upstream and around the reservoir 110
        6.4.2 Downstream of the reservoir 111
    6.5 Changes in reservoir and catchment usage 111
        6.5.1 Reservoir 111
        6.5.2 Catchment 112

7	FLOODS AND WAVES		113
	7.1	Flood assessment	113
		7.1.1 Introduction	113
		7.1.2 Flood producing rainfall	114
		7.1.3 Catchment area, water surface areas and areas of rapid runoff	115
		7.1.4 Stream and catchment slopes	119
		7.1.5 Soil and geology	121
		7.1.6 Vegetation	125
		7.1.7 Maximum flood	126
		7.1.8 Other factors	126
		7.1.9 Further advice	128
	7.2	Passing design flood	128
		7.2.1 Overflow assessment	128
		7.2.2 Flood attenuation	132
		7.2.3 Overflow arrangements	133
		7.2.4 Main overflow	136
		7.2.5 Auxiliary overflow	141
		7.2.6 Other aspects	143
	7.3	Waves and freeboard	144
	7.4	Flood event procedures	145
		7.4.1 Emergency procedures	145
		7.4.2 Procedure following substantial flow	145
		7.4.3 Other meteorological events	146
	7.5	Floods during construction	147
8	SITE INVESTIGATION		148
	8.1	Introduction	148
		8.1.1 General	148
		8.1.2 Site investigation approach	148
	8.2	Information study	150
		8.2.1 Approach	150
		8.2.2 Sources of information	151
		8.2.3 Buried and other services	151
	8.3	Site survey	152
		8.3.1 Approach	152
		8.3.2 Levelling survey	152
		8.3.3 Information to record from site walkover	152
	8.4	Ground investigation	154
		8.4.1 Approach	154
		8.4.2 Trial holes	156
		8.4.3 Soil augers	156
		8.4.4 Boreholes	157
		8.4.5 Information Required	159
		8.4.6 Laboratory testing	160

	8.5	Geotechnical assessment	168
		8.5.1 Ground conditions	168
		8.5.2 Construction materials	171
		8.5.3 Permeability assessment	174
		8.5.4 Embankment fill appraisal	174
		8.5.5 Design aspects	175

9 DESIGN ASPECTS 176

9.1 Reservoir design 176
- 9.1.1 General 176
- 9.1.2 Slope stability 176
- 9.1.3 Reservoir perimeter 176
- 9.1.4 Erosion 176
- 9.1.5 Islands 177
- 9.1.6 Sedimentation 177

9.2 Embankment design 177
- 9.2.1 General 177
- 9.2.2 Design requirements 180
- 9.2.3 Design options 181
- 9.2.4 Homogeneous design 182
- 9.2.5 Zoned design 183
- 9.2.6 Diaphragm design 183
- 9.2.7 Lined design 184
- 9.2.8 Vegetation on the embankment 187

9.3 Earthworks 190
- 9.3.1 General 190
- 9.3.2 Embankment foundations 190
- 9.3.3 Seepage control measures 191
- 9.3.4 Slope and crest design 193
- 9.3.5 Slope protection and freeboard 194
- 9.3.6 Crest protection 198
- 9.3.7 Drainage 198
- 9.3.8 Borrow areas 200
- 9.3.9 Core materials 200
- 9.3.10 Shoulder fill materials 199

9.4 Reservoir lining materials 201
- 9.4.1 General 201
- 9.4.2 Geomembranes 201
- 9.4.3 Clay 201
- 9.4.4 Bentonite 202

9.5 Geotextiles 202
- 9.5.1 General 202
- 9.5.2 Geotextiles for drainage applications 203
- 9.5.3 Geotextiles for erosion control 203

9.6 Pipework 203
- 9.6.1 General 203
- 9.6.2 Types of pipe 204
- 9.6.3 Pipe bedding and surround 204

9.7	Drainage materials		206
9.8	Inlet structures		207
	9.8.1	Gravity supply	207
	9.8.2	Feeder channels or pipes	207
	9.8.3	Watercourse control weirs	208
	9.8.4	Silt traps	210
	9.8.5	Pumped supply	211
9.9	Drawoff works		216
	9.9.1	General	216
	9.9.2	Drawoffs for gravity discharge	216
	9.9.3	Low-level drawoffs	216
	9.9.4	Other drawoff pipes	218
	9.9.5	Pumping arrangements	219
9.10	Overflow works		220
	9.10.1	General	220
	9.10.2	Main overflow – impounding reservoirs	220
	9.10.3	Auxiliary overflow – impounding reservoirs	228
	9.10.4	Overflow works – non-impounding reservoirs	230
9.11	Ancillary structures		234
	9.11.1	Bywash channels	234
	9.11.2	Fish passes	234
9.12	Concrete		234
	9.12.1	General	234
	9.12.2	Design of structures	234
	9.12.3	Concrete mix specification	235
	9.12.4	Aggregates	235
	9.12.5	Water-cement ratio	236
	9.12.6	Admixtures	236
	9.12.7	Crack control and joints	237
	9.12.8	Reinforcement and cover	241
9.13	Masonry		241
	9.13.1	General	241
	9.13.2	Concrete blocks	242
	9.13.3	Bricks	242
	9.13.4	Mortars	242
	9.13.5	Reinforcement	242
	9.13.6	Durability and impermeability	242
	9.13.7	Drainage	243
9.14	Environmental considerations		243
	9.14.1	Siting	243
	9.14.2	Shoreline	244
	9.14.3	Bank shape	246
	9.14.4	Water depth	246
	9.14.5	Islands	248
	9.14.6	Aquatic and shoreline planting	248
	9.14.7	Vegetation	252
	9.14.8	Tree planting	255
	9.14.9	Habitat creation for breeding wildfowl	256

	9.14.10	Habitat creation for fish	256
	9.14.11	Habitat creation for other wildlife	257
	9.14.12	Recreational and sporting use	257
	9.14.13	Management for fishing	257
	9.14.14	Management for shooting	258
	9.14.15	Examples of design	258
	9.14.16	Burrowing animals and other pest control	259
9.15	Safety considerations		262
	9.15.1	Public and operational safety	262
	9.15.2	Emergency water supply for fire fighting purposes	264
	9.15.3	Livestock	265
9.16	Design records		266
	9.16.1	General	266
	9.16.2	Subsequent records	266

PART 3 CONSTRUCTION 267

10 CONSTRUCTION ASPECTS 267

10.1	General		267
	10.1.1	Introduction	267
	10.1.2	Weather considerations	267
	10.1.3	Working limits and setting out	268
	10.1.4	Construction access and temporary works	269
	10.1.5	Construction monitoring	270
10.2	Buried and other services		271
	10.2.1	General	271
	10.2.2	Services	271
	10.2.3	Inspection holes	271
	10.2.4	Diversion of services	271
10.3	Earthworks		271
	10.3.1	General	271
	10.3.2	Control of surface water and groundwater	272
	10.3.3	Construction plant and equipment	274
	10.3.4	Site preparation	274
	10.3.5	Foundation preparation	277
	10.3.6	Unsuitable material	277
	10.3.7	Trenches and excavations	278
	10.3.8	Excavation of fill materials from the borrow pit	278
	10.3.9	Reservoir works	279
	10.3.10	Embankment construction	280
	10.3.11	Fill placing and compaction	281
	10.3.12	Damage to soil and vegetation by construction plant and equipment	286
	10.3.13	Stockpiling materials	289
10.4	Seepage control measures		290
	10.4.1	Cutoff trench	290
	10.4.2	Upstream clay blanket	290
	10.4.3	Anti-seepage collars	291

10.5 Drainage 291
    10.5.1   Toe and other trench drains...291
    10.5.2   Manholes...292
    10.5.3   Drainage blankets...292
    10.5.4   Drainage associated with structures.................................294
    10.5.5   Headwalls..294
    10.5.6   Geotextiles 295
10.6 Slope and crest protection 295
    10.6.1   Upstream slope...295
    10.6.2   Crest 295
    10.6.3   Downstream slope..296
10.7 Reservoir liners 296
10.8 Concrete and masonry...296
    10.8.1   General..296
    10.8.2   Reinforcement..297
    10.8.3   Formwork..297
    10.8.4   Waterstops..298
    10.8.5   Concrete placing and compaction....................................298
    10.8.6   Joint formation and concrete finishes 299
    10.8.7   Masonry construction...299
10.9 Drawoff and other pipework 300
    10.9.1   General..300
    10.9.2   Drawoff pipework 301
    10.9.3   Thrust blocks..301
    10.9.4   Outlets 301
10.10 Overflow works..302
    10.10.1   Construction of drop inlet overflow weirs......................302
    10.10.2   Construction of crest weir overflow works....................303
    10.10.3   Construction of auxiliary overflow works 303
10.11 Environmental/conservation work 305
    10.11.1   Shallows and islands...305
    10.11.2   Aquatic planting...305
    10.11.3   Terrestrial planting...305
10.12 Construction safety..306
    10.12.1   General..306
    10.12.2   Safety related to the construction of earthworks 307
    10.12.3   Safety related to construction plant 307

11 CONTRACTUAL ARRANGEMENTS 309
11.1 Contract requirements 309
11.2 Contract options 311
    11.2.1   General..311
    11.2.2   Non-standard form of contract...312
    11.2.3   Standard form of contract 313
11.3 Specification...314
11.4 Drawings 314

	11.5	Financial arrangements	315
		11.5.1 General	315
		11.5.2 Clarification of terms	315
	11.6	Statutory provisions	316
	11.7	Insurance	316
		11.7.1 General	316
		11.7.2 Public liability	317
		11.7.3 Employees of contractor	317
		11.7.4 Insurance of the works	317
	11.8	Supervision	317

PART 4 POST-CONSTRUCTION 319

12 OBSERVATIONS AND MONITORING 319

- 12.1 Introduction 319
- 12.2 Reservoir level and flow measurement 320
- 12.3 Identification of defects 320
- 12.4 Problem monitoring 321
- 12.5 Reservoir margin areas 326
- 12.6 Embankment areas 327
  - 12.6.1 General 328
  - 12.6.2 Embankment crest and upstream shoulder 329
  - 12.6.3 Downstream shoulder and embankment toe 333
- 12.7 Inlet, overflow, drawoff and ancillary structures 335
- 12.8 Downstream areas 337
  - 12.8.1 General 343
  - 12.8.2 Seepage 344
  - 12.8.3 Instability 344
- 12.9 Other aspects 344
  - 12.9.1 Public safety 344
  - 12.9.2 New development 345
  - 12.9.3 Appraisal of visual observations and monitoring records 345

13 MAINTENANCE WORKS 346

- 13.1 Introduction 346
- 13.2 Embankment areas 346
  - 13.2.1 General 346
  - 13.2.2 Internal erosion and deterioration 347
  - 13.2.3 Surface erosion 347
  - 13.2.4 Surface deformations and shallow instability 348
- 13.3 Reservoir perimeter 348
- 13.4 Geomembranes and geotextiles 348
  - 13.4.1 Geomembranes 348
  - 13.4.2 Geotextiles 349
- 13.5 Inlet structures 349
- 13.6 Drawoff works 349
- 13.7 Overflow works 349
- 13.8 Ancillary structures and drainage 350

13.9	Burrowing animals and pest control	350
13.10	Vegetation management	351
	13.10.1 General	351
	13.10.2 Aquatic vegetation	351
	13.10.3 Grassland	352
	13.10.4 Scrub and trees	352
	13.10.5 Timing of management work	353
13.11	Safety	354

14 REMEDIAL WORKS ... 355

14.1	Introduction	355
14.2	Embankment areas	355
	14.2.1 General	355
	14.2.2 Leakage reduction and control	356
	14.2.3 Internal erosion	356
	14.2.4 Stability improvement	357
	14.2.5 Slope protection	357
	14.2.6 Surface erosion	358
14.3	Inlet, drawoff, overflow and ancillary structures	359
14.4	Geomembranes and geotextiles	359
	14.4.1 Geomembranes	359
	14.4.2 Geotextiles	359
14.5	Vegetation	360
14.6	Reservoir perimeter	360
	14.6.1 General	360
	14.6.2 Wave erosion and slope protection	361
	14.6.3 Springs and surface water flow	361
	14.6.4 Stability improvement	362
14.7	Changes in storage capacity/water area	362
14.8	Dismantling reservoir structures	363

References		365
Appendix A	Reservoirs Act 1975	375
Appendix B	Capacity and maximum depth of water	383
Appendix C	General requirements for various recreational and sporting purposes	390
Appendix D	List of useful addresses	392
Appendix E	Waterborne diseases	402
Appendix F	Flow gauging (after Ref 2)	403
Appendix G	Estimation of stream flow (after Ref 2)	407
Appendix H	Typical flood assessment	411
Appendix I	Effects of water surface areas	413
Appendix J	Sources of information	417
Appendix K	Typical field sketch	421
Appendix L	Reservoir liners	422
Appendix M	Geotextiles	431
Appendix N	Concrete	436
Appendix O	Typical bill of quantities	440
Appendix P	Reference gauge post for reservoir level monitoring	443
Appendix Q	Surveillance indicators of possible defects	444

Figures

Figure 1.1	Reservoir and dam terminology	40
Figure 1.2	Embankment terminology	41
Figure 2.1	Typical non-impounding irrigation reservoir	44
Figure 2.2	Reservoir associated with commercial development	44
Figure 2.3	Ornamental/landscaping reservoir	45
Figure 2.4	Environmental/conservation reservoir	45
Figure 3.1	Main steps in the planning process in England and Wales	65
Figure 3.2	Main steps in the licensing and consent process in England and Wales	69
Figure 4.1	The main aquatic plant zones	86
Figure 4.2	Features of a well-designed reservoir for wildlife	87
Figure 4.3	Features of a poorly designed reservoir for wildlife	88
Figure 6.1	Average annual rainfall (1941–1970) (Ref 52)	98
Figure 6.2	Potential evapotranspiration (Based on average evapotranspiration losses (mm) (1930–1949)) (Ref 53)	101
Figure 7.1	RSMD (mm) in Great Britain (Ref 54)	115
Figure 7.2	Assessment of peak flood flow	116
Figure 7.3	Flood peak intensity for small reservoir catchments (After Ref 54)	117
Figure 7.4	Definition of catchment area	118
Figure 7.5	Assessment of catchment area, water surface areas and areas of rapid runoff	120
Figure 7.6	Assessment of stream slope factor	121
Figure 7.7	Assessment of soil factor	125
Figure 7.8	Assessment of vegetation factor	127
Figure 7.9	Assessment of maximum flood	128
Figure 7.10	High reservoir level threatening overtopping	129
Figure 7.11	Damage as a result of overtopping	129
Figure 7.12	Concrete weir chamber (with provision for adjustable weir boards)	134
Figure 7.13	Vertical pipe with bellmouth (operating at top water level)	135
Figure 7.14	Simple concrete main overflow	136
Figure 7.15	Stone-faced overflow with fishpass	136
Figure 7.16	Blockwork overflow with dropweir	137
Figure 7.17	Side entry overflow	137
Figure 7.18	Grassed auxiliary overflow	138
Figure 7.19	Gabion-lined auxiliary overflow	138
Figure 7.20	Basic arrangement of main and auxiliary overflows	139
Figure 7.21	Sizing overflow structures	143
Figure 8.2	Inadequate support to trial holes	156
Figure 8.3	Selection of soil augers (after Ref 58)	158
Figure 8.4	Cable percussive boring rig	159
Figure 8.5	Assessment of clay content by hand texturing (Ref 60)	163

Figure 8.6	Initial assessment of particle size from settlement test (Ref 2) 164
Figure 8.7	Example of planar polished surfaces 167
Figure 8.8	Grading curves for coarse-grained soils 169
Figure 9.1	Slope protection of boards and stakes to reservoir perimeter 178
Figure 9.2	Slope protection of logs and stakes to reservoir perimeter 178
Figure 9.3	Slope protection of stone near top water level around reservoir perimeter 179
Figure 9.4	Slope protection of vegetation near top water level around reservoir perimeter 179
Figure 9.5	Typical designs for a homogeneous dam (a) on impermeable foundation (b) shallow depth of permeable foundation 184
Figure 9.6	Typical design for a zoned dam (a) on impermeable foundation (b) on shallow depth of permeable foundation 185
Figure 9.7	Typical design for a dam incorporating a central diaphragm 186
Figure 9.8a	Typical design for a lined dam (a) incorporating a clay liner (b) incorporating a geomembrane 188
Figure 9.9	Typical design of anti-seepage collars 193
Figure 9.10a	Bank protection methods (Ref 64) 196
Figure 9.10b	Bank protection methods (Ref 64) 197
Figure 9.11	Drainage details 199
Figure 9.12	Typical arrangement for gravity filling (After Ref 2) 208
Figure 9.13	Alternative protection at inlet using gabions 210
Figure 9.14	Typical arrangement of inlet structure 211
Figure 9.15	Excavated silt trap 212
Figure 9.16	Typical chamber silt trap details 213
Figure 9.17	Typical pump layout (after Ref 2) 214
Figure 9.18	Typical arrangement of a pumping chamber 217
Figure 9.19	Typical pump and screen 217
Figure 9.20	Screened gravity drawoff 218
Figure 9.21	Gravity drawoff with penstock control 219
Figure 9.22	Typical weir chamber arrangement (after Ref 2) 221
Figure 9.23	Typical concrete sill and channel arrangement 224
Figure 9.24	Stepped overflow channel 226
Figure 9.25	Landscaped stepped overflow 227
Figure 9.26	Timber stoplogs in a brickwork overflow and channel structure 228
Figure 9.27	Typical auxiliary overflow arrangement for flow over a steep downstream slope 230
Figure 9.28	Typical auxiliary overflow arrangement for flow around the dam with a gently sloping channel 231

Figure 9.29	Limiting velocities for flow over plain and reinforced grass (Ref 70)	232
Figure 9.30	Effects of water/cement ratio on concrete permeability (Ref 37)	236
Figure 9.31	Concrete joint details (Ref 37)	239
Figure 9.32	Waterstops in typical construction joints (Ref 37)	240
Figure 9.33	Good and bad examples of shoreline design (Ref 44)	244
Figure 9.34	Variable shoreline (mainly vegetated)	245
Figure 9.35	Variable shoreline (mainly open)	245
Figure 9.36	Gently sloping variable shoreline	246
Figure 9.37	Water-depth variation to maximise wildlife benefit	247
Figure 9.38	Good and bad examples of island profiles (Ref 44)	249
Figure 9.39	Islands created by excavation	250
Figure 9.40	Islands created by filling	250
Figure 9.41	Typical design of a simple floating raft	253
Figure 9.42	Typical design of a timber raft	253
Figure 9.43	Floating island showing access ramps	254
Figure 9.44	Features of a well designed reservoir for wildlife	259
Figure 9.45	Features of a poorly designed reservoir for wildlife	260
Figure 9.46	Environmentally poor reservoir design	260
Figure 9.47	Environmentally good reservoir design	261
Figure 9.48	Typical warning sign	263
Figure 9.49	Typical lifebelt facility	264
Figure 10.1	Effects of ill-considered access after wet weather	268
Figure 10.2	Hydraulic excavator	276
Figure 10.3	Tractor and rib roller	276
Figure 10.4	Large dozer with towed vibrating roller	282
Figure 10.5	Crawler loader with towed vibrating roller	283
Figure 10.6	Self-propelled roller	283
Figure 10.7	Sheepsfoot roller	284
Figure 10.8	Relationship of moisture content, dry density, compaction and permeability (Ref 37)	286
Figure 10.9	Use of plant to break up dry fill	287
Figure 10.10	Addition of water to dry fill	288
Figure 10.11	Toe drain construction using sightline	292
Figure 10.12	Simple manhole arrangement	293
Figure 10.13	Anchorage of geotextiles for erosion protection	304
Figure 11.1	Contract requirements	312
Figure 12.1	Typical long-term seepage area at toe of a dam	323
Figure 12.2	Embankment instability	323
Figure 12.3	Surface movements at toe of dam	324
Figure 12.4	Instability of upstream slope following reservoir lowering	324
Figure 12.5	Seepage around reservoir perimeter	328
Figure 12.6	Instability around reservoir perimeter	329
Figure 12.7	Erosion around reservoir perimeter	330
Figure 12.8	Unhealthy vegetation	331

Figure 12.9	Tree becoming established adjacent to overflow	333
Figure 12.10	Damage to channel by trees	334
Figure 12.11	Slope erosion	335
Figure 12.12	Severe crest erosion	337
Figure 12.13	Damage to geotextile slope protection	337
Figure 12.14	Erosion to upstream slope between roots	338
Figure 12.15	Erosion to upstream slope undercutting vegetation	339
Figure 12.16	Severe upstream erosion	340
Figure 12.17	Damage to exposed liner	340
Figure 12.18	Erosion damage to drain outlet	341
Figure 12.19	Blockage of channel restricting flow	342
Figure 12.20	Settlement and distress of overflow structure resulting from erosion beneath	343
Figure 13.1	Sandbag protection to upstream slope	347
Figure 14.1	Geotextile wrapped drain to control seepage flows	358
Figure 14.2	Use of gabions to control wave erosion	361
Figure F1	Installation of 90° vee-notch weir flow gauge	404
Figure F2	Installation of rectangular weir flow gauge	406
Figure I1	Flood attenuation procedure	415
Figure I2	Determination of flood attenuation ratio, R (Ref 54)	416
Figure L1	Anchor trench details	428
Figure M1	Soil grading curves – cohesive soil	434
Figure M2	Soil grading curves – granular soil	435

Boxes

Box 3.1	The Town and Country Planning General Development Order 1988	61
Box 3.2	Countryside protection	62
Box 3.3	Environmental assessment	63
Box 3.4	Minerals and waste	64
Box 3.5	Impounding licence	66
Box 3.6	Water abstraction licence	67
Box 3.7	Consent for works	68
Box 3.8	Fish farming	71
Box 3.9	The Salmon and Freshwater Fisheries Act 1975	72
Box 3.10	Fishing rights	73
Box 3.11	Wildlife and Countryside Act 1981	74
Box 3.12	Forestry Act 1967 and tree felling	75
Box 3.13	Pollution control legislation	76
Box 4.1	The site and its environment	81
Box 4.2	Environmental effects of reservoir creation	82
Box 6.1	Factors affecting the runoff of surface water into a reservoir	97
Box 6.2	Soil infiltration and evapotranspiration	99
Box 7.1	Selection of appropriate multiplier	132
Box 8.1	Features to note during a site walkover survey	153

Box 8.2	Sampling methods and frequency	155
Box 8.3	Excavation of trial holes	157
Box 8.4	Estimation of insitu permeability by simple infiltration test	166
Box 8.5	Engineering behaviour of fine (cohesive) soils	169
Box 8.6	Engineering behaviour of coarse (granular) soils	170
Box 8.7	Properties of main soil types	171
Box 8.8	Unsuitable embankment foundation	172
Box 9.1	Erosion control	195
Box 9.2	Types of pipe	205
Box 9.3	Pipe bedding and surround criteria	207
Box 12.1	Problem identification – maintenance works	321
Box 12.2	Problem identification – remedial works	321

Tables

Table 7.1	Stream slope factors	121
Table 7.2	Classification of soil properties – (Ref 56)	122
Table 7.3	Classification of soils by runoff potential – (Ref 56)	123
Table 7.4	Description of typical soil classes – (Ref 56)	124
Table 7.5	Soil factors	124
Table 7.6	Forestry index	126
Table 7.7	Reservoir classification – (based on Ref 54)	131
Table 7.8	Length of weir to pass design discharge	140
Table 8.1	Mass of soil sample required for laboratory tests	155
Table 8.2	Field identification and description of cohesive and organic soils	161
Table 8.3	Field identification and description of granular soils	162
Table 8.4	Estimation of soil compactness/strength from simple field tests and observations	165
Table 8.5	Soil permeability	166
Table 9.1	Minimum trench widths for pipes (Ref 71)	204
Table 9.2	Maximum channel velocities and slopes for feeder channels (Ref 2)	208
Table 9.3	Friction losses in pumped suction and delivery pipes (Ref 2)	215
Table 9.4	Weir chamber – minimum pipe diameter and fall (Ref 2)	222
Table 9.5	Maximum velocity of flow over auxiliary overflow	233
Table 9.6	Concrete cover to reinforcement	242
Table 9.7	Common plant species of the main aquatic plant zones	251
Table 10.1	Implications of weather conditions	268
Table 10.2	Excavators and transportation	274
Table 10.3	Recommended compaction plant for construction of small embankment-type reservoirs	284
Table 10.4	Optimum moisture content: typical values for soil types	286
Table 10.5	Compaction requirements for typical soil types	286
Table 10.6	Assessment of pipe thrusts	301

Table 12.1	Action on discovering instability or seepage	325
Table 12.2	Vegetation indicating poor drainage – (Ref 123)	327
Table L1	Geomembranes – butyl rubber-based products	423
Table L2	Geomembranes – polyethylene products	423
Table L3	Geomembranes – polyvinyl chloride (PVC)-based products	424
Table L4	Bentonite liners – bentonite mats	424
Table L5	Geomembranes – other lining products	424
Table M1	Geotextiles	431

Glossary

Abutment	The valley sides forming the foundation to the embankment. Depending on the width of the valley floor, the abutments may extend over a relative short length or much of the embankment length
Agrichemicals	Chemicals used in agriculture
Alluvium	Material which is transported by a river and deposited, ranging in size from alluvial clay to alluvial gravel
Amenity use	A use which is of benefit for the local environment and community
Aquifer	A permeable soil or fissured rock capable of yielding water in economic quantities
Artesian water	Groundwater confined between impermeable strata which is under sufficient pressure to raise the water above ground level
Attenuation ratio	The ratio between the reduced depth of flood flow which passes over the overflow of a reservoir and the depth which would have passed if the reservoir was absent. The reduction results from the presence of the large surface area of the reservoir which allows temporary storage of some of the flood
Auger	A hand-operated soil sampler
Auxiliary overflow	A secondary overflow designed to operate only during extreme flood flows
Average flow	The flow in a watercourse resulting from the average annual rainfall minus losses averaged throughout the year
Base flow	The flow in a watercourse during dry weather periods which originates from water bearing strata and not from surface runoff
Bearing capacity	The capacity of the ground to sustain load
Bentonite	A fine clay which has considerable swelling properties in the presence of water. It is often used to reduce seepage in reservoir embankments
Borehole	A small diameter exploratory hole in the ground formed by an auger or shell and auger methods for investigating the sub-surface ground conditions

Borrow pit	An excavation to provide material for building an embankment or other construction works
Breaching	The removal of a section of the embankment due to erosion, foundation movement or excessive flood
Bywash channel	A channel which leads water around a reservoir and embankment to prevent entry into the reservoir
Catchment	The area of land from which precipitation, normally rainwater or snow melt, drains into a reservoir, pond, lake or stream. It is usually expressed in hectares or square kilometres
Clay	Strictly a constituent of soil the grain size of which is predominantly less than 0.002 mm, but the term is used colloquially to describe a soil which is cohesive
Cohesive soil	Soil whose major constituent is clay, which tends to shrink on drying, expands on wetting and gives up water when compressed. Cohesive soils possess a measurable plastic limit
Compensation flow	The minimum flow that must be released from an impounding reservoir into the watercourse downstream of the dam.
Creep	The slow downhill movement of soil which occurs close to the ground surface
Crest	Top of an embankment
Dam	A combined structure including an embankment, overflow(s), drawoff works, pipelines, etc.
Dead storage (dead water)	The water in a reservoir which is not available for use as it lies below the level of the sump or drawoff works
Desiccated surface	The surface of soil which has dried out due to lack of rainfall to form a hard surface, usually associated with cracking of the surface
Design flood	The flood used for assessing the size of the overflows; based on the maximum flood and the reservoir classification
Design mix	Concrete mix for which the purchaser is responsible for specifying the required performance and the producer is responsible for selecting the mix proportions to produce the required performance
Dilatancy	The property can be recognised by moulding a lump of soil in the hand and then squeezing it – any surface moisture will be seen to recede

Direct abstraction	The removal of groundwater or water from watercourses for use in direct supply
Drainage class	A class which gives a measure of the rate at which precipitation can infiltrate the ground. It is divided into 'rapid', 'medium' and 'slow' and is dependent on the soil type, e.g. whether it is gravel-like or clay-like
Drainage divide	A topographical boundary from which runoff or infiltration flows in different directions. It marks the limit of a catchment area or watershed
Drainage medium	The material used in a drain (usually crushed stone) through which water can pass easily
Drawdown	The rapid lowering of the water level of a reservoir
Drift map	Geological map which shows superficial deposits overlying rock formations
Drought conditions	Continuous dry weather without significant rainfall
Dry slope	Downstream or outer slope of a dam
Effective mean soil moisture deficit	The quantity of water, expressed in millimetres, required to restore a soil to the point at which free drainage occurs
Embankment	Man-made bank generally constructed of soil to retain water. Also includes associated slope protection, drainage, etc.
Erosion	The removal of soil by natural agencies such as rainfall, river and flood flows, undermining or gravity, or by human or animal activities
Evaporation	The conversion of water to water vapour by solar radiation
Evaporite	Material deposited by evaporation and normally found as a geological stratum: often contains potentially aggressive constituents. Typical evaporites include halite (rock salt), gypsum and hydrite
Evapotranspiration	Water returned to the atmosphere by evaporation and transpiration
Exploratory hole	Hole into the ground to examine the nature of the subsurface soil material, e.g. borehole, auger hole or trial pit
Fetch	The length, measured in a straight line, along a reservoir over which winds can build up waves
Fill	Soil, rock or waste material placed in an excavation or other area to infill or raise the surface elevation
First filling	The initial filling of water into a reservoir

Flood attenuation	The reduction of a flood flow as it enters an impounding reservoir caused by the reservoir or other water area temporarily accommodating the flood water
Flood event	A period of flood significantly above the dry weather flow for the time of year. The term does not imply flooding
Flood flow	The flow in a watercourse resulting from a period of heavy rainfall
Flood routing	The process for estimating how an impounding reservoir will affect the behaviour of flood flows in watercourses
Flood surcharge	The maximum rise of still water level above reservoir top water level during a design flood. (Surcharge water is not retained in the reservoir but is discharged until the normal retention level is reached)
Freeboard	The vertical height from top water level to the top of the embankment. (Freeboard is required to contain flood surcharge plus wave-splash of specified severity)
Gabions	Rectangular or tubular baskets made from steel wire or polymer mesh and subsequently filled with large stones or rock fragments
Geomembrane	Synthetic impermeable material used for reservoir lining
Geotextile	Synthetic permeable fabric used in conjunction with soil and vegetation; principally for erosion control, filtration, soil reinforcement and drainage
Graded surface	A ground surface which has a uniform level or inclination
Granular soil	Soil whose major constituent is sand and/or gravel, which tends to have negligible volume changes with changes in moisture content and no measurable plastic limit
Groundwater	Water that occurs below ground level in rock or soil below the water table
Gulleying	The process whereby fine material is eroded away by runoff, resulting in a series of downslope steep-sided channels
Hard protection	Collective term for bank protection using materials such as steel, concrete, etc.; as distinct from protection using natural soft materials such as vegetation and timber
Hydration	The swelling of a soil caused by taking water into the soil structure

Impounding reservoir	Reservoir formed by damming a watercourse
Infiltration	The flow of water from the land surface into and through the upper soil layers
Inflow	Water entering a reservoir
Lake	Body of water stored above or below ground, the capacity of which is greater than 2500 m^3 or depth of water is greater than three metres
Liquid limit	The moisture content at which soil begins to behave in a liquid manner
Made ground	See 'Fill'
Main overflow	A primary overflow structure composed of non-erodible material, e.g. concrete or steel pipes, to pass low floodflow. In an impounding reservoir the flow is likely to be continuous
Maximum flood	The maximum volume of water that can enter a reservoir depending on various climatic and catchment characteristics
Moisture content	The ratio expressed as a percentage of the weight of water in a given soil mass to the weight of dry solid particles (see 'Optimum moisture content')
Non-impounding reservoir	Reservoir with inflow received from pumped or gravity fed supply from nearby stream or borehole, and not from any significant natural inflow
Normal retention level	See 'Top water level'
Operating level	The normal level of water in a reservoir which may be variable in an abstraction reservoir or essentially fixed at a constant level ignoring any flood rise
Optimum moisture content	The water content in a soil at which the greatest compaction can be theoretically achieved
Outlet works	The structural parts of the dam which permit water to be released from the reservoir, e.g. drawoff pipes
Overburden	The unsuitable material overlying potentially usable soil for earthworks
Overflow	A structure for passing flow safely over the embankment to discharge downstream; a dam normally has a main overflow of non-erodible material to pass continuous flow and frequently an auxiliary overflow of grass or reinforced grass to pass occasional extreme flows

Particle size distribution	A laboratory test to determine the composition, in terms of particle size, of a soil giving the percentages of:
	• boulders > 200 mm
	• cobbles 60 – 200 mm
	• gravel 2 – 60 mm
	• sand 0.06 – 2 mm
	• silt 0.002 – 0.06 mm
	• clay < 0.002 mm
Pass	A single directional movement of a compactive roller
Peak flood flow	The maximum rate at which water that can arise from a catchment, dependent only on catchment characteristics and size and the RSMD
Peat	Soil largely made up of organic remains
Penstock	A means of controlling, varying or stopping the flow in a pipeline, normally located on the end of the pipeline
Percolation	The downward flow of water below the subsurface under the influence of gravity until it reaches the water table
Perennial flow	A stream which flows continuously throughout the year, being fed by groundwater during seasons of low rainfall and by runoff during seasons of high rainfall
Permeability	Measure of the rate at which groundwater can flow through the soil or along discontinuities in rock
Piping	An erosion process which generally occurs in fine sand or silt when a flow of water removes material adjacent to the flow channel thereby steadily increasing the size of the channel
Plasticity index	The difference between the liquid limit and the plastic limit
Plastic limit	The moisture content at which a soil begins to behave in a plastic manner
Pond	Body of water stored above or below ground whose capacity is less than 2500 m^3 or depth of water is less than three metres
Pore space	Any voids in a soil whether air or water filled. When the voids are fully filled with water, the soil is said to be saturated
Potable water	Water used for public consumption
Precipitation	The sum of water which falls as rain, snow, hail, fog and dew, and expressed in millimetres over a specified period

Public supply	Potable water taken directly from mains supply
Raised reservoir	A reservoir designed to hold or capable of holding water above the natural level of any part of the land adjoining the reservoir
Reservoir	An artificial body of stored water (see Section 1.3)
Reservoir losses	Water stored in a reservoir lost through evaporation or seepage through the foundation and dam
Return period	A measure of the expected time (in terms of probability rather than forecasting) between floods equal to or greater than a stated magnitude
Riparian owner	Owner of the land alongside a watercourse. In the UK, ownership normally extends to the centreline of the channel
Rock	Component constituents which are cemented together. Is generally considerable stronger and harder than soil
Roller	Compaction cylinder in contact with soil attached to compaction plant
RSMD	A factor used for assessing the maximum peak flood. Also the 'Flood Studies Report' index for flood-producing rainfall which is defined as the one day's rainfall of a five year return period less the effective mean soil moisture deficit
Run-up	The maximum vertical height attained by a wave running up a dam face, referred to the steady water level without wind action
Saturated	Soils in the groundwater zone and within which all pore spaces have been filled with water
Sedimentation	The deposition of sand, silt and clay carried by a watercourse as a result of the reduction in velocity of the stream/river as it enters an impounding reservoir
Seepage	Movement of water through soil or rock
Setting out	Marking out the position on site of the proposed embankment and any associated structures
Shear strength	The stress at which a soil fails by shearing
Sill	The element of an overflow controlling the discharge level of the reservoir over the overflow

Site	The area of land where the proposed works are to be constructed, including the borrow area, plus any adjacent areas which may be required for construction or access purposes
Slope protection	Protective measure to prevent erosion and general degradation of an embankment by the reservoir water on the upstream surface of the embankment (see Stone pitching)
Soft protection	Opposite of hard protection, normally employing natural materials, for bank protection
Soil	Component constituents which are not cemented together but the coherent structure of which depends on the degree of interlocking between grains and cohesive bonding
Soil class	A classification of soils by their runoff potential; Soil Class 1 has the lowest runoff potential, whilst Soil Class 5 has the greatest runoff
Soil profile	Profile of the sub-surface soil as revealed from an exploratory hole
Solid map	Geological map which does not show superficial deposits overlying rock formations
Specialist adviser	Competent person able to advise on specific technical, legal or other matters
Stone pitching	A facing of stones on the upstream slope of an embankment as a protection against erosion and wave action
Storm losses	That part of precipitation which does not become runoff within the flood period because it has evaporated, infiltrated, been retained in the soil or temporarily ponded on the catchment surface
Subsoil	A layer of weathered soil beneath the topsoil usually containing minerals leached from the topsoil and many roots (denoted as the B horizon by soil scientists)
Superficial Materials	Materials, often soft or weak, which have been redeposited by the action of rivers, glaciers, landslides or other movement, processed or produced as a result of weathering action
Surface water	Water which flows in watercourses either permanently or seasonally, and stored in ponds, lakes or man-made containment structures
Topography	The general lie of the land which may be either artificial or natural

Topsoil	A layer of soil which can support vegetation. It usually comprises the uppermost 150 mm of soil (denoted as the A-horizon by soil scientists)
Top water level	(a) For a reservoir with a fixed overflow sill, the lowest crest level of that sill (b) For a reservoir from which the overflow is controlled wholly or partly by movable gates, siphons or other means, the maximum level to which water may be stored exclusive of any provision for flood storage
Transpiration	The emission of water vapour into the atmosphere by living plants
Trial hole	Small pit generally dug (using an excavator) to examine the sub-surface soil profile
Undertaker	Persons using or intending to use a reservoir. Where the reservoir is not used or intended to be used, the owner or the lessee occupying the land including the reservoir. Lessees of a reservoir, e.g. a fishing club, may be classed as joint undertakers, together with the owners of the reservoir. Other owners of part of the dam or reservoir may also be undertakers (see also Appendix A)
Urban developments	Includes areas of hardstanding and impermeable surfaces which offer little resistance to runoff
Valve	A means of controlling, varying or stopping flow in a pipeline
Vegetation	An assemblage of living plants comprising: • grass — Member of a distinct non-woody, botanical family widely utilised by man. The term does not include rushes, sedges and reeds • herbs — generally non-woody flowering plant. The term which excludes grasses, rushes and sedges • shrub — woody plant branching abundantly from the base and not reaching a large size (< 4 m) • tree — woody plant, usually with a single stem (trunk), bearing lateral branches and often reaching a considerable size

Watercourse	The path taken by flowing water on the ground surface, e.g. a river, stream or ditch; the watercourse may have been altered by previous human operations. A watercourse may contain perennial or seasonal flow or may normally be dry
Waterlogged	Soil at the ground surface which has become fully saturated
Water quality	Quality of water in terms of bacteriological and chemical composition
Waterstops	Specially-shaped strips, generally of nominal width range (150 mm to 325 mm) and made of polyvinyl chloride (PVC) or rubber, which are incorporated into concrete construction at expansion and contraction joints to maintain watertightness.
Water table	The level below which the soil or rock is fully saturated
Wave surcharge	The vertical rise of water against a dam created solely by the run-up of waves of specified probability
Wave surcharge allowance	The theoretical wave freeboard sufficient to prevent overtopping and threatening the embankment crest for the particular type and design of embankment under consideration
Weir	Structure placed across a watercourse to measure the rate of flow
Weir chamber	Structure used to allow the flow of water from a reservoir through to the outlet pipe beneath the embankment, also functions as the controlling level on the retained water
Well	A hole formed for the specific purpose of abstracting groundwater
Wet slope	Upstream or inner slope of dam
Winter rain acceptance potential	A measure of the rate at which precipitation can infiltrate the ground. It is divided into five categories from 'very high' to 'very low' based on the degree of runoff, e.g. if runoff is very high, the winter rain acceptance is very low (equivalent to Soil Class 5)
Works	The temporary and permanent works required for the construction of a reservoir
Yield	Generally refers to the average rate of water abstraction from a reservoir during a season of use

Notation

A	catchment area factor (ha)
A_C	area of catchment (ha)
A_f	forested area within catchment (ha)
A_R	area of rapid runoff (e.g. urban area) within catchment (ha)
A_W	area of surface water within catchment (ha)
AAR	average annual rainfall (mm)
B	length of overflow (m)
B_A	length of auxiliary overflow (m)
B_m	length of main overflow (m)
D_{15}	particle size of the in-situ soil where 15% of the soil particles are smaller by size (mm) (derived from particle size distribution test)
D_{85}	particle size of the in-situ soil where 85% of the soil particles are smaller by size (mm) (derived from particle size distribution test)
F	forestry index
G	stream slope factor
H	discharge head over weir (m or mm)
H_A	discharge head over auxiliary overflow weir (m or mm)
H_D	discharge head over main overflow to pass design discharge of $Q_m/10$ (m or mm)
H_m	discharge head over main overflow weir (m or mm)
L	Breadth of flow measuring weir (m)
Losses	losses expressed as percentage of average annual rainfall (%)
M	flood multiple dependent on reservoir classification
MSL	length of the major watercourse in the catchment (km)
O_{90}	pore size of the geotextile where 90% of the pores are smaller by size (mm)
Q_i	design flood (m³/s)
Q_m	peak flood flow (m³/s)
Q_{max}	maximum flood (m³/s)
Q_o	attenuated design flood (m³/s)
R	attenuation ratio
RSMD	regional soil moisture deficit (mm)
S	storage ratio
$S_1 - S_5$	percent areas of soil classes
S1085	mean stream slope (m/km)
SOIL	soil index
V	soil factor

V_r	reservoir volume (m³)	
W	vegetation factor	
X	slope factor	
Z	embankment height above stripped ground level (m)	
a	reservoir surface area at a height of $h_u/2$ above top water level (m²)	
d	maximum depth of water in reservoir, measured at dam (m)	
g	mean catchment slope (expressed as 1V:gH)	
h	head over flow-measuring notch or gauge (m)	
h_u	maximum unattenuated flood rise (m)	
l_e	embankment length, measured along centreline (m)	
l_p	reservoir perimeter (m)	
l_r	length of reservoir, measured as a straight line from the dam to the upstream end of the reservoir (m)	
n	valley side slope (expressed as 1V:nH)	
s	valley bed slope (expressed as 1V:sH)	

Abbreviations

ACE	Association of Consulting Engineers
ADAS	Agricultural Development and Advisory Service
AONB	Area of Outstanding Natural Beauty
BALI	British Association of Landscape Industries
BASC	British Association for Shooting and Conservation
BIAC	British Institute of Agricultural Consultants
BTCV	British Trust for Conservation Volunteers
CIRIA	Construction Industry Research and Information Association
DoE	Department of the Environment
FWAG	Farming Wildlife Advisory Group
GDO	General Development Order
HSE	Health and Safety Executive
ICE	Institution of Civil Engineers
IFM	Institute of Fisheries Management
LPA	Local Planning Authority
MAFF	Ministry of Agriculture Fisheries and Food
NRA	National Rivers Authority
PDR	Permitted Development Rights
QSRMC	Quality Scheme for Ready Mixed Concrete
RLA	Relevant Licensing Authority
SSSI	Site of Special Scientific Interest

Conversion factors

CONVERSION FACTORS
CORRECT TO FOUR SIGNIFICANT FIGURES

	SI Unit	× Conversion Factor	= Imperial Unit	× Conversion Factor	= SI Unit Symbol	
Length	kilometre	0.6214	mile	1.609	km	
	metre	1.094	yard	0.9144 *	m	*exact conversion factors
	metre	3.281	foot	0.3048 *	m	
	millimetre	0.03937	inch	25.40 *	mm	
Area	hectare (10,000m^2)	2.471	acre	0.4047	ha	
	square metre	1.196	sq yd	0.8361	m^2	
	square metre	10.76	sq ft	0.09290	m^2	
	square millimetre	0.001550	sq in	645.2	mm^2	
Volume/Capacity	cubic metre	1.308	cu yd	0.7646	m^3	
	cubic metre	35.31	cu ft	0.02832	m^3	
	litre (10^{-3} m^3)	0.03531	cu ft	28.32	l	
	litre	0.2200	gallon	4.546	l	
Mass	tonne	0.9842	ton	1.016	t	
	kilogramme	2.205	pound	0.4536	kg	
Density	tonne per cubic metre	62.43	lb/cu ft	0.01602	t/m^3	
	kilogramme per cubic metre	1.686	lb/cu yd	0.5932	kg/m^3	
Force	kilonewton	0.1004	ton f	9.964	kN	
	kilonewton	224.8	lbf	0.004448	kN	
	kilonewton	102.0	kgf	0.009807	kN	
Stress/Pressure or Bearing Capacity	kilonewton per square metre	0.009324	ton f/sq ft	107.3	kN/m^2	
	kilonewton per square metre	20.89	lbf/sq ft	0.04788	kN/m^2	
	kilonewton per square metre	0.1450	lbf/sq in	6.895	kN/m^2	
	newton per square millimetre	145.0	lbf/sq in	0.006895	N/mm^2	

Acceleration due to gravity (g) 9.807 m/sec^2

Part 1 Feasibility stage

1 Introduction

1.1 BACKGROUND

Some 2500 reservoirs come within the scope of British reservoir safety legislation, with perhaps a similar number of small water storage reservoirs not subject to this control. Where the reservoir has a capacity in excess of 25 000 m^3 above the natural level of any part of the land adjoining the reservoir, it is classed as a 'large raised reservoir' under the Reservoirs Act 1975[1], which places certain legal obligations on the owners of the reservoir and other parties. Such reservoirs may only be designed and constructed under the control of a civil engineer approved by the DoE. Whilst few large reservoirs for public water supply, river regulation, flood alleviation or hydro-electric power are likely to be constructed in the future, it is estimated that a substantial number of small water storage reservoirs will be created annually over the next few years.

In the past, small reservoirs were constructed primarily for irrigation and other farm purposes. Many were also created for power generation and other industrial purposes. More recently, a wider use has been evident, with functions as diverse as fish farming, fishing, fire fighting, conservation and amenity use. Increasingly, small reservoirs have been included as part of commercial, industrial and leisure developments to provide a water feature. A significant number of small reservoirs fulfil a dual or multiple purpose role. Many owners have found that an amenity lake increases the value of their property. In some instances, particularly in south east England, the enhancement may approach the cost of construction.

Reservoirs designed, built and maintained for the Water Service Companies (formerly the Water Authorities), British Waterways Board and many other large concerns have generally been substantial and handled by experienced in-house staff or specialist consulting engineers. Many of the smaller reservoirs have been developed by people unfamiliar with accepted techniques of dam and water engineering and the standards of design and construction have sometimes been far from adequate. Planning and design has often been undertaken by inexperienced contractors, landscape architects, or other practitioners with limited specialist knowledge of dam and water engineering. Construction has sometimes been carried out by contractors with little experience in reservoir construction. Earthworks and fill placing have been treated as an uncontrolled muck-shifting exercise without any effort being made to achieve adequate fill selection, placing and compaction. Standards of supervision have often been limited or absent whilst contractual arrangements have frequently left much to be desired. Therefore well-intentioned efforts to obtain a low-cost reservoir may result in high construction costs, operation and maintenance problems, environmental insensitivity or legal problems.

This guide is provided for all aspects of small water reservoir creation from conception, through the feasibility, design and construction stages, to maintenance and monitoring. A previous guide entitled 'Water for Irrigation' (Bulletin No. 202)[2], produced by the Ministry of Agriculture, Fisheries and Food (MAFF) in 1967, concentrated essentially on farm irrigation reservoirs. This was last revised in 1977, but has now been superseded by current practice and the need to consider the increased usage and wider functions of these small water reservoirs.

Design and construction of these small reservoirs can be affected by most of the problems influencing larger reservoir construction, albeit on a reduced scale. On the other hand certain features that need particular consideration for a small reservoir development would often be less critical for a larger reservoir. These features which might have a significant effect on design and construction have not always been appreciated at a sufficiently early stage, leading to significant difficulties and increased costs at a later date. Thus clear indication of those areas where further specialist advice must be sought is an important aspect of any guide.

Construction of a reservoir is likely to require planning permission and it is prudent to discuss this and other statutory aspects with the relevant authorities at an early stage. Certain licences and consents are also normally required from the relevant licensing authority in England and Wales for river abstraction or for impounding and obstructing a water course and these must be obtained prior to construction. Requirements will often differ in Scotland and Northern Ireland and specific advice should be sought in these areas.

Environmental aspects must be fully studied during the early stages of planning and design to allow the full effects of the reservoir to be assessed. Although some change to the existing environment must be accepted, there is often considerable opportunity for landscape enhancement and the creation of habitats for flora and fauna. It is essential that allowances for enhancement, and, if necessary, mitigation should be incorporated in the design and construction.

The best intentions may not always be successful and so comment and advice on remedial works is included towards the end of this guide. This is prepared primarily for problems with new works, but may be extended, with caution, to problems with existing reservoirs. Many were adequately planned and designed in their day, but need to be maintained in a safe state and uprated to take account of changes and good practice in small reservoir engineering.

1.2 SCOPE

This guide is aimed at persons who have limited experience in the field of small reservoir construction. It attempts to explain the procedures and principles which form good practice through all the phases of feasibility, planning, design, construction, maintenance and remedial works. In particular, the guide seeks to:

- identify the engineering, economic and environmental factors to be taken into account in the planning and design stage, and to describe good design practice. Careful consideration is given to the safety of the structure, protection of the public, reliability of use and environmental acceptability
- promote good standards of construction
- draw attention to problems which have detracted from the effective functioning of low-cost water storage reservoirs in the past and, where appropriate, to indicate remedial measures for new and existing embankments
- set out a practical and cost-effective approach to monitoring and maintenance.

The major criteria to be considered when assessing the feasibility of a small water reservoir development are given in Section 2. Thereafter, the guide addresses specific requirements and illustrates these, where possible, by using figures and photographs. More basic information is given at an early stage and developed throughout the remainder of the section. As good practice involves judgement, experience, a guide of this nature can only draw attention to aspects which need consideration or on which specialist expert advice should be sought.

A bibliography and list of references are included for more detailed information on aspects mentioned in the guide. Other information is given in appendices, together with sources of information and independent advice. Some terms differ from those in conventional use and so the glossary defines key technical terms used in this guide. In general, technical terms and notation conform to civil engineering convention.

Some of the aspects in this guide are also applicable to water retained in natural or man-made depressions in the ground whilst certain operational and maintenance aspects may also be useful for existing reservoirs.

The guide does not, however, deal with the following aspects:

- reservoirs with a capacity in excess of 25 000 m^3
- flood attenuation reservoirs
- masonry and concrete dams or service reservoirs
- mining tailings, farming or industrial slurry lagoons or liquid waste tips
- river training embankments and sea defences
- hazards to downstream areas as a result of controlled or uncontrolled discharge from the reservoir.

Whilst many of the aspects covered in this guide are applicable to some of the above structures, certain features require a different approach (e.g. wave protection on river training embankments) and specialist advice should be sought.

The Construction (Design & Management) Regulations 1994 (the CDM Regulations) will impose statutory duties on clients, designers and contractors and introduce the roles of planning supervisor and principal contractor. The Regulations provide the framework for managing the issues of health and safety during the construction, repair, maintenance and demolition of civil engineering, building and engineering construction works.

One of the features of the Regulations and the associated Approved Code of Practice (ACOP) is the fresh emphasis they lay on the importance of considering health and safety aspects of construction during the various phases of designing a project. As the regulations are a post-study development they are not considered in detail in this report. (see CIRIA Report 145: CDM Regulations – Case Study Guidance for Designers: An interim report).

1.3 DEFINITIONS

A reservoir, within the scope of this guide, is a body of water retained by means of a man-made structure and includes the dam and all other appurtenant structures and features. Where the water is retained above ground level, it is strictly a raised reservoir, but is generally referred to as a reservoir for simplicity throughout this guide. A reservoir is distinct from a naturally occurring body of water which is stored in a depression below adjacent ground level. Either of these may form a pond if they are of a small size, or a lake if they are larger. No formal distinction between pond and lake exists, but everyday usage readily places a water area into one or other category. For the purposes of this guide, a pond can be taken to be a body of water up to three metres depth. A reasonable upper limit for the total capacity of a pond is 2500 m^3.

Reservoirs within the scope of this guide will be either impounding or non-impounding defined as:

- non-impounding reservoir – has no significant natural inflow and is filled artificially by pumping, or by a gravity-fed supply from an independent water source. Also known as an offstream reservoir
- impounding reservoir – is sited across an existing watercourse and receives inflow mainly from the watercourse and by direct inflow. Also known as an onstream reservoir.

In the past, the terms 'embankment' and 'dam' have been used interchangeably to describe the water-retaining earth structure of a non-impounding and impounding reservoir. In MAFF Bulletin 202 these terms were taken to be synonymous. For the purpose of this guide, however, an embankment refers to a man-made bank to retain water, but includes the slope protection and drainage works. A dam refers to the whole water-retaining structure including the embankment, overflows, drawoff works and pipelines.

The following terms have also been adopted to define the various slopes of the dam:

- upstream — refers strictly to an impounding dam, but considered to include the inner slope of a non-impounding dam. Also known as the wet slope
- downstream — refers strictly to an impounding dam, but considered to include the outer slope of a non impounding dam. Also known as the dry slope.

The main reservoir and dam terminology is shown on the diagrammatic plan in Figure 1.1, whilst the key embankment terminology is shown on the cross-section in Figure 1.2.

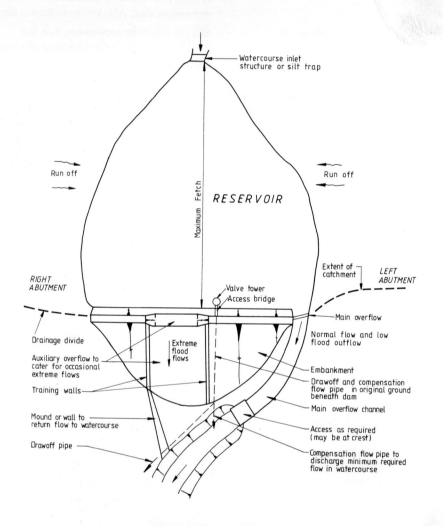

DIAGRAMMATIC PLAN

Figure 1.1 Reservoir and dam terminology

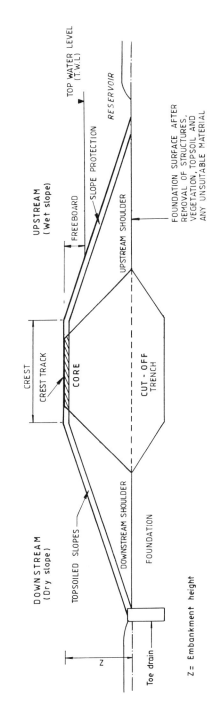

Figure 1.2 Embankment terminology

2 Reservoir feasibility

2.1 INITIAL ASSESSMENT

Selecting a reservoir site will depend on many factors. These include:
- reservoir purpose
- water requirements
- source and amount of available water
- available land
- location of water usage (where applicable)
- reservoir situation
- influence on or from existing water or other features
- topography
- land use (including services)
- ground conditions
- floods and waves
- available construction materials
- planning and licensing consents
- environmental constraints and considerations
- cost
- access and existing rights of way
- visual appearance
- maintenance and monitoring requirements.

The cost may be minimised by utilising existing ponds, lakes, gravel or clay pits and low-lying land adjacent to existing water bodies or by creating the reservoir in conjunction with mineral extraction operations.

The would-be reservoir creator/owner (the 'undertaker') must consider all the various factors and arrive at a location and a design which may involve some compromise; rarely can all aspects be fully satisfied. Reservoir safety, however, is paramount and must never be compromised in any way.

Once a possible site has been selected, it would normally be prudent to seek specialist comment and advice on the outline proposals as detailed in Section 2.16. This will allow aspects which may have been overlooked by the undertaker to be considered and may also offer scope for various technical improvements to the proposed reservoir before the undertaker has proceeded too far with the initial design. There is also scope for potential cost savings from this approach.

Options for the siting of the reservoir and embankment should also be identified, if possible, at this early stage. Time and money may then be saved at a later date if the selected site proves to be unsuitable as a result of unforeseen ground or other conditions revealed during the site survey and ground investigation.

2.2 RESERVOIR PURPOSE

A reservoir for water storage may be created for many purposes, but these tend to fall into two groupings:

Water abstraction
- irrigation or other farm use
- industrial use
- other water supply purpose
- fire fighting.

The volume of water is normally the main requirement for water abstraction reservoirs, although the depth may be a factor in some instances.

Amenity and other uses
- environmental conservation
- landscaping / ornamental
- recreation
- sporting
- shooting
- fishing
- golf course
- fish farming
- storage of polluted water or slurry
- water areas associated with commercial and industrial development.

A reservoir may fulfil a dual or multiple role, e.g. a golf course may require water storage reservoirs for not only both landscaping and recreational reasons but also as a source of irrigation water. The surface area of non-abstraction reservoirs may often be the key requirement although the depth may also be important.

Figures 2.1 to 2.4 show typical small reservoirs within the scope of this guide.

Figure 2.1 Typical non-impounding irrigation reservoir

Figure 2.2 Reservoir associated with commercial development

Figure 2.3 Ornamental/landscaping reservoir

Figure 2.4 Environmental/conservation reservoir

2.3 WATER REQUIREMENTS AND SIZE OF RESERVOIR

The storage capacity of a reservoir, normally expressed in cubic metres, is of major concern at the feasibility stage because it dictates whether the proposed reservoir will fall within the scope of the Reservoirs Act 1975[1]. If it has a capacity in excess of 25 000 m^3 above the natural level of any part of the land adjoining the reservoir, it is classified as a large raised reservoir within the Act and subject to specific legal obligations. It should be noted that the reference point from which the level is taken is the lowest ground level adjacent to the reservoir; in the case of an impounding reservoir it is normally the lowest bed level of the watercourse immediately downstream of the proposed position of the downstream toe of the dam. The capacity is measured to top water level and any temporary flood increase above this level is not included. The Act places certain legal obligations on the undertaker and other parties, and such reservoirs may only be designed and constructed under the control of a DoE approved civil engineer as defined in the Act. The main aspects of the Act are summarised in Appendix A. The Act also places a continuing obligation on the undertaker during the lifetime of the reservoir. In the context of the Act, the term 'undertaker' is used to denote the person using or intending to use the reservoir and this term has also been employed throughout this guide.

As the Act is based on volume and not on a hazard rating, many smaller reservoirs within the scope of this guide could pose a greater risk to life or property downstream than size alone could intimate. Consequently, the framework for responsible supervision and maintenance contained in the Act is often adopted for smaller reservoirs as well. Construction of a reservoir below the 25 000 m^3 limit to exclude it from the provision of the Reservoirs Act 1975[1] may appear to be an attractive option to minimise future costs, but the need for continued inspection and maintenance is still present, and the lack of a regular statutory inspection by a specialist potentially puts a greater burden on the undertakers of the reservoir. Whether the reservoir falls within this Act or not, a need for continual monitoring is present, as discussed in Section 12, to ensure that risks to third parties are not allowed to develop and the reservoir is maintained in a safe condition.

A preliminary appraisal of the required size of a reservoir for a given volume can be assessed from Appendix B. This appendix shows how a preliminary estimate can be made at an early stage which should be sufficiently accurate for initial feasibility purposes. It is generalised by necessity, but it is possible to appoximate most potential reservoirs to fall within the terms of the Appendix. The constraints on the periods for abstraction/filling and the probable need for dead storage should not be forgotten. Certain reservoir usages required a minimum area and the Appendix also shows how the volumes and maximum depth of these can be assessed. Various examples are given in the Appendix. Where possible, it is desirable to minimise the volume of earthworks and maximise the water-to-earth ratio, i.e. the volume of water storage made available by the movement of a given volume of earth. This will potentially allow the reservoir to be constructed at the minimum cost, subject to the other constraints on the site.

In a non-impounding reservoir, the storage capacity is dependent on the shape of the reservoir, the ground slope and the depth of stored water. Some increase in storage will be gained from the excavation within the reservoir, but this is not likely to be substantial and is generally ignored in a preliminary assessment. The water-to-earth ratio increases as the depth of water decreases for a reservoir of a given volume to a practical minimum of about two metres. In most situations, the depth of water is determined by the volume required and the amount of land available. In this way the maximum ratio is fixed by site conditions and is achieved by excavating most of the fill material from the reservoir floor area to construct the perimeter embankment with any shortfall provided by fill from outside the reservoir area. A reservoir constructed in this manner on a level site has a final floor level below and an embankment crest level above the original ground level. Large shallow reservoirs may appear economic in terms of the amount of earth moving required, but are likely to need extensive works for the control of excessive weed growth and maintenance of the long lengths of embankment, high evaporation losses, a large proportion of dead storage and a likely surplus of topsoil and unsuitable materials from the foundation preparation works.

The method of estimation of the storage capacity of an impounding reservoir has been prepared on the basis of uniform ground slopes, assuming the valley is approximately parabolic in cross section and the materials used for dam construction are excavated from within the reservoir area. The water to earth ratio for an impounding reservoir depends on the longitudinal bed slope of the valley and the height of the dam. It is less dependent on valley side slopes, provided they are of uniform gradient, although gently sloping valleys will store more water for a given impounding height than steeper sloping valleys. Various combinations of dam height, fill volume and water-to-earth ratio can be considered for any given capacity (or surface water area, by calculation) and the most desirable combination found. The Appendix shows a typical example of uses for both water abstraction and amenity purposes where the criteria are the available storage and surface area respectively.

Although Appendix B has been prepared as a guide to determine the capacity and maximum depth of water for a reservoir, it should be noted that reservoir sites are rarely uniformly sloping and so the methods in these appendices must be considered as indicative only. The final estimate of volume, once more definite arrangements have been formulated, should be assessed accurately, following levelling of the proposed site.

Determination of the amount of storage or surface water area required by the undertaker may depend on several factors that are outside the scope of this guide. Irrigation or fire fighting needs, for example, will require specified rates of flow at certain times of the year, whilst fish farming may require a specific surface area. Other usages, particularly sporting or recreation, will necessitate minimum areas of open water and controls on the depth and extent of shallow areas. Appendix C details the general requirements for various recreational and sporting purposes. Many reservoirs will be limited by existing topographical, environmental or other constraints, which will effectively dictate the maximum size or depth of water, whilst reservoirs with substantial areas deeper than about two metres may have little environmental value over their more shallow boundaries.

Where specific information is required, specialist advice must be sought in the relevant field, e.g. ADAS or from one of the private sector consultancies through the British Institute of Agricultural Consultants (for irrigation advice), the British Association for Shooting and Conservation (for guidance on shooting and allied interests, etc). The contact addresses of the relevant bodies and organisations are given in Appendix D.

2.4 WATER SOURCES

The source of water required to supply a reservoir depends upon the type of reservoir. The principal water sources for the main types of reservoir are as follows:

- Non-impounding reservoir

Non-impounding reservoirs have no catchment and water entering the reservoir naturally is limited to rainfall onto the reservoir itself. Therefore water to fill the reservoir must be obtained either by abstraction from groundwater sources, or more usually from a nearby stream. It may be possible to divert the water flow from a watercourse by gravity along a pipeline or open channel and save the cost of pumping. Any weirs or similar structures, which are required to allow the level to be raised in a stream to let the water be abstracted or diverted, would require an impounding licence as well as an abstraction licence.

The adequacy of the supply, in terms of rate and volume of water flow, must be established and proven. Sufficient storage must be provided to allow for times of drought when demand for the stored water is likely to be greatest and replenishment of water into the reservoir is least. Where the requirement of a non-impounding reservoir is to maintain a certain surface area, the source of supply needs to be sufficient to balance the losses from the reservoir after the first filling.

- Impounding reservoir

The majority of the inflow into an impounding reservoir will be the flow in the watercourse with a small contribution from direct valleyside runoff into the reservoir and rainfall directly into the reservoir.

The quantity and rate of inflow into an impounding reservoir is affected by climatic and topographic influences and the effects of land use in the catchment. Any watercourse with a perennial flow should be adequate to fill and maintain the required water area for a reservoir within the scope of this guide.

Flow from an adjacent river catchment can be added via pipelines or open channels or by means of pumping. Water can also be extracted from groundwater sources to supplement the direct inflow or, occasionally, by constructing near horizontal adits into the surrounding ground to intercept the groundwater table. Such methods are normally only suitable to the larger reservoirs and are not likely to be viable for the smaller sizes.

Runoff from local ditches, land drains and drainage systems may be sufficient to fill a very small reservoir. Drainage from roofs, glasshouses, paved areas etc., may also be utilised in appropriate instances although such sources would normally form only a contributory part of the water entering a reservoir.

Section 6 discusses the means of assessing the actual amount of water available for storage and the water losses from a reservoir. The quality of the water should be considered to ensure, where relevant, that it is fit for the intended purpose.

2.5 AVAILABLE LAND

It is important that there is sufficient land available for a reservoir of the required storage capacity or surface area. This should also include consideration of:

- suitable topography
- the existing land use and its ecological value
- ground conditions and available construction materials
- proximity of the stored water to its intended use, if appropriate
- ease of access to reservoir
- liability of land to flooding
- proximity of reservoir to land which may adversely affect the intended use of the reservoir, e.g. nitrate-rich runoff from fertilised fields
- land ownership and rights of way
- effects on adjacent structures and natural features
- buried or other services
- archaeological or environmental constraints or areas subject to special planning controls and land use management policies.

2.6 RESERVOIR SITUATION

The situation of the reservoir relative to existing or proposed features outside the immediate reservoir area should be considered at an early stage. These may influence the design, construction or operation of the reservoir or be affected by the reservoir creation. These aspects are discussed throughout the guide and potential problems highlighted, but include:

- a small reservoir situated in the flood plain of a large river
- effects of other reservoirs or other water surface areas upstream of the proposed site

- two or more reservoirs in close proximity, or effectively in sequence, will require particular consideration of the arrangements to pass flood flows safely
- effects of areas of urban development or areas where rainfall is likely to runoff rapidly upstream of the proposed site. These would include extensive areas of hardstanding, greenhouses, plastic sheeting (may be present seasonally)
- influence of the reservoir creation on existing or proposed developments, including reservoirs or water bodies, downstream of the proposed location.

It should not be forgotten that the construction of a reservoir may cause problems for an owner or resident adjacent to or downstream of a reservoir. Depending on the problem, this may put the owner or resident in a position to bring an action for nuisance.

2.7 TOPOGRAPHY AND LAND USE

The initial selection of a potential site for a reservoir will often depend on knowledge of the area, backed up by a study of available information and maps. Aerial photographs may offer some assistance in selecting potential sites. These aspects are discussed in Section 8.

The following topographic features, if present, will help to maximise storage for an impounding reservoir:

- where one or more tributaries meet the main watercourse
- where a narrow section for the dam itself is present, with a wider basin for the reservoir
- where topographic features result in change in the gradient of the stream bed (from flatter to steeper downstream).

These features tend to occur where a harder band of material is present along a watercourse. This need not be a different foundation type, but a relatively narrow harder band within a softer material may be advantageous. Many of the very stiff clays of south east England tend to have such features.

Topographic features have less influence on the siting of a non- or partially-impounding reservoir, but siting the reservoir on gently sloping ground will tend to maximise the storage volume and minimise the excavation.

A potential site should be walked over and features checked which could influence the reservoir, including:

- topographic features
- visual geological features and ground conditions
- land use
- unstable or potentially unstable ground
- afforestation

- overflow positions
- sources of fill materials
- possible constraints on the top water level
- surface and sub-surface services
- archaeological features.

A simple level survey along the proposed dam centre-line (and up the centre of the valley for an impounding reservoir) should enable an initial assessment of capacity and/or water area as described in Section 2.3.

A more detailed levelling survey will be required to produce a contoured plan of the chosen reservoir and the immediate surrounding areas as discussed in Section 8. The plan can then be used to show the dam and the reservoir accurately, and enable estimation of the stored water volume and required fill quantities. Subsequent changes to the reservoir and records of maintenance and remedial works should also be recorded on the plan.

2.8 GROUND CONDITIONS

The ground conditions at the reservoir site will influence the feasibility and cost of the reservoir. A preliminary indication of the conditions can be gained by a study of available information (local geological maps and guides, information from adjacent areas, etc.) and a visual appraisal of potential foundation conditions, groundwater levels, existing slopes, areas of soft or unstable ground, changes in vegetation type, etc. These aspects will influence the design, construction and operation of a reservoir and are detailed extensively in Section 8.

Difficult conditions or unfavourable strata to a substantial depth require extensive site investigation and specialist advice and will necessitate more complex design and construction procedures. The cost of the extra investigation and works may approach the original cost of the reservoir and it may be prudent to abandon the site if there are other options.

Geomembranes or bentonite or imported clay linings to form an impermeable barrier are expensive, but may make an otherwise unsuitable site useable, if cost is not a deciding factor.

2.9 FLOODS AND WAVES

After intense rainfall, the runoff from the catchment will increase substantially. Flood flows entering the reservoir must be passed safely and discharged downstream of the embankment by means of a suitable overflow structure or structures. Failure to provide adequate arrangement to pass the flows will lead to discharge, threatening the integrity of the embankment. Floods during construction pose a major threat to the partly completed embankment and are discussed further in Section 7.5.

Section 7 details the method to assess the flood inflow into the reservoir and describes the procedure for determining the design flood to be passed over the dam. The design flood selected for sizing the overflow works is partly dependent on the downstream land use. Habitation or development in this area will result in larger design floods.

An impounding reservoir will usually require a main overflow of non-erodible material to pass low flood flows, with an auxiliary overflow to allow larger flows to pass on a much less frequent basis. The main overflow should be constructed of concrete or other material with a high resistance to long-term erosion, whilst the auxiliary will only be used on very infrequent occasions for short durations and will normally be an earth channel of shallow depth, protected by a dense cover of grass. The main overflow must be located on undisturbed ground; consequently its position and length will be limited by topographical constraints. Existing vegetation, particularly trees, or structures may also limit the position and extent of the main overflow. The auxiliary overflow can rarely be sited on original ground without substantial excavation, but in many instances it can be located on the embankment itself. This will require the extent of trees and woody shrubs on the slope downstream of the overflow to be limited and the crest to be maintained in a sufficiently level and satisfactory condition.

Particular flood problems are likely if the reservoir is to be created on a watercourse with a large or steep catchment. The size and rate of runoff will be large and so the design outflow will necessitate major works to safely pass the flood. This may result in overflow works forming the major part of the dam. Such dams are beyond the scope of this guide and require specialist advice.

Non-impounding reservoirs have a negligible natural inflow and so the overflow facilities can be much reduced compared to impounding reservoirs. There is a need, however, to provide some means of overflow, to control both the reservoir levels following very extensive rain and to allow for any over pumping or introduction of water into the reservoir.

Waves result from wind action on the reservoir and sufficient freeboard between flood surcharge level and the top of the embankment must be provided to prevent water overtopping the dam. Methods to determine the height of the freeboard are given in Section 7. The provision of shelter belts of trees may help reduce the wind action in addition to any environmental benefits.

2.10 AVAILABLE CONSTRUCTION MATERIALS

Locally available clays, sands and gravels are normally the most economical materials for the construction of the dam. Concrete will generally be too costly unless the storage is for hundreds rather than thousands of cubic metres and is likely to be unsympathetic in a rural environment.

Fill materials should be excavated from borrow pits ideally within the reservoir limits to avoid potentially undesirable excavation which would be visible after filling and to increase the storage capacity, especially in small reservoir sites. Borrow pits should increase storage, but avoid creating areas of dead water which are unavailable for use or unattractive to wildlife. They should also avoid causing instability of adjacent ground.

Available materials and their suitability for use as embankment fill are discussed in Sections 8, 9 and 10.

Measures are often necessary to select and rework locally available fill materials, although these may be able to be placed as excavated. Where potentially unsuitable materials are present and no suitable fill is available, it is normally more economic to incorporate such materials in preference to bringing in better material from locations off-site. Specialist advice is likely to be needed in such cases to produce a satisfactory and economic design.

2.11 PLANNING, LICENCES AND CONSENTS

Constructing a reservoir normally requires planning approval, although the situation with regard to their construction as part of agricultural operations on farmland is less clear. All works on farmland, however, have to be notified to the local planning authority (LPA) and any change from previous agricultural use would require permission. Certain areas of countryside are subject to special planning controls and land use management policies and permission for reservoir creation is not likely to be given (e.g. National Parks, Areas of Outstanding Natural Beauty, areas with archaeological or environmental constraints, SSSI, etc.).

A planning consent is likely to contain specific conditions attached to the construction, which may put constraints on its use and also affect the feasibility. Certain environmental details, e.g. tree planting, together with access and road safety requirements, may also be included.

Early consultations with the Local Planning Authority are recommended prior to any expenditure on survey and ground investigation. A site meeting is often helpful to avoid misunderstandings.

Licences and consents are also required from the Relevant Licensing Authorities (RLAs) before any work is commenced and early consultations with site meetings are often helpful. The RLA is the Environment Agency in England and Wales (previously the NRA until April 1996). The RLA in Scotland is the Scottish Environment Protection Agency and in Northern Ireland, the Environment and National Heritage Agency.

A water abstraction licence must be obtained whether the water is used directly or stored in a non-impounding reservoir. The issue of a licence can depend on the effect of a reservoir scheme on existing licence holders and water users, environmental factors, river quality and the existing uses of the water. A licence, therefore, should offer protection against future loss of water to others. The licence will contain conditions, which may require abstraction to cease or to be reduced under certain drought conditions; normally these will be the times that water demand is greatest and sufficient storage must be provided. It should be noted that the issue of an abstraction licence is no guarantee of the quantity or quality of the water source.

An impounding licence is required where any works are proposed which would impound, impede or divert the flow in a watercourse (defined as a river, stream or other watercourses, whether natural or artificial and whether tidal or not). Impounding also includes the construction of a weir across the line of a watercourse to raise the water level locally to allow abstraction upstream of the weir. It is also likely that the RLAs will require a specified minimum flow of water to be maintained in the watercourse downstream of the weir or dam at all times. Restrictions on filling are likely to be included in the licence which may limit both the time of year when filling can take place and the time when flows at a specific position elsewhere on the watercourse are below a minimum stated amount.

The construction of any structure which affects the flow of water or the drainage capabilities of a watercourse requires a consent for works. The consent by RLAs is generally given solely on the basis of river shape, flow, environment and flood defence criteria and does not consider the design and suitability of the proposed works other than its impact on flows and the effects on the watercourse and flood plain. The RLAs may have a right of access along the banks of the watercourse and so may seek to limit development and vegetation within a certain distance of the bank of the watercourse and require channels to be bridged to allow access.

More detailed aspects of these various planning approvals and licences and consents are given in Section 3. Comments on other consents, legislation and legal responsibilities which may affect the feasibility or the purpose of the proposed reservoir are also included in this section. The various statutory controls are liable to change and reference should be made to the new statutory legislation and European Directives which may appear after the preparation of this guide. It should be appreciated that statutory controls apply whether a reservoir is to be newly constructed, extended from an existing feature, adopted, extended, brought back into use, changed in use, abandoned or whatever.

2.12 ENVIRONMENTAL CONSTRAINTS AND CONSIDERATIONS

A reservoir cannot be located without adequate regard for its impact on the surrounding area. Figure 2.1 shows how reservoirs have frequently been constructed in the past, with no real regard for their situation or imposition into the surrounding area. Although some loss of the existing environment must be accepted, there is considerable opportunity for potential landscape enhancement and the creation of supplementary habitats for flora and fauna. Section 4 details the typical environmental aspects that need to be considered when planning the creation of a reservoir, together with possible mitigating effects. Not all the points mentioned will be relevant for every reservoir proposed, but many schemes would benefit if they are considered during the early stages. Where tree planting is to extend onto the embankment, the types and locations must be considered carefully as detailed in Section 9.2.8.

2.13 COSTS

The costs of any proposed reservoir clearly affect both the feasibility and economic viability of the scheme, as discussed in Section 2.1. Estimates should be prepared for a number of reservoir locations and sizes. Costs for alternatives may be useful, if, for any reason, the initially selected site becomes unavailable or is found to be unsuitable.

When assessing the economics of reservoir construction, account should be taken of all survey and site investigation costs and fees for specialist advisers in addition to the actual design and construction costs. The costs of a reservoir do not stop following construction; continuing expenditure on monitoring and maintenance work keeps the reservoir safe. It is also prudent to make provision for remedial works during the life of the reservoir. Section 5 gives further information on costs.

Grants may be available from sources such as the Countryside Commission, Local Authorities or English Nature if the reservoir is to be created for landscaping or conservation.

2.14 INSPECTION, MONITORING AND MAINTENANCE CONSIDERATIONS

Early consideration of these aspects allows the long-term costs of monitoring and maintenance to be reduced as well as maximising many features of the design and environmental aspects. These considerations are discussed in Sections 12 and 13.

They detail good practice that should minimise both plant and labour input costs and yet are sympathetic to the likely rural nature of reservoirs within the scope of this guide. In particular, careful planning, planting and maintenance of vegetation can help minimise other construction and maintenance costs in addition to enhancing the general environmental.

2.15 ALTERATIONS/EXTENSIONS TO EXISTING RESERVOIRS

An existing reservoir may require alteration and/or extension to fulfil a new purpose or an intensification of use. This may require the volume or surface area to change or some other feature of the reservoir or dam to be amended.

If the required changes are limited to the reservoir itself, this should not prove particularly difficult but potential changes must be considered carefully. An increase in volume may also bring the reservoir within the scope of the Reservoirs Act 1975, as described briefly in Section 2.3.

Changes to the dam are potentially more complex. An existing embankment dam may not be adequate to retain an increased head of water. If detailed records exist, the design of the alterations may be relatively straightforward, but a conservative approach should be taken. Where an old dam is to be altered, particular care has to be taken. Any works which involve an increase in water level will not be straightforward and are likely to fall outside the scope of this guide.

2.16 SPECIALIST ADVICE

A guide of this nature can only present a straightforward and conservative approach to reservoir creation and give guidance based on reasonable site conditions. No two sites are identical and every location must be assessed individually. The guide identifies potentially difficult conditions; where the situation is complex or uncertain, it is essential that competent specialist advice is sought. This may involve legal matters, but is more likely to include specific technical aspects associated with the design, construction and operation of the dam and allied works. In in many instances, an experienced and competent water engineer must be consulted and in situations where reservoirs are on difficult sites specialist advice may be required throughout the design and construction process.

The undertaker or his advisers should be aware of the areas of potential problems and uncertainties that may arise from his particular proposals for small reservoir creation. It is the duty of the undertaker to ensure that significant or possibly significant matters are referred to a specialist adviser. This should minimise the risks and potential expenditure at all stages and maximise the benefits of the reservoir. This approach also reduces the risks of future action from other bodies.

Such expertise will clearly involve additional costs in the short term, but should lead to reduced construction and maintenance costs and help to avoid operating difficulties. Failure to seek advice when needed may result in an inappropriate design, leading to significant extra costs and delays during construction and increased maintenance difficulties and costs. Operating difficulties may also result. Ultimately, major remedial works or complete reconstruction may be necessary.

Specialist advice may also be necessary during construction if unforeseen conditions are revealed or as a result of external influences which necessitate changes to the design. The onset of problems during reservoir operation, particularly at the time of the first filling, is likely to require advice to control and deal with these matters and allow the reservoir to function as intended. The implications of changes to the reservoir or the surrounding areas may also need to be referred to a specialist adviser.

Throughout the creation and use of the reservoir, the risks and the resultant implications of embankment failure and the consequent liabilities for the escape of water must not be forgotten. Specialist advice at the appropriate stages during and following the onset of problems or changes will help to minimise these. A periodic visit and review of the reservoir by a specialist is also recommended.

The undertaker should ensure, as far as possible, that any adviser is competent and adequately experienced for the type and scale of advice needed. This may not be straightforward, but any potential advisers should be asked to provide details of their experience to-date and, if necessary, details of references which can be taken up. The names of private sector consultants specialising in dam creation may be obtained from the Association of Consulting Engineers, the Institution of Civil Engineers and the Institution of Water and Environmental Management or from other appropriate organisations. Useful contact addresses are given in Appendix D. These include the majority of the organisations representing the specialist advisers, to whom reference should be made in the first instance.

Where the reservoir is to be created for agricultural purposes, preliminary advice and information is readily available from the local ADAS office. Under the Agriculture Act 1986[3], the Minster (and hence ADAS adviser) have a responsibility to give balanced advice with due regard to both agricultural and environmental interests. Advice on a wide range of questions associated with reservoir creation and management and other relevant aspects (e.g. irrigation) can be obtained from the local ADAS office through the ADAS main address given in Appendix D. A cost would normally be associated with this service. Advice is also available from private sector consultants through the British Institute of Agricultural Consultants.

3 Licences and statutory requirements

3.1 INTRODUCTION

Reservoir creation and management is subject to legal, planning and operational constraints. Reservoirs within the scope of the Reservoir Act 1975[1], with a capacity of 25 000 m^3 of water above the natural level of any part of the adjacent land, have significant controls on their design, construction and maintenance as detailed in Appendix A. Smaller reservoirs below this capacity are not subject to the requirements of this Act, but cannot normally be built or extended without the appropriate licences and planning consents. All reservoirs will be subjected to statutory Health and Safety legislation during construction. Where the reservoir forms part of an undertaking which involves the reservoir as part of that undertaking (e.g. fish farming, irrigation supply, etc.), the Health and Safety legislation will continue to apply.

This section summarises the most important statutory controls that affect reservoir creation. These controls are embodied within separate Acts of Parliament and the Statutory Instruments (or Regulations) made under the provisions of those Acts. Acts of Parliament, Statutory Instruments and other legislative documents are sometimes couched in legal language that is difficult for the non-specialist to understand. Interpretation of the requirements is sometimes far from clear, even to specialists. While this section attempts to describe, in general terms, the main provisions of the relevant legislation for the non-specialist, and to provide a measure of interpretation, it cannot cover specific situations. The guidance given here should help non-specialists to get the most benefit from those discussions and to avoid what can be expensive legal pitfalls. Difficult or particularly contentious cases require appropriately qualified specialists to advise on the best approach including multiple or parallel planning applications. It should be stressed that the guidance on legislation given in this document is advisery as only the courts can interpret legislation with authority.

Licences and statutory requirements, together with any liabilities and responsibilities of the undertaker which might arise to other parties, may differ in Scotland and Northern Ireland. The application of certain Acts is detailed in the relevant section, but it is recommended that specific advice and clarification should be sought in these areas.

The following summary of the legislation is not exhaustive, but it is designed to provide guidance as to which authorities have a role in the planning, construction and management of reservoirs. Environmental considerations can also affect or be affected by the creation of a reservoir and these must be taken into account, together with any liabilities and responsibilities of the undertaker which might arise to other parties during the construction and operation of the reservoir. Licences and statutory requirements may change with time and this guide can only detail the legislation at the time of preparation; the current situation must be confirmed with the relevant authorities.

Planning controls and considerations in England and Wales are administered by the Local Planning Authority (LPA). This normally comprises the district planning department for planning approval although works involving mineral extraction are handled at a county level. This situation may change or vary locally as a result of the proposed local government restructuring. In Scotland, the LPA is currently the district council. In Northern Ireland, the DoE are the LPA and consult with the district councils. In England and Wales, the local office of the Environment Agency should currently be contacted for the issue of an impounding licence and consent for works in a watercourse, together with abstraction and discharge licences if required. This body may be superseded by another or a replacement statutory authority controlling these licences and consents in due course and thus the types and details may change in the future. In Scotland the situation is different as there are no main rivers and the riparian owners have greater freedom than in England and Wales to carry out any works. Where required, the appropriate licences can be obtained from the River Purification Board. Fishing rights are often separated from the riparian owners in Scotland, but the appropriate agreements must still be obtained. In Northern Ireland, the Department of Environment deals with licensing excepting drainage matters, which are handled by the Department of Agriculture. If there is any doubt as to whether any particular proposal requires a licence or permission or is affected by any other statutory provision or riparian owner, advice must be taken from the relevant authority. These authorities are referred to as the 'Relevant Licensing Authorities (RLAs)' throughout this guide.

Appendix D lists the addresses of the main offices of each of the current RLAs. The LPA can be contacted through the local district council offices. An early approach is recommended, accompanied, where possible, with a plan of the proposed location, grid reference and outline proposals which will allow the other parties to assess more readily the impact of the proposed reservoir. A site meeting at an early stage is often helpful to avoid misunderstandings and take account of any particular problems or aspects that may concern the undertaker or the relevant LPA or RLA.

3.2 PLANNING CONSIDERATIONS

Reservoir creation is normally subject to the laws and regulations affecting the siting, design and operation of engineering works and the development of land. These controls are concerned with the effect that a development will have on the area, particularly for neighbours and adjacent land, and may vary depending on the size, location and purpose of the reservoir. The creation of a very small water area, such as a garden pond, for example, is not normally subject to any planning constraints.

The Town and Country Planning Act 1990[4] sets up the framework for planning control by the LPA in England and Wales, covering all aspects of development, including reservoirs and such diverse issues as quarrying and listed buildings consent. In principle, all development requires planning permission. Certain types of development, however, are categorised as 'permitted development' for which individual planning permission is unnecessary to lessen the burden on the LPA and to avoid unwarranted delays. Some agricultural development is also given 'permitted development rights' (PDRs) to encourage agricultural production. The rules covering permitted development are set out in the Town and Country Planning General Development Order 1988[5]. Various amendments have modified the Order, but Amendment Nos 2[6] and 3[7] have significantly altered and restricted the PDRs that applied previously to agricultural development. Box 3.1 details the main features of this Order and its amendments. It is strongly recommended that no assumption be made with regard to PDRs without first taking specialist advice and/or consulting the LPA.

Whilst the need for planning permission for non-agricultural operations is readily apparent, the situation regarding planning approval of a reservoir for agricultural use is not entirely clear and appears to be open to interpretation by each LPA. The controls and their interpretation may also change with time. Where there is doubt whether a reservoir requires permission, the LPA may be requested to give a formal comment, known as a 'Section 64 Determination' to rule on whether planning permission is required. An informal discussion at an early stage, to discuss the outline proposals of the planned reservoir appears sensible and should enable the planner's views to be obtained before potentially unproductive work is carried out. In the past, some landscaping and conservation ponds on farmland have been permitted as permitted development, but these have been small (less than 0.25 ha) and formed by machinery available on the farm and hence have been classified as 'agricultural'. The change of use of previous agricultural land would constitute development and permission would be needed. Agricultural usage of waters might include irrigation, the supply of water for livestock, water for fire fighting and possibly fish farming. Any proposed usage other than these would constitute a change of use and thus be subject to planning procedures. The construction of a reservoir for non-agricultural use would also clearly require permission. Where the proposal would result in the loss of 20 hectares or more of higher grade farmland (Grade 1 to 3A), MAFF must also be consulted by the LPA.

The planning situation and controls in Scotland and Northern Ireland are likely to vary. The specific requirements should be ascertained by discussions with the LPA and specialist advisers.

The planning approval is likely to contain specific conditions attached to the construction of the reservoir and may put constraints on its use, such as limiting the number of rods allowed at any time on a fishing lake. Environmental constraints are increasing and these may require a landscaping plan for the reservoir to be approved by the LPA prior to construction. A time limit may also be put on the commencement of the reservoir, whilst additional requirements on access and road safety matters may be included.

Box 3.1 The Town and Country Planning General Development Order 1988

Permitted development

Agricultural works (including reservoir construction) is deemed a permitted development (i.e. not requiring specific planning permission) so long as specified conditions, including the following, relevant to reservoir creation, apply:

- The works must be 'reasonably necessary for the purposes of agriculture within that unit'. An agricultural unit is defined in the GDO as land which is occupied as a unit for the purposes of agriculture. Generally, farm holdings, or units, of less than five hectares are subject to full development control. In addition, an individual parcel of land below one hectare which forms part of a unit of five hectares or more, but is separated from it (for example, by a public road or land in different ownership), is also subject to full development control. A farm may consist of several non-contiguous parcels of land, each attracting PDRs. It is unclear whether PDRs apply to works carried out on one parcel of land to meet the needs of others, even though the parcels may be part of the same farm holding. For example, an irrigation reservoir on one parcel of land may not qualify as a permitted development if it was intended to store water for use on a non-contiguous parcel of land. In practice, some LPAs accept that as long as at least some of the output relates to the parcel of land on which it is sited, PDRs still apply.

- Works on agricultural units of less than five hectares have only PDRs in respect of the repair or dredging of reservoirs and ponds for agricultural purposes under the GDO.

- The works should occupy an area not exceeding 465 m^2. Any development, of which any part is within 90 m of the proposed development, carried out within the preceding two years counts towards the 465 m^2 limit. The limit includes not only the works themselves, but any associated hardstanding areas, tracks, roads or other allied works.

- The works must be further than 25 m from a metalled road.

- In the case of slurry storage, the works must be further than 400 m from the curtilage of a 'protected' building. (A 'protected' building includes buildings normally occupied by people but excludes residences occupied by farmworkers, whether or not they are employed on the agricultural unit for which the development is planned).

- The works must be less than 12 metres in height.

- Development for fish farming in certain areas of landscape interest is not a PDR under Amendment No 3 to the GDO[7].

- The works should not be within 2 km of a SSSI.

Prior approval

PDRs cannot be exercised when excavation or engineering operations are proposed which are reasonably necessary for agricultural purposes within the unit unless the farmer or developer has applied to the LPA for a determination as to whether their prior approval is required in respect of certain details of the development, design and appearance.

A written description of the siting, design and external appearance of the building or works must be submitted for the LPA's prior approval. (This requirement applied previously only in the case of National Parks and certain other designated areas.) Work on site is not allowed to start unless or until:

- the applicant receives notice from the LPA that prior approval is not required,
- where prior approval is required, the giving of such approval,
- the expiry of 28 days following the application receipt date if the LPA has not communicated the need or otherwise for prior approval to the applicant.

It should be noted that there is a restriction on prior notification for excavations in excess of 0.5 hectares.

Differing planning legislation is in force in Scotland and Northern Ireland.

The LPA has a duty to confer with other relevant bodies and to make the application available for public inspection. Objections to the application must be reviewed by the LPA and, where appropriate, taken into account when determining the application. Other proposed developments which might affect the reservoir, or vice versa, must be taken into account by the LPA.

Reservoirs in certain areas, as detailed in Box 3.2, will be subject to special constraints and an early approach to the LPA is necessary. Other bodies will be involved in many instances and this will tend to extend the planning stage of any proposal. The controls on Scotland and Northern Ireland may differ in many respects from those in England and Wales and the LPA should be consulted for the detailed requirements.

Box 3.2 Countryside protection

Special planning controls

Certain areas of countryside in England and Wales are subject to special planning controls and land use management policies. These include:

- national parks – (National Parks and Access to the Countryside Act 1949[8]),
- national nature reserves – (notified by English Nature under the Wildlife and Countryside Act 1981[9]),
- sites of special scientific interest – (English Nature will have provided the landowners and occupiers with a list of potentially damaging operations. Prior to any of these being carried out, English Nature must be notified in writing so that the effect on the scientific interest can be assessed),
- country parks – (Countryside Act 1988[10]),
- local nature resources – (designated in County Development Plans),
- areas of outstanding natural beauty – (designated by Countryside Commission and responsibility of LPA),
- natural scenic areas – (applicable to Scotland),
- areas of high landscape value,
- protected woodlands,
- ancient woodland – (appear on pre AD 1600 maps as woodland),
- environmentally sensitive areas,
- listed buildings and scheduled ancient monuments,
- conservation areas.

The effects of a development on the environment now receive considerably more comment and study than in the past. A recent EEC directive[11] that has been incorporated into the Planning Acts aims to ensure that permission for larger development projects is not granted until its effects on the environment have been fully assessed. The term 'environmental assessment' describes the technique and process by which information about the environmental effects is collected from many sources by the developer and taken into account by the LPA in forming their judgement on whether the development should go ahead. The technique and process are described in Box 3.3. Whilst this procedure is intended primarily for larger projects, the guidelines could be taken to be applicable to any structure retaining water on a long-term basis and, in theory, the LPA could request that an environmental assessment be provided for a small reservoir. No definite comments on the likely need for environmental assessment can be made, but it is considered that most small reservoirs which fall within the scope of this guide should not be subject to this procedure. The various matters that need to be considered in assessing the influence of the reservoir construction on the environment and the possible enhancement and mitigation measures are detailed in Sections 4.2 to 4.4.

Additional planning constraints and requirements are applied if the removal of minerals or the deposition of waste is involved, as described in Box 3.4, and require consultation with the LPA, and possibly the RLA, at an early stage.

Where borrow areas for fill, or features, such as access roads, form part of the proposed works, these must be included in any planning application.

An existing reservoir may be listed and/or may require planning permission if changes to the water area or top water level are envisaged. The agreement of the RLA is also likely to be needed. Dismantling of the dam (see Section 14.8) will also require a similar approach. An existing water feature may also be subject to planning controls or statutory protection which may limit the extent of any proposed changes.

Box 3.3 Environmental assessment

Statutory controls and regulations

The European Community Directive on 'The Assessment of the Effects of Certain Public and Private Projects on the Environment' (85/337/EEC)[11] came into effect in July 1985 and was incorporated into consent procedures in the UK as the Town and Country Planning (Assessment of Environmental Effects) Regulations 1988[12] and its subsequent amendments and allied regulations. The Regulations are explained and the intended policy is detailed in the DoE Circular 15/88[13].

Differing planning legislation is in force in Scotland and Northern Ireland.

Application of Regulations

The Regulations are designed to ensure environmental assessment is carried out before development consent is granted for certain types of major projects which are judged likely to have significant environmental effects. A list of major projects where environmental assessment is required is included in the DoE/Welsh Office publication 'Environmental Assessment; A Guide to the Procedures'[14] and includes 'Dams and other installations designed to hold water or store it on a long-term basis'. Whilst the Regulations were issued to cover major projects, an LPA may request an assessment from the applicant for small reservoirs. One LPA in southern England has stated that it considers an assessment could be required for reservoir creation of any reasonable size. Loss of agricultural land in excess of 20 hectares of Grade 1, 2 or 3a is also included, as well as developments within 250 metres of land which has been used for waste disposal in certain circumstances in the past 30 years. Salmon hatcheries or salmon rearing installations beyond a specified size are also subject to the Directive. Development in areas subject to special countryside protection may also be included.

Where developments outside the reservoir area form part of the proposed works (e.g. possible borrow area for fill materials, access, etc.), these will need to be included within the scope of any Environmental Assessment.

Box 3.4 Minerals and waste

Removal or deposition of material

Further restrictions apply to the construction of water areas if the removal of minerals or the deposit of waste is involved. The minerals cannot be moved off the land (but see 'Fish Farming') nor waste brought onto the land without express separate planning permission. This may be an important consideration in the construction of large irrigation reservoirs in sand and gravel bearing strata or where imported material is needed to replace unsuitable material for the construction of the dam.

The main steps in the planning process are summarised in Figure 3.1.

Obtaining planning permission is often a difficult and time-consuming activity. A period of eight weeks from the receipt of the submitted application for determination is normal, but this may be extended by agreement in writing. Different LPAs adopt individual procedures and time scales may extend to several months before a decision is reached. Where special planning controls are in place, the time scale may be extended considerably when various other parties, such as English Nature, are consulted. In many instances, planning permission for reservoirs which fall within one of the areas described in Box 3.2 may be virtually impossible to obtain. Development undertaken in breach of the conditions imposed either by the Act, the PDRs or the LPA, may be the subject of enforcement action. Developments not considered by the LPA to fall within the scope of PDR are subject to normal planning procedures. Detailed advice on applying for planning permission in England and Wales can be found in References 15 and 16.

A fee is chargeable for all planning applications which will vary with the location size and type of proposed development. LPAs also have a policy of charging for change of use, and some for engineering; usually based on an area rate.

3.3 LICENCES AND CONSENTS

Permission is required from the local office of the RLA for the following in England and Wales:

- impounding licence
- water abstraction licence
- consent for works
- land drainage consent
- discharge consent.

These licences and consents are required under the Water Resources Act 1991(17), together with the Water Resources (Licences) Regulations 1965(18) in the case of a water abstraction licence. The Land Drainage Act 1991(19), in addition to the Water Resources Act 1991, governs the duties of the RLAs in relation to flood drainage in England and Wales.

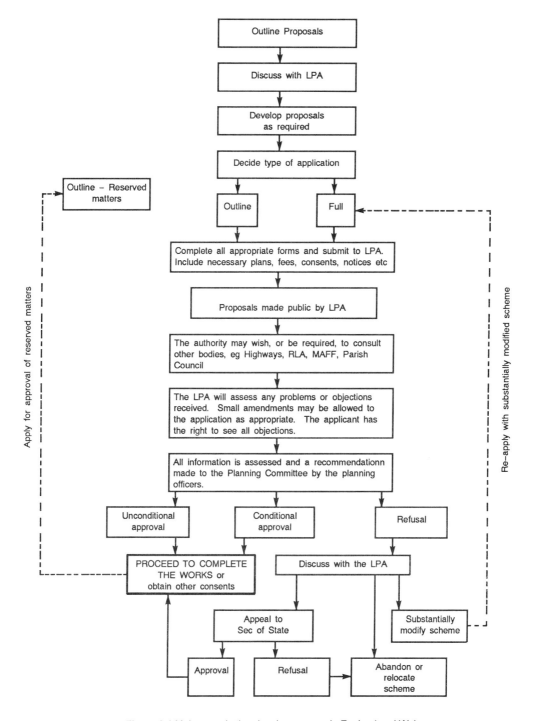

Figure 3.1 Main steps in the planning process in England and Wales

A licence is required before constructing or allowing any works for impounding, impeding or diverting the flow in a watercourse (defined as a river, stream or other water course whether natural or artificial and whether tidal or not). Impounding also includes the construction of a weir across the line of a watercourse to raise the water level locally to allow abstraction upstream of the weir. It also includes the construction of a dam across the line of the watercourse for the purpose of creating an impounding reservoir. A non-impounding reservoir which involves works for diverting the flow from a watercourse to feed the reservoir will also require an impounding licence. Box 3.5 describes the various aspects relating to the issue of an impounding licence.

Box 3.5 Impounding licence

Method of application

The procedure to apply for an impounding licence is similar to that for an abstraction licence.

Constraints

The licence specifies the place where the impounding can take place; the retention level of the impounded water and the residual (compensation) flow of water, if any, which must be maintained in the watercourse downstream of the weir or dam. The RLA will normally stipulate that the initial filling and any subsequent refilling is restricted to certain times of the year (normally winter) and may limit filling when flows as measured at certain locations elsewhere on the drainage system fall below certain specified levels. It is generally preferable that a reservoir is filled over a long period at a slow rate and restrictions may also be included in the licence in this respect. The RLA may require the installation of a measuring device to monitor inflow and discharge during impounding and also subsequently if water is to be abstracted.

A water abstraction licence is required before any abstraction occurs from a body of water, watercourse, well, borehole, surface excavation or groundwater source. Box 3.6 describes the means of obtaining a licence and the various aspects and constraints affecting the issue of a licence.

If a licence is obtained to abstract water from any source and that water is then stored in an offstream reservoir, then any subsequent abstraction from the reservoir is not subject to control, i.e., it can be used as and when needed. An exception may occur, however, under certain Drought Order provisions. As water is usually abstracted during the winter months for storage in an offstream reservoir, the possibility of an imposed reduction in the licensed quantities by the RLA during the winter months is remote and the licence holder is reasonably assured of an adequate supply.

The situation is different in the case of water stored in an impounding reservoir. The impounding licence permits the storage of a specified amount of water but this is not a licence to abstract water from the reservoir. Abstraction from an impounding reservoir must be licensed separately and will usually take place in the summer months; it should be recognised in periods of drought that there could be a possibility of an imposed reduction in the licensed quantities. If no water is abstracted from a reservoir, i.e. the reservoir is an amenity feature, fishing lake, etc., then an abstraction licence is not required.

Box 3.6 Water abstraction licence

Method of application

Applications can be made only by the occupier of land adjacent to the water or by an individual with right of access to the land. The making and determination of an application is a legal process and there are certain requirements which must be fulfilled by the applicant. A Notice must be published in the London Gazette and for two successive weeks in a newspaper with a circulation in the locality of the proposed abstraction. The notice gives information on the proposed abstraction and a place within the locality where a copy of the application may be viewed by the public. Any member of the public may make written representation to the RLA who must consider all the various points before deciding whether or not to grant the licence; this will also depend on the availability of water at the particular location.

Constraints

The RLA must not normally grant a licence which would worsen the conditions of existing licence holders or persons with an existing right to the water, whether surface flows or groundwater. A licence, therefore, should give protection against future loss of the waters to others. The RLA must take account of environmental factors, river quality objective, and the existing uses of the water.

Where there is considerable demand or a severe drought in some area, a licence may be issued which requires abstraction to cease when flows or levels fall below a predetermined value. The licence will also normally specify the point of abstraction; the land on which the water is to be used; the purpose for which the water is to be used; the quantity of water which can be abstracted during specified periods; the means of abstraction and the basis of measuring or assessing the amount of water actually abstracted. In the event of a severe drought, the RLA may also serve notices in certain instances, reducing the amount of water that may be abstracted for a specific time period. Non-impounding reservoirs which are for retaining water for use will require to be filled from an adjacent stream, water source or borehole. If they are created without impeding or obstructing the flow in a stream or watercourse they will require only an abstraction licence, possibly requiring a residual flow at the intake, but no impounding licence. It should be noted that the issue of an abstraction licence is no guarantee of the quantity or the quality of the water source.

A consent for works is required for the erection, construction or alteration of any of the following, as described in Box 3.7:

- a dam or other similar obstruction to flow in a watercourse
- a culvert that would be likely to affect the flow in any watercourse
- any other structure in, over, or under, a watercourse
- any structure or building on, or near (usually within eight metres), the bank of a watercourse
- any structure (including earthworks) in the flood plain of a main river unless the planning permission is in existence.

Where the watercourse lies within the area of an Internal Drainage Board and is not a main river, the RLA is then the Internal Drainage Board for issue of the consent. Information on the status of a watercourse or an Internal Drainage Board can be obtained from the RLA.

Box 3.7 Consent for works

Method of application

Application for the consent on the appropriate form must be accompanied by a plan of the proposed reservoir and a detailed drawing at a larger scale showing the proposed works. Sufficient drawings are required to show the proposed works adequately. The application and drawings must be submitted in duplicate. Calculations may be requested in some instances. A prior discussion with RLA staff before submission of the application is recommended to ensure the proper details have been included. A guide for the submission of the application is usually available from the RLA, detailing the required scales of drawings and matters to be included in the application.

Constraints

The consent will briefly describe the proposed reservoir and its intended purpose and relate these to the drawings showing the works. One set of the drawings will normally be returned together with a series of notices to be submitted at certain key times. The consent by the RLA is given solely on the river shape, flow, environment and flood defence criteria and does not give approval to the design and suitability of the proposed works other than its impact on flows and effects on the watercourse or floodplain. In practice, however, the RLA would be likely to comment on any feature which it considers unsatisfactory or poor practice. The RLA may have a right of access along the banks of watercourses for maintenance purposes and thus may seek to limit development or vegetation within a certain distance from the watercourse and require channels to be bridged to allow free access.

Earthworks in or around the reservoir may be carried out as part of the works for landscaping purposes or deepening parts of the reservoir. Similarly, surplus or suitable material may also be deposited around or in the reservoir or possibly downstream of the embankment. The RLA will wish to ensure that this is not likely to affect the drainage or flood storage capacity; any plans for materials deposition should be clearly shown in the proposals.

Further consents may be necessary from the RLA, or Internal Drainage Board if relevant, in respect of Flood Defence Byelaws, or under the Salmon and Freshwater Fisheries Act 1975[20] in respect of fishpasses. Existing drainage consents may be affected by the reservoir construction and the consent for works, and the licences may have constraints, or possibly be refused, on these grounds. Advice should be obtained from the RLA in the first instance. Other consents are also required for fisheries as subsequently discussed.

Compensation flow and outflow through the overflow works from an impounding reservoir do not need a discharge consent, but other discharges from both impounding or non-impounding reservoirs are likely to require the appropriate consent which may contain conditions on quantity, rate, water quality etc. A discharge consent will be required under pollution control aspects of the Water Resources Act 1991[17] if any discharge of water (including reservoir emptying) is to be carried out to a watercourse or to an underground strata. Further comments on pollution are given in Section 3.5.3

The main steps in the licensing and consent process are summarised in Figure 3.2.

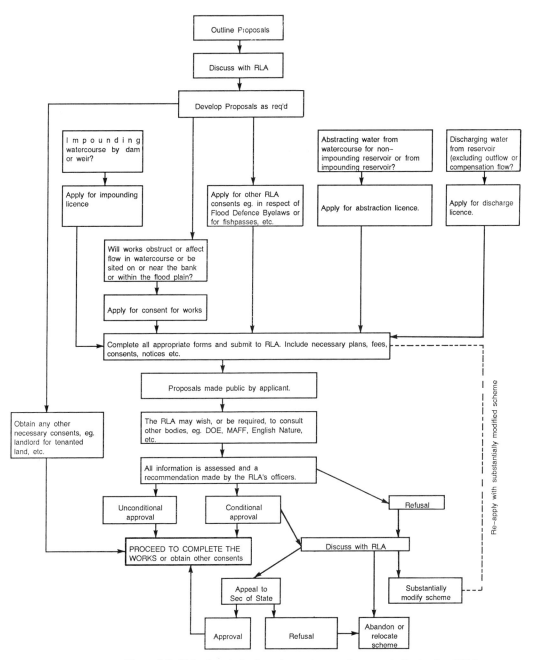

Figure 3.2 Main steps in the licensing and consent process in England and Wales

The specific requirements for the individual licences and consents from the RLA are detailed within Boxes 3.5 to 3.7, but some comments may be made regarding other more general aspects of obtaining consents. An application fee is normally charged in relation to the licences and consents required for each structure. These fees are payable towards the cost of examining the proposals and may vary. Upon receipt of a full and proper application, the RLA has a specified period in which to grant or refuse a proposal. They are obliged to grant consent to reasonable proposals which will not endanger the flow of the watercourse or possibly cause any additional risk of flooding. They also have a duty under the Water Resources Act 1991[17] and the Land Drainage Act 1991[19] to refuse consent if the proposals might prove detrimental to the environment and to further and enhance conservation interests. The former Act also requires due regard to be given to the interests of fisheries which may include consideration of the water quality for fish and the effects of the development on existing fish in or using the watercourse. If notification either refusing or consenting to the proposals is not received within the specified period from the receipt of a full and proper application, then the RLA are deemed to have given consent. A right to appeal exists if consent is believed to have been withheld unreasonably and normally an arbitrator will be appointed to settle the matter if agreement cannot be reached. If arbitration is not accepted by the applicant, the matter will be referred to and determined by the relevant Minister.

Where works are carried out without formal written consent the RLA has the power to serve a notice requiring the nuisance to be abated within a specified time. If the undertaker fails to comply with the notice, the matter can be taken to court and a fine imposed. Any further failure to comply may result in an additional fine which can be levied on a daily basis. Under the Land Drainage Act 1991 (19), the RLA may remove, alter or demolish any unauthorised work and recover the costs incurred.

Documentary evidence is required before works are carried out if the consent or approval of a third party is required.

Where a reservoir is to be created by a tenant on a farm, The Agriculture Act 1986[3] requires the consent of the landlord to be obtained for the construction. This applies to any form of water area for whatever purpose. Clearly permission would normally be required for any tenanted property before a reservoir could be constructed. Advice on the law and practice involving tenancies is available primarily from Chartered Surveyors (Rural Practice Division).

Various water byelaws may be relevant to reservoir construction. In particular, Water Supply Byelaw 47[21] requires that every pond, fountain or pool, the capacity of which exceeds 10 m^3 and which is filled or supplied with water from a mains supply, shall have an impervious lining or membrane to prevent the leakage or seepage of water.

The requirements in Scotland and Northern Ireland vary and specific advice should be sought from the RLA.

3.4 FISHERIES

Subject to certain controls detailed in Box 3.8, the construction of reservoirs for the rearing of fish in England and Wales is given permitted development rights under Town and Country Planning Act 1990[4].

Box 3.8 Fish farming

Constraints

The business must be registered as a fish farm with the Ministry of Agriculture, Fisheries and Food under the Diseases of Fish Act 1983[22] and the Registration of Fish Farming and Shellfish Farming Business Order 1985[23]. Notification to the Health and Safety Executive may also be required. The following constraints apply to the creation of a reservoir for this purpose:

- The area of the site where operations would be carried out should not exceed 2 ha.
- No works must be carried out within 25 m of any metalled road.
- If minerals (e.g. sand and gravel) have been extracted, no excavation must be deeper than 2.5 m nor greater in area than 2 ha (including other excavations on the land during the preceding two years).
- Certain other constraints may apply which may exclude construction of an initial pond or a development for a new business.
- The development will not be considered as a PDR in certain areas (Box 3.1) nor if the agricultural unit is less than 5 ha.

Under the provisions of the Salmon Act 1986[24], it may now be no longer necessary in certain instances for the proprietors of registered fish farms to obtain the consent of the RLA prior to stocking ponds once the original stocking of the fish farm is complete. This aspect should be clarified with the RLA at the time of the original stocking (which requires consent), but in all other cases, consent is required for all fish movements.

It is important that the reader checks the situation in Scotland and Northern Ireland.

Various Regulations apply to the movement and stocking of fish in a reservoir. In particular, Section 30 of the Salmon and Freshwater Fisheries Act 1975[20] requires written consent from the RLA for the introduction of fish or spawn in inland waters. This includes fish to be reared for food as well as ornamental purposes. A licence will also be required from MAFF to introduce into the wild any animal (including any fish, shellfish or spawn) not ordinarily resident in Great Britain or which is listed in Schedule 9 of the Wildlife and Countryside Act 1981[9]. Shellfish are not included within the Salmon and Freshwater Fisheries Act 1975.

Two other Acts are relevant to fisheries in inland waters; these are the Diseases of Fish Act 1937 as amended 1983[22], together with the Fish Health Regulations 1992[25] as amended, and the Theft Act 1968[26]. The purpose of this legislation is to prevent the introduction and spread of fish diseases; the latter Act deals with the taking or destruction of fish from private property. This applies to any kind of fish and it should be noted that 'taking' includes the catching of and attempting to catch fish with the intention of returning them to the water.

The establishment and operation of fish farming, fish movement and stocking and the rights associated with fishing can be seen to be potentially complex, as described in Boxes 3.9 and 3.10 and the requirements may differ in Scotland and Northern Ireland. Unless an undertaker wishes to establish fishing facilities at a reservoir for his individual use, it is suggested that specialist advice is sought. This is particularly necessary regarding the leasing or sale of fishing rights.

Box 3.9 The Salmon and Freshwater Fisheries Act 1975

> **Relevant provisions of the Acts**
>
> The Salmon and Freshwater Fisheries Act 1975[20] contains provisions to protect fish by prohibiting the use of certain implements and of explosives, poisons and electrical devices for taking fish. Such substances and devices may, however, be used for scientific purposes or for protecting or replacing stocks of fish, with the permission in writing of the RLA. The Act also lays down minimum close seasons for fishing although these may be altered or dispensed with by byelaw. It requires that those fishing for salmon, trout, freshwater fish and eels in England and Wales must hold a rod licence issued by the RLA. Any person or association entitled to an exclusive right of fishing in any inland waters may be granted a general licence to fish in those waters subject to any conditions agreed between the RLA and the licensee. The Act also prohibits the introduction of fish or spawn into an inland water unless written consent to do so is first obtained from the RLA.
>
> Under the terms of the Water Resources Act 1991[17], the RLA is empowered to make byelaws regulating fishing. These may, apart from altering or dispensing with the close seasons for fishing laid down in the 1975 Act, prohibit the taking of fish without lawful authority, regulate the size of fish that may be taken and where they may be taken, and specify the instruments and lures that may be used. Copies of local fisheries byelaws may be obtained from the relevant regional office of the RLA.
>
> It is important that the reader checks the situation in Scotland and Northern Ireland.

3.5 ENVIRONMENTAL ASPECTS

3.5.1 Species protection

The Wildlife and Countryside Act 1981[9] and the Water Resources Act 1991[18] place great emphasis on species protection and thus are likely to be of particular importance in reservoir creation. A summary of the former Act is given in Box 3.11.

Other legislation may be applicable in Scotland and Northern Ireland and specialist advice should be sought.

Box 3.10 Fishing rights

Summary of fishing rights

The right to fish in an inland water generally belongs to the owner or occupier of the land beneath the water. Usually this is the same as the owner of the bank of the watercourse (the riparian owner) and extends to the middle of the watercourse. The right of fishing may have been separated from ownership of the bank of the watercourse, so that only the owner of the fishery has the right to the fishing. Where a reservoir is owned by a number of parties, the right is divided amongst them, and if a reservoir belongs to one owner, the fishing rights belong to that owner unless in either case, it can be shown that the right of fishing belongs to another person. A fishery owner is not entitled, as a matter of course, to use the embankment or elsewhere around reservoir perimeter unless he has acquired the right expressly or by implication as an ancillary right to his right to fish or has the permission of the owner; a fishing lease would normally include the right to use these areas. The public have no right to use any part of a watercourse bank, embankment or reservoir perimeter which is privately owned, even though they may be entitled to fish in the watercourse or reservoir, nor can they fish from an adjacent path or highway. The rights of fishing may be bought, sold or let and can be transferred with or without ownership of the soil beneath. Lease of land will pass the fishing rights with it, unless the lease expressly reserves such rights for the landlord. If such rights are leased, the lease should contain all relevant constraints and limits. Those wishing to fish in inland waters may only do so with the permission of the owner. In England and Wales they will also need a fishing licence from the RLA.

It is important that the reader checks the situation in Scotland and Northern Ireland.

3.5.2 Felling consents and tree preservation orders

The felling of trees in Great Britain (excluding Northern Ireland) is controlled by the Forestry Commission under the Forestry Act 1967[27]. A summary of the requirements of the Act and other aspects related to felling trees is given in Box 3.12. A licence from the Forestry Authority, as part of the Forestry Commission, is normally required to fell growing trees, but a small amount of timber may be felled by an occupier without a licence. It is likely that woodland management operations as part of reservoir creation or maintenance would require a licence.

Where the trees are growing in a Site of Special Scientific Interest (SSSI), in a conservation area, in protected woodlands or are subject to a Tree Preservation Order, special considerations apply and permission for felling is less likely to be given. Strict liability applies, and the owner must check the statutory controls and be satisfied that all necessary permissions have been obtained before any felling or other work is carried out. Trees in a SSSI require English Nature or the relevant county agency to be notified and agreement gained, whilst for the remainder, the district council must be consulted before any work is carried out. Similar constraints apply in Scotland and Northern Ireland.

Box 3.11 Wildlife and Countryside Act 1981

Birds

Part 1, Section 1 of the Act states that it is an offence to take, damage or destroy the nest of any wild bird whilst the nest is being built or is in use. Thus, the construction and management operations for a reservoir should take account of the nesting period and only be carried out at the appropriate times. A defence that the damage was an accidental result of an otherwise lawful action and could not reasonably have been avoided (e.g. occurred during normal agricultural practice) may offer some protection; however, it has been shown to be unsuccessful in the management of vegetation. Felling of trees may result in the destruction of nests and the disturbance of birds on or near the nest. Certain bird species are given further protection, principally relating to disturbance, as detailed in Schedule 1, whilst others, as listed in Schedule 2, may be taken outside the close season, or for some birds at all times by an authorised person. This is defined as the owner or occupier of the land, or any person authorised by the owner or occupier, on which the authorised action is taken. The close seasons are detailed in a separate order and vary between species.

Animals

The legal protection of animals under this Act is less comprehensive than that afforded to birds. Only certain species are listed in Schedule 5, and Section 9 of the Act makes it an offence to take, kill or injure any of these animals, which include certain butterflies, insects and reptiles. It is also an offence to damage, destroy or obstruct any place used by these animals for shelter, or to disturb the animals in that place. A defence of normal agricultural practice, as for birds, may be an adequate defence against prosecution, but would be unlikely to be satisfactory for other operations. The illegal use of poisons, which are often agricultural pesticides, is an issue which has yet to be resolved by the Act. Recent amendments to the Act to make landowners more responsible for what happens on their land and, with more effective enforcement of existing pesticide laws, should reduce this practice.

Plants

Plant protection under the Act is split into two groups comprising the rarer species, as noted in Schedule 8, and the remainder. Under Section 13, a person commits an offence if they intentionally pick, uproot, or destroy any wild plant in Schedule 8, or not being an authorised person as defined above, intentionally uproot any other wild plant. Thus no plant may be dug up and moved to a reservoir without the owner's permission and certain rarer plants may not be uprooted. The only defence is if it can be shown that the act was an incidental result of a lawful operation and could not reasonably have been avoided. Thus, permitted reservoir creation should be acceptable, but the presence of rare plants may affect the initial planning approval.

Release of non-native animals and plants

It is also an offence under Section 14 to release, or allow to escape into the wild, any animal which is not ordinarily resident or a visitor to Great Britain in a wild state or to plant or to cause certain plants to grow in the wild. All reasonable steps must be taken to ensure animals and plants do not escape into the wild.

3.5.3 Pollution

It is an offence under the Water Resources Act 1991[17] to allow a noxious substance to enter a watercourse or body of water. Trade or sewage effluent, fuel oils, pesticides, slurry silage effluent, daily washing and runoff from yards or hardstanding, sheep dip fluids, and similar are likely to be considered as noxious substances. They must be retained in tanks, compounds or lagoons for disposal elsewhere or cleaning prior to discharge.

Box 3.12 Forestry Act 1967 and tree felling

Exemptions to felling licence

An occupier may fell 5 m^3 of growing timber in any calendar quarter year provided that not more than 2 m^3 are sold. A number of exemptions may be relevant to reservoir construction and these might include:

- trees below 8 cm in diameter, measured 1.3 m from the ground, or, in the case of thinning, below 10 cm in diameter
- coppice or underwood below 15 cm in diameter
- dead or dangerous trees or those causing a nuisance
- where tree removal is included within a proposal and planning permission has been granted.

Method of application

Where trees are to be felled, the Forestry Authority should be approached for information and guidance. An application for a licence must be made by the owner of the land on which the trees are growing or by a tenant whose lease entitles him to fell the trees or by an agent acting on behalf of the owner or tenant. Applications cannot be accepted from timber merchants unless they come into one of the above categories. Timber merchants should nevertheless assure themselves that a licence has been issued before undertaking any felling, since they are liable to prosecution for illegal felling if it transpires that no licence exists.

Applications should be made on Form PW11 available from the Forestry Authority. Applications should be submitted along with a plan of the area at least three months before felling is due to commence to allow time for inspection and any necessary consultations. The explanatory notes on the back of form PW11 give guidance on how it should be completed. It should be noted that felling must not be commenced until a licence has been issued.

The regulations are enforced by the RLA, at present the National Rivers Authority (NRA) in England and Wales, which has powers to serve notice where facilities present 'a significant risk of pollution'.

The various Acts and statutory codes controlling pollution which are relevant to reservoir creation in England and Wales are detailed briefly in Box 3.13. This is a complex area and one which is changing continuously. It is therefore suggested that specialist advice is sought. Farmers can get free general advice from ADAS on preventing pollution.

'The Code of Good Agricultural Practice for the Protection of Water'[31] and 'The Code of Good Agricultural Practice for the Protection of Soil'[36] both issued by MAFF, and the CIRIA publication 'Farm Waste Storage – Guidelines for Construction'[37] detail current good practice.

The NRA, as the RLA in England and Wales, have recently issued a code of practice for the protection of groundwater[38] and will consider this when considering applications for licences and consents.

Box 3.13 Pollution control legislation

This box summarises the main legislation in Wales and England. It should be noted that the situation may vary in Scotland and Northern Ireland.

Water Resources Act 1991

The Act[17] is the principal Act affecting water resources, quality and pollution. The Act repeals the Water Act 1989[28] which itself replaced Part 2 of the Control of Pollution Act 1984 (COPA)[29].

Section 85 of the Act defines 'pollution'. In essence it is an offence, without proper authority, to cause or knowingly permit discharge of poisonous, noxious or polluting matter or any solid waste matter to any controlled waters. A defence against such an offence would be either a waste-disposal licence issued under Part 1 of the Control of Pollution Act 1974 (COPA)[30], a discharge consent given under the Act or other defence in Sections 88 or 89 of the Act. However, this defence is largely academic, as it is unlikely that the Waste Disposal Authority or the NRA would grant consent for an activity which was likely to cause pollution. 'Controlled waters' comprise underground waters, ponds, lakes, ditches, streams, rivers and coastal waters extending for up to three miles out to sea.

The Act provides for the following functions:

- imposing a general prohibition on discharges and deposits into waters

- granting consents for discharges to conditions; these conditions relate mainly to the quantity and character of the discharges

- providing for the establishment of a comprehensive scheme of administrative controls to monitor and enforce the control of pollution of water and water supply.

There are in addition certain precautionary preventive measures and provisions for the restoration of waters.

Section 90 refers to the offence of removal of deposits and vegetation in inland freshwater without the consent of the Authority.

The Act places the responsibility with the NRA for monitoring, controlling and remedying pollution. It also has the power to prosecute offenders.

Section 92 provides for requirements to take precautions against pollution and, in certain circumstances, for the associated expenses to be recovered.

Sections 93 to 96 provide for special controls in water protection zones or nitrate sensitive areas.

Section 97 provides for Codes of Good Agricultural Practice to be issued as guidelines and to promote desirable practices

The Code of Good Agricultural Practice for the Protection of Water

The Water Code[31] is a Statutory Code. Whilst contravention of the Code will not itself give rise to criminal or civil liability, failure to comply could be taken into account in any legal proceedings. The Code describes the main risks of water pollution from different agricultural sources.

In each section, good agricultural practice is defined in a way that balances the need to minimise the likelihood of water pollution while enabling economic agricultural practice to continue. For each class of potential agricultural pollutant (e.g. dirty water, silage, slurry, fertiliser) the Code sets out the principles to be adopted to ensure safe storage, treatment and/or disposal.

The Control of Pollution (Silage, Slurry and Agricultural Fuel Oil) Regulations 1991

The Regulations[32], known as the Farm Waste Regulations. They have a major effect on the design and construction of farm waste storage facilities to minimise the risk of pollution. The Regulations are enforced by the NRA. The Regulations cover three activities of silage making and storage, slurry and dirty water storage and agricultural fuel oil storage. They do not cover solid wastes (e.g. farmyard manure), but run-off from solid manure stores in the farmyard is considered to be slurry.

The aim of the Farm Waste Regulations is to ensure that storage facilities are designed and constructed so as to minimise the risk of unintended escape of silage effluent, slurry (including dirty water) and fuel oil. This is achieved by prescribing performance criteria that the facilities must satisfy, together with a number of detailed technical requirements.

The Environmental Protection Act 1990

The Act[33] affects a range of pollution issues, including odours and smells. It repeals, amends and replaces parts of the previous legislation relating to statutory nuisances contained in the Public Health Act 1936[34], the Public Health (Recurring Nuisance) Act 1969[35] and the Control of Pollution Act 1974[30].

The effect, generally, of the Act is to focus attention on pollution issues. It provides authorities with additional powers to curb pollution and makes higher penalties available. This Act, together with the relevant provisions of the Water Resources Act 1991[17] and the Farm Waste Regulations[32] places a very clear and direct responsibility on the users, designers and builders of reservoirs and waste storage structures to ensure that facilities minimise the risk of water or air pollution by waste emissions.

The Control of Pollution Act 1974 (COPA)

Part 1 of the Act[30] made it an offence to disposal of 'uncontrolled' waste which is poisonous, noxious or polluting, on any land in such a way as is likely to give rise to environmental hazard. This includes both solid or liquid waste. Waste disposal authorities are empowered to survey all wastes arising in their area. Part 1 of COPA has been replaced by Part II of the Environmental Protection Act 1990[33]. Much of this part of the EPA has not yet come into effect and COPA[30] controls still apply. Part 2 of COPA was repealed and incorporated into the Water Act 1989[29] and subsequently into the Water Resources Act 1991[17].

3.6 RESPONSIBILITIES

3.6.1 Owner's liability

The principle of strict liability applies (i.e. no defence can be offered), and the owner and occupier of the land are liable for any action of the escaping water from the reservoir. Water in quantity is, or is capable of being, dangerous; the risks are clearly known, and no negligence need be proved. The rule applies not only to escapes of water, but also to seepage and pollution. The owner and occupier are liable not only for the acts of employees, but also for the acts of their independent contractors or advisers, i.e. those who design and build the reservoir. Recourse against them should then be available provided the contractual arrangements, as discussed in Section 11, are properly set up. An injunction would be obtainable against the undertaker in an appropriate case, e.g. imminent threat of collapse or other danger.

A possible defence to strict liability could be if the damage occurred as a result of a totally unforeseen event; a possibility not recognised by human prudence. Extremes of weather, even very severe wind and storms, could hardly come within such a category. Recent weather events have demonstrated the variability of the climate and any defence along these lines would be very difficult to prove. The act of a third party which could not be foreseen and over whom no control could be exercised is likely to be a defence. However, experience and contemporary knowledge would suggest that possible trouble upstream or downstream should be considered by undertakers and their advisers. If the victim caused the damage himself, such as opening a sluice pipe, it would be considered his own fault and no defence would be necessary. The law of nuisance also applies, i.e. the unreasonable use of the land by an owner or occupier so as to cause damage to his neighbour's property by escape of water, etc.

It is envisaged that most reservoirs within the scope of this guide will be in sole ownership of the undertaker. Where a dual or multiple ownership is planned, the liabilities and responsibilities of the various parties shall be agreed clearly prior to construction. Such agreement should also include the responsibilities and liabilities for the monitoring of the completed works and the necessary maintenance and remedial works.

A carefully maintained record of the design and construction of the reservoir and the subsequent monitoring, maintenance and remedial works (Section 9.16) would be of benefit in the event of any action by other parties arising from creation of the reservoir.

The owner of premises or land has a liability, under the Occupier's Liability Act 1957[39] in England and Wales, to ensure the safety of visitors using the premises for the purposes of their visit. This may require secure fences around reservoirs with notices warning of deep water, soft ground, etc. and buoyancy aids where visitors are anticipated. Some warning signs and means of exclusion should be provided on private property where visitors are not expected or permitted to guard against unauthorised entry. Other liabilities may also apply.

Construction of an impounding reservoir will affect the flow conditions further downstream regarding both the size and pattern of typical and flood flows. These may affect operations which are being carried on prior to and as a result of the reservoir creation and thus certain liabilities may be present. The impounding will also affect groundwater levels in the vicinity of the reservoir and alter the flow pattern of the river just upstream of the reservoir, causing possible deposition of silt, development of vegetation or other changes. These effects may lead to liabilities to other parties.

3.6.2 Employer's liability

Under the Health and Safety at Work Act 1974[40], the employment of persons will require the Employer, so far as is reasonably practicable, to ensure the health, safety and welfare of himself and all employees and third parties affected by his work activities. The Act also gives inspectors, appointed under this Act, enforcement powers and allows them to request any information relating to the reservoir. Other liabilities may also be relevant, particularly with regard to hazardous substances[41]. The Management of Health and Safety at Work Regulations 1992[42] requires Employers to carry out additional measures, relating primarily to risk assessment and other matters.

In cases where there appears to be a danger, but the Health and Safety at Work Act does not apply, an inspector may advise the local authority with a view to possible action under the Local Government Act 1972[43] or other local authority enforced legislation. Local authorities are empowered, where an emergency or disaster involving destruction or danger to life or property is imminent or there are reasonable grounds for apprehending such a situation, to incur expenditure to avert, alleviate or eliminate the potential effects. An inspector of the Health and Safety Executive could also seek specialist advice from within the Executive or outside consultants.

Equivalent legislation is in force in Scotland and may vary from that detailed for England and Wales.

3.6.3 Liability during construction

Other Acts and associated regulations relate to the carrying out of construction works and the necessary plant involved in addition to the Health and Safety at Work Act[40]. Whilst these will be the prime responsibility of the contractor carrying out the construction work, some responsibility will fall on the undertaker. It should be noted that a risk of pollution will be present as a result of accidents or vandalism and thus the various constraints will apply which are within the Control of Pollution Act 1974[30], the Land Drainage Act 1991[19] and the Water Resources Act 1991[17]. Various other liabilities may also apply and these are discussed further in Section 11.

The risk of liability for the discharge of the large flood through a partially completed dam must be borne in mind during the construction period.

3.6.4 Liability of others

The principal liabilities of any specialist advisers should be adequately covered in the contractual arrangements. In England and Wales, they will also be liable in tort, i.e. a civil wrong for which the remedy is a common law action for unliquidated damages which is not exclusively the breach of a contract or trust or other reasonable obligation.

The specialist adviser has a duty to exercise reasonable care and skill. He should be practical and competent and have sufficient and suitable expertise to advise on the aspects required. Failure to follow existing guides and practice may invite a strong case for negligence.

The contractor will be liable under the principles of contract and tort, and the standard construction contracts, if these are used.

The LPA, the RLA and other authorities and undertakings may be liable in certain instances. An employee may act negligently in the course of his duties and vicarious liability would apply. Similarly, if incorrect or negligent procedures are followed, the authority or undertakings may be liable. Certain other liabilities are applicable to the authorities and undertakings and these may be sufficient to invite action by the undertaker.

4 Environmental considerations

4.1 GENERAL

Wetlands and water bodies of all types are a declining habitat, under pressure from detrimental activities including pollution, drainage and infilling. When new reservoirs are created, for whatever reason, some thought should be given to preserving existing wetlands and enhancing wildlife, conservation and other environmental benefits.

Environmental considerations are addressed throughout this guide. In the sections on design, construction and maintenance, guidelines for wildlife and nature conservation are included.

This section lists environmental issues that need consideration at an early stage and outlines the effects that a new water body may have on the existing environment. Many of the points listed will not be applicable in all cases, but any reservoir creation will have some effect on the environment. Whatever the reason for construction, the opportunity exists to create a reservoir that also has wildlife and conservation benefits. These may be aimed at specific groups such as waterfowl, fish, amphibians or dragonflies. Alternatively, it may be desired to create an attractive environment with general habitat value. Positive measures that improve the wildlife and environmental value of the reservoir and associated wetland can also be included to provide a varied habitat, recreate an earlier lost environment or mitigate adverse features. Whatever the objectives, they can be maximised by considering ecological issues at the outset and incorporating them into the design, construction and management procedures.

This section does not discuss the environment assessment which may be a statutory requirement, and which is described fully in Section 3.2. It details the various matters to be considered, however, whether a formal environment statement is prepared or not. If the existing habitat is shown to be valuable, then preservation should be considered and the reservoir located elsewhere.

4.2 THE SITE AND ITS ENVIRONMENT

Before the effects of a proposed reservoir creation on the existing environment can be assessed, it is necessary to consider the existing site and its environment at an early stage. Box 4.1 outlines some of the aspects that may need to be considered for a particular scheme. Many of the points listed will not be significant, but should be considered in a preliminary appraisal.

Some of these considerations relating to population and development, may also be useful in assessing flood risk and reservoir safety in Section 7.

Box 4.1 The site and its environment

Population and development	• proximity of reservoir site to local population and development
	• amount of local population
	• present land use and development of reservoir site
	• present land use and development in areas downstream of reservoir
	• existing planning controls and special countryside protection
	• proposed changes to existing use or development in areas downstream of the reservoir
	• reference to site and areas further downstream in local development plans
Recreation	• existing recreational use of the land and areas further downstream
	• proposed changes to recreational use downstream of the reservoir
	• recreational demand and requirements
Watercourses, drainage and water quality	• location of existing aquifers, watercourses, springs, water areas, marshes, etc.
	• source and quality of water for initial and subsequent filling
	• chemical composition of water (affecting type of flora and fauna)
	• potential and actual sources of pollution, suspended solids and nutrient loading
Flora and fauna	• existing habitats, particularly protected habitats
	• plant and animal species, particularly protected and rare species
	• migrating animal species that make use of the land
	• predator/prey interactions and their ecological balance
Soil and geology	• agricultural or horticultural value of land
	• soils of conservation value, e.g. natural peat beds
	• site geology
	• land forms and land surface features
Landscape and topography	• physical landscape, topography, ground slopes and levels, etc.
	• soft landscape, trees, hedgerows, etc.
	• aesthetic landscape, visual quality of scenery
	• protected/designated areas, AONB, National Park, SSSI, etc.
Cultural features	• architectural, historical, heritage and archaeological features within area affected by reservoir (possibly given statutory protection (listed) by the LPA, of interest to a conservation society or known to be present)
	• an existing reservoir to be enlarged may be listed by the LPA
Climate	• precipitation (rain, snow, hail, sleet, fog or dew)
	• typical precipitation patterns (duration, local distribution pattern, etc.)
	• season variation of climate (summer, drought, etc.)
	• site orientation
	• countrywide position (more western areas generally wetter)
	• local climatic influence of existing topography and vegetation
	• air quality.
Other relevant environmental features	• footpaths, rights of way, etc.
	• other longstanding rights.

Box 4.2 Environmental effects of reservoir creation

Population	•	the effect the reservoir is likely to have on the local community and wider population, in economic and social terms for example
	•	the numbers of population affected, whether directly or indirectly, by the reservoir
Recreation	•	restriction of existing recreation pursuits in area of reservoir
	•	new recreational activities as a result of the reservoir creation
	•	effects on features and facilities associated with recreational activities
Watercourses, drainage and water quality	•	deterioration in waterside ecology due to diversion or realignment of watercourses
	•	sedimentation and scouring due to damming or other obstructions to water flow
	•	pollution, such as runoff, agrochemicals, chemicals or a change in water chemistry
	•	disruption to natural drainage system and effects on groundwater
	•	waterborne diseases as detailed in Appendix E
Flora and fauna	•	damage to existing habitats
	•	disturbance and damage to animals
	•	restriction to the movement of migratory fish
	•	addition of exotic species in restoration work can upset natural ecological balance
	•	changes in groundwater and surface flows can often be to the detriment of existing animal and plant communities
Soil and geology	•	reduction in agricultural or horticultural value of soil
	•	sterilisation of mineral resources
	•	physical damage to soil structure
	•	chemical damage to soil through pollution
	•	erosion of light soils
	•	loss or damage to geological, palaeontological (fossil) and surface landform features
Landscape and topography	•	visual effect on surrounding area and landscape
	•	physical effect by removing vegetation or changing topography
	•	physical effect of increasing groundwater levels adjacent to the reservoir
	•	effect of earth moving on stability of existing slopes and structures adjacent to the reservoir
	•	effects of erosion by water around reservoir perimeter
Climatic effects	•	removal or creation of shelter by removing or planting trees
	•	change in microclimate for invertebrates and plants
	•	increased incidence of mist as a result of reservoir
Traffic and access	•	disturbance caused by construction traffic on local roadways
	•	traffic hazard to pedestrians
	•	compaction of soil and destruction of vegetation at access to site
	•	parking and traffic problems
	•	diversion of footpaths and other rights of way
Noise, discharges and residues	•	pollution by whatever means downstream and adjacent to the reservoir
	•	dust deposition on nearby vegetation and buildings
	•	vehicle exhaust fumes
	•	noise disturbance to local residents

4.3 ENVIRONMENTAL EFFECTS OF CONSTRUCTING THE RESERVOIR

When planning a new reservoir, the effects on the existing environment need to be assessed at an early stage. Box 4.2 outlines some of the environmental effects that may result from such a scheme. Most of the points listed will not be applicable in many cases, but any reservoir creation will have some effect on the environment. Differing effects may be apparent during the construction phase and these may be more significant, although of shorter term, than those during use of the completed reservoir.

4.4 ENHANCEMENT AND MITIGATING MEASURES

4.4.1 General

Mitigating measures are those required to reduce any adverse environmental impacts of the development. This could include creating a habitat where another has been lost or planting trees to screen an embankment. This section also outlines the positive measures that can be taken to improve the value of a new reservoir for wildlife; these can normally be included at little or no cost if considered during the design stage. Section 9.14 details the particular features to be considered at the design stage, whilst Section 10.11 describes measures to be taken during construction to maximise the ecological benefit of the reservoir. Specialist advice on appropriate nature conservation measures can be obtained from English Nature, the appropriate County Wildlife Trust, ADAS or FWAG, whilst the Forestry Commission or private sector forestry consultants can advise on tree planting. The RLA may be also able to offer advice on certain aspects. Some useful, more detailed comments are contained in References 44 to 46.

There may be specific wildlife requirements that need to be catered for in the design when designing a new reservoir. If the reservoir is to be used for fishing, the focus will be on habitat creation for this purpose. Similarly, if the objective is wildfowl shooting then design features that would encourage the breeding of relevant species would be incorporated. Advice on such specific requirements is contained in Sections 9 and 10 and the information produced by specialist organisations should also be consulted, e.g. the British Association for Shooting and Conservation[47], the Game Conservancy[48] or the Angling Foundation[49].

Although the design of the dam must focus primarily on safety and function, it is possible to incorporate environmentally sympathetic measures. The shape of the embankment can be varied by introducing flatter slopes or a wider crest in places. Crest level can also be varied, provided this does not affect any flood and wave retention ability. Tree and shrub planting can be extended to the embankment itself and to maximise the ecological benefit of a new reservoir, should be considered as an integral part of the design. Provided vegetation is sited with care and certain critical areas are kept clear and unsuitable species avoided, it will not only increase the wildlife value but can also improve the safe functioning of the dam.

4.4.2 Preservation of existing habitats

When planning a new reservoir, the existing habitat at the proposed site should be considered, not only in terms of the best location but, more importantly, where not to locate it. Creation of a new water body will necessarily destroy the existing habitat of the site; to achieve maximum ecological benefit it is essential that the existing habitat is not in itself of high conservation or wildlife value. Statutory designations may exist to prevent any form of damaging activity, for example the site may be a Site of Special Scientific Interest (SSSI) or may harbour plant or animal species protected under the Wildlife and Countryside Act 1981[9]. Even if the site is not of designated importance, it may still have a present wildlife value that is greater than the proposed replacement. In general, semi-natural or areas of interest should not be destroyed, especially those with poor or slow regenerative abilities. The importance of all existing habitats should be assessed fully and they should be retained if they are of particular value.

If the new reservoir is to replace existing wetland an area of the former habitat should be retained immediately adjacent to the water body. This not only adds interest and diversity to the shoreline, but may provide a good source of aquatic material for colonisation of the new area.

In ecological terms it is preferable to locate a new reservoir near to existing wetland habitat. This not only increases the size of the habitat for wildlife but also provides a close source of flora and fauna for colonisation of the new area. If this is not a possibility, then the new reservoir should be located close to other semi-natural habitat such as grassland, hedgerows or scrub, and in quiet, secluded locations to minimise disturbance to wildlife. These measures will enhance the value of the new reservoir rather than creating a separate element in an already fragmented semi-natural landscape.

Dams alter drainage patterns, often to the extent that could be detrimental to existing communities of wetland animals and plants downstream. The new reservoir could have an adverse effect on an existing natural wetland, and thus the habitat loss may be greater than the potential gain.

4.4.3 Habitat creation

It is important that habitat provision is carefully planned to maximise the ecological benefit of a new reservoir so that the best conditions for the development of ecological zones are created. Different species of plants and animals vary in their habitat requirements and it is important to provide plenty of different habitat features. Ideally, there should be space created to allow a succession of grassland to nearby scrub and woodland. The maximum benefit for wildlife will only be achieved if this habitat diversity is created. Indeed, if nature conservation is the primary objective, it is often better to create a slightly smaller water body with surrounding space for associated terrestrial habitats.

The likely water quality of the reservoir is important to any long-term plans and the effects of any pollutant, suspended solids and nutrient loading in the inflow must be considered. These will affect the habitat and preclude or inhibit the establishment of some species or cause others to flourish and take over to the detriment of less intrusive species. Seasonal variations may also be as important as any long term changes, as well as any sudden effects from livestock contamination, road drainage or sprays and fertilisers.

4.4.4 Aquatic and shoreline planting

Where a new reservoir is supplied by a river or stream, there will be a supply of plant material for natural colonisation, although not all of this may be suitable for the new static water conditions. Planting will speed the establishment and assist in the development of an aquatic community and improve the appearance of the new water body. Care must be taken to introduce only those species which are naturally occurring in the local area. Unusual or exotic species should be avoided as they may upset the natural ecological balance and rapidly become dominant to the detriment of the natural species. If different species are planted in groups, competition leads to the most vigorous species becoming dominant. It is advisable, therefore, to plant the reservoir margins with separate blocks of species, allowing each to become established and develop a varied waterside flora.

For maximum ecological benefit, a complete planting programme should be planned that is based on the natural gradation from dry land plants, through marginals and emergents, to submerged and floating species. These zones are based on the position the plants occupy in relation to the water margin and are illustrated in Figure 4.1. Species can be selected for planting that have a particular food value for fish or wildfowl, are visually attractive or that simply provide general habitat value.

Where the water level will vary substantially as a result of the reservoir usage, the planting will require careful planning and consideration. Few plants will be able to flourish in the zone between the maximum and minimum water levels. Many species, however, are tolerant of short-term changes of limited extent and will be able to become established despite occasional periods of flooding or lowered water levels.

4.4.5 Terrestrial vegetation

Terrestrial vegetation should be planned so that a variety of habitats adjacent to the reservoir are created or extended and structural diversity of vegetation is provided. It is desirable to introduce only native species as these also tend to have a greater wildlife value.

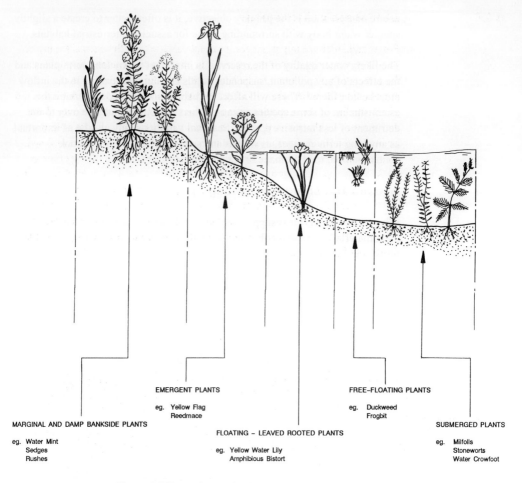

Figure 4.1 The main aquatic plant zones

Tree species should be selected with care and mixtures of species with different growth rates and habitats should be avoided. Native species in keeping with existing woodland should be selected to create a 'natural' area in most instances and 'ornamental' species should be restricted generally to the non-rural settings. Examination of existing shrubs and trees in and adjacent to a proposed reservoir site will provide an indication of the species that are likely to establish successfully.

Tree planting may continue onto the dam, providing the trees are located in acceptable positions and are maintained in a healthy condition. Vegetation on the embankment is discussed more fully in Section 9.2.8.

Tree planting may also provide shelter to the reservoir as discussed in Section 9.1; this will also help to diversify the nature of the water surface and surrounding areas, and will encourage the establishment of a wider range of habitats and species.

4.4.6 Landscape enhancement

Ponds and lakes are important features that contribute to the natural beauty of the countryside and thus sensitively designed and well managed reservoirs can form an attractive addition to the landscape. Variability in the vegetation surrounding the water with woodland, scrub and grassland greatly improves the visual quality of the reservoir setting and increases landscape diversity. Trees and shrubs can also be used to screen an embankment which might otherwise be intrusive. The reservoir itself will be more attractive if a varied shoreline is created with islands, spits and bays, and varying steep and shallow perimeter banks. This type of reservoir becomes more a part of the natural landscape than one with a uniform shape and constant slopes, as can be seen in Figures 4.2 and 4.3. Such features may be confined to the reservoir perimeter, but there is normally no reason why many cannot be continued onto the upstream face of the embankment, provided they are compatible with the reservoir operation and other possible constraints.

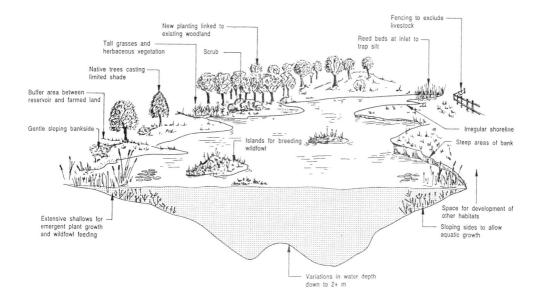

Figure 4.2 Features of a well-designed reservoir for wildlife

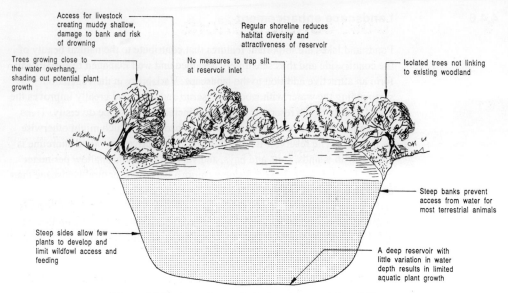

Figure 4.3 Features of a poorly designed reservoir for wildlife

5 Cost considerations

5.1 INTRODUCTION

The cost of any small embankment-type reservoir scheme will depend primarily on the topographic and ground conditions at the site, the type and size of reservoir and particular requirements dependent on the intended use of the reservoir. The costs are therefore site-specific and likely to bear little relationship to 'typical' values. Reservoirs to be used for irrigation purposes may be eligible for financial aid towards the cost of the scheme. Further information on the grants available can be obtained from a local ADAS adviser. Additional grants may be available for other purposes, e.g. conservation, etc.

The creation of a reservoir is likely to add additional value beyond that of the land value alone. This will vary substantially depending on the size of the undertaker's land and of the reservoir, the reservoir purpose and situation and location in the country. In the most favourable circumstances, the enhanced value may approach the cost of constructing the reservoir and so a scheme may become virtually self financing.

Outline cost estimates should be prepared for a number of options in terms of the following:

- size of reservoir
- secondary use
- reservoir sites
- type of dam.

This should provide an economic basis for site selection whilst other considerations must also be taken into account. These include site constraints, requirements of the appropriate licensing and planning authorities, safety and environmental requirements. This approach also provides an indication of costs for alternatives which may be useful if for any reason the initial site becomes unsuitable. The need for and cost of specialist advice should also be assessed at each stage of a scheme.

Monitoring and maintenance costs should be included in these outline estimates with provision for possible remedial measures.

5.2 SITE INVESTIGATION

The technical suitability of the selected site or sites needs to be established before significant expenditure has been incurred. Most of this information can be obtained relatively inexpensively, with the undertaker carrying out much of the work himself by following the guidelines in Section 8. As a general rule of thumb, the site investigation cost should be between three and five percent of the total project cost unless boreholes by a specialist contractor and extensive laboratory testing are necessary. More complex ground investigation alone may cost up to 15% of the total project cost, or more on particularly complex sites. Although potentially expensive, the ground investigation should not be skimped on as inadequate investigatory work is likely to prove considerably more expensive at a later date if adverse conditions are revealed during construction.

Topographical and geological survey maps, published memoirs and other information can be obtained for relatively small costs from HMSO outlets and other sources listed in Appendix D. Information searches are usually free of charge from these sources except for the cost of photocopying.

A close visual examination of the prospective sites, including taking photographs, will help the undertaker. This entails minimum cost and could be useful in selecting the best site for the proposed works. At this stage the undertaker should have established where the reservoir is to be constructed; this initial study helping to narrow down the possible sites and minimising the cost of a survey and ground investigation.

A survey of the site should be carried out and a plan prepared by a qualified surveyor. For a small reservoir this is a relatively rapid inexpensive item that can normally be completed in two or three days. An adequate survey and an accurate plan will, in the long term, save time and cost at all stages of the scheme. The survey should be used initially to assist in the ground investigation stage, and subsequently in the preparation of drawings for the project for planning and licence purposes and later for the design and construction stages.

A simple ground investigation can be carried out relatively cheaply if the equipment required is available to the undertaker. If not, the undertaker should hire an excavator, and possibly an auger, at an agreed rate from a locally based supplier. A simple ground investigation should be completed within a few days for a straightforward site. Advance preparation will be necessary, however, to gain access and mark out the positions of the exploratory holes, and to purchase bags, ties and labels for the collection of soil samples.

If more advanced work is required it will greatly increase the cost of the site investigation. The undertaker should seek specialist advice and assess the benefits before any work commences. The cost of an adequate investigation for a potentially difficult site may be a significant amount of the total construction cost and other sites, if available, may prove more attractive.

5.3 PRE-CONSTRUCTION STAGE

If the site investigation has shown the site to be feasible, preliminary proposals should be prepared for discussion with the LPA and RLA. These may be prepared by the undertaker without undue costs, but specialist advice may be required to clarify certain aspects. Budget costs (see Section 11.5) for the proposals can be obtained from potential contractors. Costs will be involved in the submission of the applications.

Once an outline scheme has been shown to be viable and likely to meet with the requirements of the statutory authorities, a detailed design can be prepared using the information in Sections 6 to 9. The costs of this stage of the work may not be significant for a straightforward site, but a potentially difficult site and/or an ambitious or complex scheme will require a substantial input from the specialist adviser. The design must be sufficient and detailed adequately in the contract documentation to allow the contractor to construct the works in the most straightforward and economic manner. In particular, the quality and number of the drawings must be sufficient to allow the requirements to be shown clearly, concisely and unambiguously. Too many, rather than too few, drawings should be prepared, with sufficient space left to include further details where this is subsequently found to be necessary. Careful planning and forethought in preparing the contract documentation is likely to prove cost effective.

Consideration of the environmental matters, together with the monitoring and maintenance requirements, during the design will help to maximise these aspects and allow suitable measures to be included at least cost. In many instances, early consideration will allow inclusion of favourable features at no or minimal cost beyond the basic price.

Once the undertaker has taken the decision to proceed with the scheme, he should consider whether the construction is to be carried out by himself or by a contractor.

If the undertaker decides to engage the services of a contractor, a standard form of contract, with a quotation with individual work items clearly defined and rated, should allow the most economic construction costs to be obtained. This approach should also minimise the likelihood of potentially expensive claims by the contractor for extra costs arising during the construction phase. The standard forms of contract are outlined in Section 11 and are inexpensive to purchase.

5.4 CONSTRUCTION STAGE

The construction stage is likely to be the most expensive phase of the reservoir creation with plant, labour and materials on site which the contractor will want to use in the most economic manner. The discovery of unknown features after an inadequate site investigation or design changes by the undertaker during construction will not be well received by the contractor, potentially resulting in extra work and possible claims for additional costs. It is essential that the site investigation is as thorough as possible to maximise the information concerning the site and that the undertaker's requirements are included in the design and contract documentation. Changes to the works agreed with the contractor should be minimised and clearly conveyed to the contractor by means of written instructions and/or amended drawings at the earliest opportunity. This will help the contractor to plan his operations and minimise the additional cost implications.

Construction costs can also be minimised by reducing constraints on the contractor's operations. The undertaker should carry out his own operations so as not to cause interference to the contractor. Where possible, the influence of third parties (e.g. unauthorised ramblers or animals) should be prevented by suitable fencing.

If the undertaker carries out the work directly, or with hired-in plant and/or labour the following points should be noted to minimise construction costs:

- The construction works should be broken down into clearly defined phases and planned in a logical sequence.
- The work should be carried out steadily and, once started, should continue without delay until completion. This is particularly true for earthworks which will form the bulk of the work.
- Allowance should be made for loss of time due to bad weather, particularly during fill placing.
- Labour, plant operators and the suppliers of hired-in plant should be informed clearly of the required work and be supervised, as appropriate.
- Excavations should be backfilled at the earliest opportunity to minimise problems.
- The possibility of large flows at any time during the construction of impounding reservoirs must not be forgotten and diversion works must be installed and functioning at the earliest opportunity.
- Any damage or unsuitable construction as a result of weather, bad workmanship or any other cause should be corrected before further work is carried out and not left.
- Specialist advice should be sought when problems are first encountered as they will tend to worsen or become more complex if not dealt with.

5.5 VISUAL OBSERVATION AND MONITORING

Regular monitoring of the dam and reservoir by the undertaker should ensure that it fulfils its safety requirements. This should include the taking of photographs at regular intervals. Such work will not incur costs to the undertaker, except for his own time. An annual inspection by a specialist adviser is also recommended to carry out a check and review the monitoring observations. Although this may seem to be a costly expense, it could save expenditure in the long term by identifying potential problems at an early stage.

The only monitoring equipment required is that recommended in Section 12 which includes the installation of means to measure the reservoir water level and flows, if required.

5.6 MAINTENANCE

When estimating the cost of a reservoir scheme, the undertaker should consider the requirements for regular maintenance to keep the reservoir in a safe condition and fully operational.

Regular maintenance to keep the reservoir in working order incurs costs. Such items as maintaining access, keeping fences and gates secure, cutting grass on the embankment and cleaning silt traps, overflow structures and channels, all require manpower. Vehicles and appropriate plant and equipment will also be needed and will incur purchase and running costs, whilst items such as pumps will need regular servicing.

Replacement materials will be required for maintenance work and may include items such as gravel, fill material and grass seed in addition to proprietary products. Safety features must be maintained carefully and items such as fencing and signs may need renewal. Allowance shall be made for these extra costs.

5.7 REMEDIAL WORKS

Further costs will be incurred to carry out remedial works. These may arise from deterioration or damage caused by any of the processes listed in Section 14. A contingency allowance should be included in the cost estimates for the project. This should allow for specialist advice and necessary remedial work. Some of these costs, however, may be covered by the undertaker's insurance, depending on the work required and the terms of the policy.

5.8 OTHER COSTS

Other costs are likely to be associated with the creation of a reservoir in addition to those discussed previously. These may include:

- additional costs associated with the specific use of the reservoir (e.g. stocking with fish)
- insurance of third party risk from the reservoir
- dealing with any pollution problems
- necessary works resulting from developments affecting the reservoir
- change or extension of use of the reservoir.

Part 2   Design

6   Yield and reservoir operation

6.1 SOURCES OF WATER

6.1.1 General

The following sources are available for water to fill reservoirs:
- surface water
- groundwater
- public supply.

Surface water includes the flow in watercourses, i.e. rivers, streams, ditches and channels, which may run permanently (perennial flow) or be seasonal, as well as water stored in ponds and lakes, whether natural or man-made, and surface water runoff adjacent to the reservoir itself. The runoff from melted snow and the flows in storm drainage systems can also be considered as surface water. Groundwater is water below ground level which may either drain into the reservoir from higher ground by means of springs or may be extracted by wells, boreholes or other means. Public supply is potable water taken directly from mains.

Recycled water from sources such as sugar-beet washing, sewerage discharge, water from farm processes etc., may be available. The quantity is not likely to be significant in most cases and this source would normally be available only in conjunction with one of those listed above. Depending on the reservoir use, the quality of the water may be doubtful in many cases.

6.1.2 Surface water

Rainfall not intercepted by vegetation or man-made devices such as drains, soakaways etc., falls on the ground and either evaporates, soaks into the ground or collects in localised depressions. When these needs are satisfied (termed 'losses'), the surplus water runs downhill and collects into streams and rivers and ultimately finds its way to the sea. When the rain is particularly intense, the surplus becomes large and the stream or river flow can increase rapidly. The flow during dry weather periods is termed the 'base flow' (which may be virtually zero), and flow in wet weather is termed a, 'flood flow', the magnitude being determined by the amount, duration and intensity of recent rainfall.

The direct runoff into a reservoir is determined by the characteristics of the catchment area, climatic factors, and the relationship between rainfall and runoff as detailed in Box 6.1.

6.1.3 Groundwater

Rainfall that has soaked into the soil and penetrated the underlying strata is termed groundwater. This process and the likely amount of infiltration is discussed in a subsequent section. Water is stored in the pores, fractures and fissures in the strata, called an aquifer; thus the quantity of water that can be accommodated depends on the available volume within such features. The water flows downward under gravity and the resistance to this movement varies widely. Materials with large pores, voids or fractures such as coarse gravel or certain rocks, such as chalk, are said to have a high permeability, whilst those such as clay have a microscopic pore size and have a low permeability with minimal flow of water. Groundwater collects until the material beneath is saturated: the top of the saturated material is said to be the groundwater table. The level is not fixed, but depends on the permeability of the material and the supply of water. In dry weather the water table falls, whilst in wet weather the level will rise and may reach the surface. The water table level is also influenced by the topography of the ground above. A layer of low permeability material will limit the downward percolation of water and may produce a perched water table above the lower main groundwater table. Where the groundwater reaches the surface, a spring will result. A natural barrier in the ground such as an impermeable fault may produce a similar effect; the flow depending on various factors which may result in seasonal variations of the spring. Water-bearing strata, overlain by low permeability material, can be under pressure. Such strata are fed from a distant source and termed a confined aquifer as the water is confined and will rise up any borehole or well which is sunk into the strata. Where the pressure is sufficient to bring water to the ground surface, it is said to be artesian, whereas a rise to below ground level would be termed sub-artesian.

Groundwater flow reaching the ground surface as a spring runs off as surface water and finds its way into the surface drainage system. Where the spring is caused by an outcrop of low permeability material, fault or similar occurrence, within an otherwise permeable strata, the spring flow may subsequently disappear back underground and become groundwater again. Such behaviour can occur frequently in chalk or limestone areas of the country. Groundwater abstraction in such areas can effectively take water from a watercourse and is likely to be subject to particularly vigorous controls.

Groundwater may be abstracted for supply from pumping wells and boreholes. The construction and maintenance of pumping boreholes is beyond the scope of this guide and specialist advice should be sought if required.

Box 6.1 Factors affecting the runoff of surface water into a reservoir

Catchment Characteristics
- shape, orientation and area of the catchment (A_c)
- slope of the catchment (expressed as 1V:gH)
- length of the major watercourse in the catchment (MSL) and mean stream slope (S1085)
- frequency of the minor streams through the catchment
- other water surface areas (A_w)
- soil type and influence of geology within and adjacent to the catchment
- land use and degree of urban development, including areas of rapid runoff (A_r)
- type and extent of vegetation both in the catchment and within the river channel
- forested area (A_f).

The area of catchment is considered to be the land surface contributing to the discharge at a particular position on a watercourse. The further downstream on the watercourse, therefore, the larger the catchment. Derivation of the catchment area, other water areas and the areas of rapid runoff are described in Section 7.1.3 whilst the assessment of stream and catchment slopes is given in Section 7.1.4. The influence of the soil types and geology within and adjacent to the catchment is discussed in greater detail in Section 7.1.5 and the effects of vegetation in Section 7.1.6. Many of these aspects are more relevant for flood assessment and will not be significant for calculations of yield. Thus the more simple methods given in Section 6.2 may be used to give an estimate of available water.

Climatic factors
- rainfall intensity and duration
- average annual rainfall
- soil moisture deficit (which is affected by rainfall in addition to evaporation and transpiration from vegetation)
- precipitation (in the form of snowfall and freezing giving rise to delayed runoff)
- altitude, which affects temperature, rainfall depth and the presence of snow in winter.

Information on the expected intensity and duration of rainfall and other precipitation at a particular location can be obtained from ADAS, the Institute of Hydrology, and other consultants, or direct from the Meteorological Office. Most sources will charge for this service. A publication by the Institute and the British Geological Society entitled 'Hydrological Summary for Great Britain'[50] (previously 'Monthly and annual totals of rainfall in the UK') is published monthly and provides useful data. A further helpful publication is 'Hydrological Data UK'[51]. From these sources it is possible to predict the quantity of rain that is likely to fall during any particular time period depending on the duration under consideration and the probability level chosen in terms of the return period (see Section 7.1.1). The average annual rainfall is usually of prime concern for reservoir yield. Average annual long-term values are given in Figure 6.1.

Relationship between rainfall and runoff

The relationship between rainfall and runoff is far from direct and may include extreme values. Complex methods for assessing flows are available, but the more simple methods that can be applied to small reservoirs are given later in the guide.

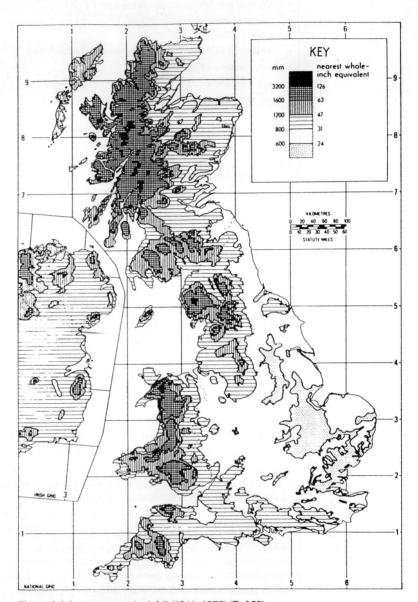

Figure 6.1 Average annual rainfall (1941–1970) (Ref 52)

6.1.4 Public supply

A source of water from public supply is only likely to be feasible for the very smallest reservoirs and is more suitable for a small pond. Certain water byelaws apply when filling ponds or reservoirs from the public supply to guard against excessive losses of water. The water quality may not be suitable for animal or plant life without further treatment. Information about using the public supply water can be obtained from the local water company who can supply details of quantity, pressure and cost of any supply which may be available and will advise on the likely availability in dry periods. The capacity of rural mains is usually insufficient without upsetting supplies elsewhere and thus only small quantities, possibly at specified times, may be available.

6.2 INFLOW AND YIELD

6.2.1 Infiltration, runoff and losses

When rain falls onto the ground, it wets all the vegetation and bare soil, and after saturating the surface, sinks into the ground or becomes runoff. Most ground surfaces, and certainly vegetated surfaces, are porous or permeable to some degree, and rainwater can soak or infiltrate into the ground. Different soils have varying permeabilities and allow different rates of infiltration.

The effects on human or animal activities may also affect infiltration, which may tend to increase or decrease it, whilst vegetation may have a significant affect on reducing or delaying the amount of runoff. These effects are more fully detailed in Box 6.2 but they tend to have less influence as the amount and duration of the runoff is increased.

The intensity and duration of rainfall that has fallen in the preceding few days has a substantial effect on the amount of rain that infiltrates into the ground or runs off. A catchment which is already wet from recent rainfall will have a greater runoff than a previously dry catchment. The amount of rainfall able to infiltrate the ground to join the existing groundwater or become runoff to join the surface water flow is further reduced by losses. Evaporation and transpiration remove moisture back to the atmosphere and seepage, drainage or other losses limit the amount of water which can flow into a watercourse or join the available groundwater.

Evaporation is the conversion of water into water vapour by solar radiation, but it is not usually possible to differentiate between losses resulting from evaporation from the land surface and losses due to vegetation. This results from water passing through the roots to the leaves, via the trunk or stem, which is then transpired into the atmosphere. The two processes are typically linked together and called evapotranspiration as detailed in Box 6.2.

Box 6.2 Soil infiltration and evapotranspiration

Soil infiltration

A gravelly or sandy soil will accept large amounts of water and most rainfall falling onto the wetted surface will tend to be accepted. Even heavy rainfall may not produce a significant overland runoff, because the water will enter the ground, although the flow may exit elsewhere and produce a flood flow. Clays, on the other hand, have little capacity to accept and pass water and thus even light rainfall may result in a water excess on the surface, although a desiccated surface after a prolonged dry spell may allow some infiltration. Where the surface is able to shed water, a substantial runoff will develop, but flatter surfaces will retain water and become waterlogged. Some rocks may be virtually impermeable, but others such as chalk or limestone, may have little effective runoff. The amount of infiltration is dependent on the soil type and the intensity of precipitation, but other factors have an influence; the ground slope, the type and extent of vegetation cover, the position of the groundwater table and seasonal effects. Increased temperature will lead to more infiltration.

The soil classification has a very significant effect on the runoff assessment. Soil maps have been prepared which define areas by soil type and hence their runoff potential. The actual basis for comparison is the reverse of the runoff potential and is termed 'the winter rain acceptance potential'. Five classes are given from Class 1 – very high, to Class 5 – very low. Where varying soil types are present in the catchment, runoff from one section may be lost elsewhere on the catchment. Alternatively, a low permeability material may bring the groundwater table to the surface and result in surface flow from springs. Thus a knowledge of the soil type and distribution over the catchment is essential. Although for smaller reservoirs the catchment is more likely to fall within one category, local minor variations may still be evident which may influence the amount of runoff.

Exposed soils can become almost impermeable when large raindrops compact the surface or wash finer material into the pores in the soil. The surface then restricts infiltration and runoff increases. Similar, but much more rapid effects may result from animal, human or vehicle activity compacting or puddling the ground surface. Dense vegetation tends to increase infiltration. Rainfall reaching the ground readily soaks into the surface because the layer of vegetation debris produces a sponge-like surface; the extensive root system prevents compaction of the ground surface; burrowing animals and insects open up the surface and the vegetation removes moisture from the soil; lowering the groundwater table and increasing the capacity of the soil to accept water.

Most surfaces in the British Isles are covered in some vegetation especially away from the more mountainous areas of northern and western Britain. The vegetation removes water from the ground whilst rainfall and, occasionally, snowfall replaces it. Where evaporation and transpiration exceed total precipitation (rainfall, snowfall, hail, fog, dew etc.), the vegetation draws on the moisture in the ground. With the reverse, the soil moisture increases and the groundwater table will ultimately rise. Over the year, there is an effective deficit of moisture, which varies throughout the

Evapotranspiration

Evaporation is virtually a continuous process that takes place throughout daylight and often during the night. The process is most active under the direct action of the sun, but will continue slowly even under increasing cloud cover. The layer of air immediately above the water surface becomes saturated and this limits further evaporation until the boundary layer is removed by wind. Thus evaporation is also a function of wind which is, itself, dependent on the position and exposure of the reservoir. The relative humidity of the air also has a significant effect on evaporation, in that a high humidity will tend to decrease the rate of evaporation. Temperature, however, has the most significant effect as more energy is available to vaporise the water. The capacity of the air to absorb water vapour also increases as its temperature rises so an increased air temperature has an enhancing effect on the rate of evaporation. In this country, the annual loss of water by evaporation is typically about 450 mm from vegetated land surfaces and slightly higher for bare soil surfaces.

The losses from land surfaces by this combined process of evapotranspiration are not of direct concern to the water already in the reservoir, but clearly have a potentially significant effect on the runoff from the catchment, reducing the amount of water which could reach the reservoir. The amount of transpiration depends on the incidence of precipitation, other climatic effects such as temperature and humidity and the type, extent and method of management of the vegetation. The amount may be increased where large deep rooted trees extract water at depths from which it would not otherwise be lost. Transpiration does not occur at night and is largely dependent on the availability of the water supply. More will be lost in evaporation if a ready supply is available and thus a distinction must be made between potential and actual evapotranspiration loss. Typical losses from evapotranspiration in the UK are shown in Figure 6.2. The information sources in Box 6.1 may be able to assist with a more detailed assessment of the actual losses at a specific location.

Forest development on a catchment can affect the amount of losses compared to those for grass cover. Where the annual rainfall is in excess of two metres (typically in the more western parts of the British Isles), there is a considerably greater loss and catchment losses are about double those for grass cover, probably because of increased interception of the rainfall in the forest canopy and subsequent evaporation. In the drier eastern parts of the country, where the annual rainfall is about 600 mm, the losses are not significantly affected by the tree cover. Mature forest also reduces the peak of the runoff compared to that from grassland, but where drainage ditches have been constructed there may be an increase and a faster rate of runoff. Immature and newly planted woodland will generally result in an increased amount and rate of runoff and this will be exacerbated by any drainage ditches. Water quality of the runoff may also be affected by the planting or removal of woodland within the catchment and this should be considered for any reservoir creation or changes in catchment use.

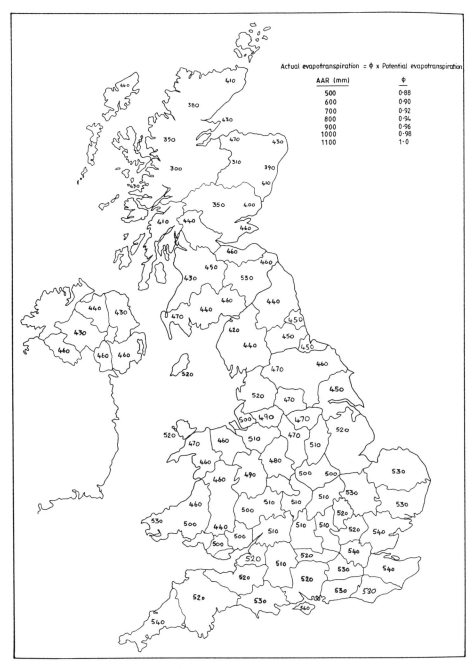

Figure 6.2 Potential evapotranspiration (Based on average evapotranspiration losses (mm) (1930–1949)) (Ref 53)

Rainfall on a catchment which is not lost by evaporation or transpiration may be removed by other processes. Geological features in the ground, such as the dip of the strata, natural impermeable barriers, or existing water passages through large voids, may control the groundwater seepage pattern. In certain instances, a flow from one valley to an adjacent one will lower the groundwater levels or decrease the runoff. Alternatively, the reverse may occur and groundwater from outside may be directed into the catchment area to increase the groundwater levels or emerge as springs. In some cases, these features may have a significant effect on seepage or drainage from the catchment, and specialist advice should be sought.

Similarly, man-made features may control both groundwater and surface water flow. Past mining activities, which now may be entirely hidden and forgotten, may have a significant effect on the drainage and seepage pattern of an area. Early industrial sites may have affected a valley, particularly where they involved water power using constructed or realigned channels. Such sites, including early water mills, may now be long abandoned but may still influence the present drainage and seepage patterns. Field drainage may also affect the amount and rate of runoff, although blocked drains may often mean that a drainage system is less effective than thought. It has also been shown that unploughed land where an extensive root system has developed can drain more readily than ploughed land with underdrainage.

Change of land use in the recent past will also tend to affect the amount of runoff and infiltration. Slopes which were once wooded or covered in dense vegetation, but which are now grassed or cropped, may still be adjusting to their new condition and thus more runoff may occur. Conversely, the afforestation of an area is likely to reduce runoff in the long-term. Any change in the vegetation is likely to change the runoff and it may take many years for new conditions to become established.

Urban developments and increased density of drainage network on the catchment usually increase the amount and speed of runoff. If the urban development exceeds about ten percent by area, the runoff is likely to be considerably greater and more rapid, and specialist advice should be sought. The position and size of the urban development is important; if it is concentrated in large areas rather than scattered pockets, or if it is located close to the reservoir, its effects can be more significant and lead to very rapid runoff following rainfall. Urban developments include areas of hardstanding and impermeable surfaces which offer little resistance to runoff. Thus developments such as car parks, glasshouses, large areas of plastic sheeting for horticultural use etc., allowing negligible seepage into the ground would fall within this category. Where the development is near the edge of the catchment, and depending on the drainage measures adopted, the runoff may be directed into the adjacent catchment. Alternatively, the runoff may be enhanced by drainage from such features just outside the catchment.

Bodies of water within the catchment area will also effect the amount and speed of runoff. Their effect is discussed in Section 7.2.2.

Where runoff is to be minimised, it may be possible to install a channel or channels to divert some of the flow away from a watercourse. Such channels must be of sufficient size and diverted around any reservoir if they are to reduce runoff into the reservoir. If this is done, the simple methods of runoff assessment given in this guide are likely to be inappropriate and specialist advice should be sought. The inadvisability of using this approach for dealing with flood producing runoff is discussed further in Section 7.1.8.

6.2.2 Estimation of available water

Many reservoirs only require a constant level as no abstraction of water is required for amenity, conservation or landscaping and most fishing and probably fish farming reservoirs. Seasonal water level variations and environmental features may also impose certain constraints. Where water is required for irrigation purposes, the maximum quantity or rate is likely to be needed in the spring and summer months when flows are at their minimum. The estimation of the required quantity or rate is outside the scope of this guide and reference should be made to the various publications by MAFF or ADAS or assistance sought from a specialist agricultural adviser. Once the demand figures are known, the size of the required storage can be assessed allowing for lost storage and drought conditions as subsequently discussed. Storage for fire fighting reservoirs should be based on the required volume plus lost storage, although the need for further water soon after the reservoir has been used should be considered.

Estimation of the quantities available from sources of water and their reliability is not an exact science. Complex methods are available, but these are not readily applicable to smaller reservoirs where the costs of extensive investigations and studies could easily approach the cost of the construction works. Experience and judgement can estimate flows based on straightforward methods of investigation and assessment.

The yield of a borehole can be readily investigated by test pumping after sinking the hole and such work is normally carried out as part of the installation process. Authority from the RLA is required to sink the borehole before work starts and to test pump the borehole. An abstraction licence is required before water is extracted on a permanent basis. Boreholes should only be sunk after suitable consideration of the geology and groundwater conditions by a specialist adviser who should also consider and interpret the pumping test to ascertain the long term yield and the possible effects during drought years. The cost of boreholes, their development and test pumping is likely to be relatively high, and thus this source of supply may not be viable for the quantity of water required in many instances.

Spring flows are difficult to forecast and require a long record of observed flows and groundwater table levels. Detective work in discovering when springs dry up, marshland becomes suitable to graze, drain flows cease etc., is also useful in assessing the flows, but verbal recollections should be treated with care and confirmatory evidence sought. A knowledge and interpretation of the geology and groundwater conditions of the area is essential and may require specialist advice unless the mechanism of the spring flow is already well understood.

A river or large stream is frequently used as a source, either for direct retention in an impounding reservoir or by abstraction into a non-impounding reservoir. The RLA should be able to advise on the amount of water that is potentially available for use and offer advice as to whether an impounding licence might contain conditions requiring abstraction to cease when flow or level fall below a given value. The conditions may result in water not being available in certain years or at certain times of the year, i.e. the source is unreliable, and these are likely to be at times when the need is greatest. Advice on the reliability of the abstraction licence should be available, possibly given as the number of years in a given time period that water would be unavailable.

Sources such as smaller streams are not likely to have detailed records or knowledge of their flows and it is then necessary to make an estimate of the flow. Frequent measurements over an extended period are the best guide. The next best estimate would be the calculated runoff based on the catchment area and its characteristics. An estimate of the summer and winter flows should be made by the most accurate method at the earliest opportunity to compare with the quantity of water required throughout the year.

Typically 80% of the total annual flow occurs during the winter months and thus water stored from this period will be available for use in the following summer. The flow can be gauged daily by means of a simple weir as described in Appendix F to assess the flow in a typical year. An estimate of the flow in a dry winter can then be made as detailed in Appendix G, which also shows a typical series of flows and the calculations involved for a low rainfall location. The flow during the summer months can also be gauged during a period of low flow following dry weather to gain some indication of the base flow in the stream. This flow arises from water-bearing strata within the ground which discharge more slowly with time unless recharged. Changes may occur slowly, and there is a lag between rainfall and runoff from these strata that may extend to days or even weeks for a larger catchment. The underlying geology will also heavily influence the rate of response which will be almost instantaneous in some limestone areas, ranging to months in many chalk areas. The base flow can cease after a drought period and the stream may dry up completely. Unless the summer flows are gauged for several years, the recorded flow will probably not include a dry year and the minimum flow will not be recorded. Direct abstraction is only possible if the minimum dry year flow exceeds the required abstraction rate by a safe margin after allowance for any flow required by the relevant licensing authority to be released from the reservoir to maintain flows downstream (termed compensation flow). Further restrictions in drought years may be invoked by the relevant licensing authority. It is likely that continuous abstraction throughout the year will not be possible unless the required abstraction rate is only a very small fraction of the river flow.

When gauging is impractical, due to lack of time or other reasons, it may be possible to assess flows based on gauged flows in a similar stream. Every catchment, however, is unique and a difference in area, slope, soil type, vegetation, rainfall or any other characteristic may substantially influence the amount and speed of the runoff and affect the river or stream flows. Records of stream flows for small catchments are generally scarce. Small streams in the upper parts of a catchment may dry up in the summer although quite substantial flows are maintained downstream; hence a correlation of flows within a catchment is not straightforward. Recourse is normally made to a theoretical estimate of the average flow based on the catchment properties.

The amount of rainfall that enters the river or stream is the average annual rainfall minus the losses in the catchment. These losses are the amount of water lost by actual evapotranspiration and as a result of geological and man-made features (see Section 6.2.1). The losses, as a result of evapotranspiration, will vary substantially between catchments and may be of the order of 25% of the average rainfall in the wetter upland areas of the west and up to 80 to 85% in parts of East Anglia in an average year. The actual percentage for a particular catchment will vary and will be strongly dependent on the soil and vegetation type and the geological and man-made features in the catchment. In the absence of any more detailed data, a preliminary estimate of the average flow throughout the year may be made from:

$$\text{Average flow (m}^3\text{/s)} = 3.2 \times 10^{-9} (\text{AAR} - \text{Losses}) A_c$$

where AAR = Average annual rainfall extrapolated from Figure 6.1 (mm)
Losses = Actual evaporation from Figure 6.2 (mm)
A_c = Catchment area (hectares)

Any reductions as a result of geological or manmade features would reduce the average flow further, the amount being dependent on the features. Specialist advice should be taken if these additional losses are considered to be significant.

In a very dry year, the flows may be only about 40% of the average flow although most of this is likely to be in the winter months. The available volume of water will be reduced accordingly. It should also be noted that some of the runoff throughout the year will arrive during or soon after a storm, and will be lost as outflow over the overflow works, further reducing the amount actually available for use. This cannot be quantified directly, but clearly could reduce the potential amount of water available for storage. Any flows downstream required by the RLA should be deducted together with an assessment for losses from the reservoir as discussed in Section 6.3.1. Considered together, these factors suggest that a watercourse is unlikely to provide a sufficiently reliable source unless the annual demand for irrigation, etc., is less than about 10% of the total rainfall on the catchment. Thus an early assessment and discussion with the RLA is necessary to assess the feasibility of a reservoir where abstraction is required.

6.3 NORMAL OPERATION

6.3.1 Reservoir losses, dead storage and compensation flow

Losses by evaporation occur throughout the year, being greatest in the summer and least in the winter. The losses are proportional to the surface area. The losses may exceed 400 mm of water depth in a dry summer, with a total annual loss of about 600 mm throughout the full year, but these may be offset against most or some of the direct rainfall on the water surface.

Seepage from a reservoir will vary with the permeability of the soil and the depth of water. It can be estimated from permeability tests on soil samples or in-situ tests on site. Some information on the permeability may possibly be obtained from other tests on the soil, but this would require careful assessment by a specialist adviser. A conservative estimate of the permeability must be assumed and seepage losses calculated on this basis. Most unlined reservoirs are founded on clay-type soils and a seepage loss of about 300 mm depth of water over an area equal to the surface area should be expected. This assumes a homogenous reservoir bottom, which is rarely the case in practice, and the actual seepage quantity may vary. Natural or man-made features on or below the reservoir floor and the embankment may allow considerable seepage. A close visual inspection of the site and past records of the area should help to locate any potential seepage paths. The features to watch out for are detailed in Section 8 for a proposed reservoir and Section 12 for an existing dam, but old land drains may offer a ready-made seepage route. Unless land drains are known beyond all reasonable doubt to be absent, it is normal to design the embankment to include measures to locate and deal with all drains as detailed in Section 9.

Seepage may also occur through the embankment fill or the fill/foundation contact if the standards of construction are poor. Adequate construction control is necessary to avoid this type of problem. Fill placing around concrete structures or pipes must also be controlled closely to minimise the risk of problems. Good construction practice should avoid this although the losses could be substantial if these are allowed to develop, which could ultimately threaten the integrity of the embankment. Seepage quantities through the fill should not normally be a significant amount.

Losses may also occur through pipes, culverts or other structures through the embankment as a result of poor construction techniques or inadequately sealing valves or penstocks.

Not all the water within a reservoir is normally available for use. The lowest level of outlet works may be limited and this will control the minimum level to which the reservoir may be drawn down. In some instances a pipe may be present at the lowest point of the reservoir to allow the reservoir to be fully drained, but this operation should only be carried out for particular maintenance or clearance works. Good operation practice is to leave not less than 500 mm depth of water over the reservoir floor if it is relatively flat. This is particularly relevant if a clay blanket is present beneath the reservoir floor. If the reservoir contains fish, a greater depth or areas of deeper pools may be necessary.

Therefore a greater depth of water than required for use is likely to be needed. Where winter abstraction only is available, with storage in an unlined non-impounding reservoir, allowance for the following should ideally be provided to give a total depth of unavailable water:

- annual evaporation   600 mm
- annual seepage   300 mm
- dead storage   500 mm

If some abstraction during the summer is possible, it may be feasible to reduce the allowances for evaporation and seepage, but a conservative approach should always be taken. A reduced allowance for evaporation may be possible if offset against summer rainfall, but this should only be considered if detailed rainfall records are available to allow assessment of reliable summer precipitation.

Flows in a watercourse are likely to continue throughout the year and thus evaporation and seepage losses will be more than adequately balanced. If the flow is not continuous, some allowance for summer evaporation and a percentage of the seepage losses should be provided. It is good practice to allow for dead storage and to take a conservative assessment for the other losses.

Compensation flow required by the RLA is generally only a percentage of the total flow, but normally must be provided throughout the year. The flow in the watercourse must be sufficient to provide this and also yield the required quantity of water when abstraction is required. If the flow is insufficient during the summer to provide the required compensation the available storage will be depleted. Depending on the actual size of the compensation flow, this may require the reservoir to be enlarged and could move the reservoir into the scope of the Reservoirs Act 1975 or alternatively make the scheme less attractive.

6.3.2 Water quality

Water quality may not be an important question for most of the landscaping or amenity reservoirs, but the quality will affect the type of vegetation, animal, fish and other fauna that can flourish in or adjacent to the reservoir. When the water is required for irrigation, fish farming etc., the water quality may be of greater significance and may limit the possible usage of the water.

Where appropriate, the quality of the water, both in bacteriological and chemical terms, must be adequate for the purpose and specialist advice may be necessary before using suspect water. Surface water may be contaminated with sewage, farm effluent, agrochemicals from the catchment, salt, petrochemicals and other chemicals, or pollutants from roads, whereas water from underground sources, particularly springs or other discharges, may contain dissolved salts.

The quality of recycled water, if used as a water source, will require careful consideration and assessment and, in most cases, specialist advice must be sought.

Water taken from public supply is satisfactory in terms of cleanliness, but the methods of treatment may result in traces of chemicals which are not satisfactory for some species of fish. If the water is left to stand in the reservoir after filling, these effects will tend to be minimised.

The presence of certain species of trees, especially conifers close to the waters' edge, may allow substantial quantities of leaves to enter the reservoir which are liable to be detrimental to the water quality.

Certain flora and fauna may be utilised to assist in improving certain aspects of water quality and ADAS, MAFF or other consultants may be able to advise further in this respect.

Measures can be installed to circulate or aerate the water in a reservoir, but these are normally not suitable for the smaller reservoirs unless a simple cost-effective method can be used.

Water quality is most likely to deteriorate during long, dry spells when the amount of water entering the reservoir is reduced or may have ceased. In recent times, algal growths have been prominent as a result of enhanced temperatures or levels of phosphate and nitrate in the water. These can prove deleterious to animals and humans, but appear to have a less serious effect on the reservoir flora and fauna. Algal growth can also develop during the first filling of a reservoir as conditions in the stream or river upstream of the reservoir adjust to the changed water levels and flow rates. Such growth normally settles down and comes into equilibrium with the other reservoir flora and fauna.

6.3.3 Operating levels and first filling

Omission of any low-level outlets will fix the operating level of a reservoir at top water level plus a rise in level to pass the flow over the overflow. Whilst such an arrangement is adequate for landscaping or amenity reservoirs and other usages which merely require the water area, the provision of a low-level outlet will greatly aid the operation and maintenance of the reservoir. An outlet pipe must be provided if a compensation flow is required, and then the operating level may drop below the overflow. A suitable arrangement would also allow the water level to be lowered for maintenance or if there were to be any emergency. Water quality may also be improved by controlled discharge and the generation of internal movement within the reservoir.

The rate of filling of the reservoir should not cause any problems so long as suitable facilities are provided for water discharge from pipes or man-made channels. Where an impounding reservoir is to be filled, this should be carried out at as slow a rate as possible consistent with the inflow and any control that may be available by the outlet pipework, if present. A specified flow must be maintained in the downstream channel during impounding and this may require temporary measures to allow this to be achieved. The flows are not likely to be prolonged and a small pump should be adequate to provide the required outflow. Alternatively, the use of a syphon may be possible to pass a small continuous temporary flow.

Operation of the reservoir at varying levels will require adequate wave protection over a range of depths on the embankment. This may need to be continued onto the reservoir perimeter or provided at certain positions if erosion is critical, i.e.

adjacent to pipework or structures. Operating at a low level may also expose unsightly stretches of mud or divide the reservoir into a number of separate water areas which may affect the flora and fauna. Where the water level is constantly at and just above top water level, the wave protection can be of limited vertical extent on the embankment, but the near constant level may result in increased erosion around the reservoir and the formation of small cliffs at the waters' edge.

Drawdown of the reservoir can lead to increased stresses within the embankment. Most drawdowns are rapid, compared to the drainage ability of the fill. Even if they occur over an extended period and lead to a lowering at the rate of one metre a week, the stability of the upstream slope is decreased although few problems with drawdown have been recorded in small embankments in the British Isles. Clearly drawdown under normal operation conditions can only take place if an outlet pipe is provided or pumping is carried out directly from the reservoir.

Under emergency conditions, pumps can be brought in to increase or provide means of drawdown if the reservoir needs to be emptied rapidly. This would normally be carried out only if the reservoir posed a serious threat of embankment instability or leakage.

6.3.4 Sedimentation

Reservoirs within the scope of this guide are likely to be located on the upper levels of catchments where the stream or river may still be eroding downwards to an equilibrium position. Thus upstream material will be eroded continuously and transported in the watercourse, particularly at times of high flow. Construction of the reservoir will decrease the speed of flow in the watercourse as it enters the reservoir; thus any material which is being carried will be deposited. Over an extended period the reservoir will be filled with material or extensive areas of very shallow water or soft silts may develop. Depending on the purpose of the reservoir, this may not be critical, but measures will ultimately be required to remove the deposited material.

Deposited material may be removed by dredging but this is likely to be expensive and will result in substantial quantities of very soft, virtually fluid, material. This will need to be disposed of in some location sufficiently far away to avoid the material washing back into the reservoir. This operation may require permission from the RLA.

If the silt load is not excessive, it may be possible to form a silt trap at the upstream end of the reservoir. This may be achieved by means of vegetation, but an adequately-sized silt trap can be readily formed by localised excavation which can be dug out as required. This aspect is discussed further in Section 9.8.4.

6.4 GROUNDWATER IMPLICATIONS

6.4.1 Upstream and around the reservoir

The presence of the reservoir will raise the level of the groundwater table in the surrounding area such that it will continue away from the reservoir as an extension of the reservoir level. Unless drainage influences the groundwater levels these are likely to rise away from an impounding reservoir to join the pre-existing groundwater table at some distance. The extent of the area affected by the change in groundwater level is dependent on several features, but the geology, topography and the previous groundwater levels, which will have reflected the topography to some extent, are the major influences. Depending on the depth of water in the reservoir and the topography, the changes may not extend for a substantial distance. Where the water depth is greater, the topography is relatively flat or the ground is moderately permeable, the effects may extend for a greater distance.

The situation for a non-impounding reservoir may be less certain. Depending on the same criteria, the previous groundwater table may have been at a moderate depth below the ground surface and thus construction of the reservoir may substantially change the flow pattern in the ground. In the worst instance, this may lead to excessive water loss and erosion problems within the foundation. Unless the groundwater surface is known to be near the surface, specialist advice should be sought before construction.

The presence of geological or man-made features beneath or adjacent to the reservoir may also affect the changes to the groundwater level. This may also require specialist advice before construction.

Groundwater changes may be reflected in many ways:

- seepages into an adjacent catchment
- temporary or continuing erosion
- existing vegetation may become waterlogged and die
- water-loving vegetation may flourish
- surrounding slopes, structures or buildings may be affected by instability or movement.

Where there is any risk of detrimental effects, these should be considered prior to construction and specialist advice sought as required.

Upstream of the reservoir, the groundwater will rise in a similar, but more restrained manner as the flow in the watercourse slows down and deepens as it approaches the reservoir. Similar effects may be evident further upstream, therefore, and these may be enhanced at times of high flows leading to more localised flooding and saturation of the ground adjacent to the watercourse.

The discharge of drain outfalls from properties or structures upstream may also be affected, particularly where the pipes have been laid to a flat gradient. Similarly, cesspits, soakaways or other features may be affected.

6.4.2 Downstream of the reservoir

Downstream of an impounding reservoir the reduced flows in the watercourse may lead to a lowering of the groundwater level. This would not normally be sufficient to cause any noticeable changes, although this could be detrimental to existing communities of wetland plants and animals. The changes upstream may cause flows around the ends of the embankment if these are significant and the foundation material is sufficiently permeable. If the flows are sufficient, this may cause difficulties both with seepage losses and possibly instability downstream of the embankment as a result of raising the groundwater table in the valley sides. Measures may be required to limit flow and allow adequate drainage. Water-loving vegetation may also develop in such cases. English Nature or the RLA should be able to advise on specific situations.

The presence of geological or man-made features beneath or around the embankment will affect the change in groundwater level and possible seepage flows. Such features could prove catastrophic for the reservoir and must be considered during the feasibility stage. If such features are suspected in any way, specialist advice must be sought at a very early stage.

Riparian owners downstream of the reservoir may have wells or boreholes which could be affected, although these are not likely to be strongly influenced by the scale of construction within this guide. The RLA will advise in this respect and will consider this aspect when issuing the impounding licence.

6.5 CHANGES IN RESERVOIR AND CATCHMENT USAGE

6.5.1 Reservoir

A reservoir will be designed and constructed for one or more specific purposes and certain features will be provided depending on these purposes. Thus, for an amenity lake where the reservoir is to operate at top water level, no low-level drawoff pipe may be provided and slope protection on the dam might be limited to a shallow range around the top water level. A subsequent change in reservoir usage may require other operating criteria to be met; water-skiing, for example, would necessitate a greater range of slope protection, irrigation would need a means of extracting water, etc. For instances, where a change of reservoir usage is envisaged, the consequences should be fully considered before the changes are implemented. This may require additional works to be carried out and may, in some cases, limit the scope of the reservoir for alternative use. In certain cases, additional or amended consents may be required, e.g. the introduction of certain types of fish farming into a reservoir.

Changes of use may also affect the environment aspects and alter the habitats which have developed as a result of the reservoir creation. This may not be popular and may require liaison and close co-operation with any interested parties prior to any changes.

6.5.2 Catchment

Changes in the catchment may increase the amount or rate of runoff. This might arise, for example, if significant areas of woodland have been felled or if the drainage system from the catchment was improved or renovated. Increased urbanisation, as defined previously, would also increase the amount and rate of runoff, particularly if this was located close to the embankment or concentrated into sizeable areas. Alternatively, a change of land usage or subsequent afforestation may decrease the amount and rate of runoff. This may be beneficial in terms of reduced runoff from major rainfall, but may also reduce the yield of the reservoir if this is relevant for abstraction purposes. Increased erosion and deposition of material in the reservoir may also occur, whilst additional floating debris is likely to find its way into the reservoir and possibly affect the outlet/overflow works.

Changes of land use in the catchment might also increase quantities of sewage effluent, agrochemical, salt, petrochemical and other chemicals entering the reservoir. The construction of a road or urban development is particularly likely to lead to a general increase in pollutants. Increased public access in any form is likely to be associated with greater vandalism, dumping and pollution.

7 Floods and waves

7.1 FLOOD ASSESSMENT

7.1.1 Introduction

Flood assessment and provision for passing flood flows are required for all impounding reservoirs whatever the size of the catchment. If a non-impounding reservoir is adopted, which has no inflow from a catchment, a suitable overflow is required to discharge direct rainfall and any continued inflow if the reservoir is full, together with any inflow from the embankment slopes themselves. A non-impounding reservoir formed by embanking on sloping ground will have a small flow from the sloping catchment, in which case appropriate facilities must be provided to pass the extra flood runoff in addition to the direct rainfall and continued inflow.

The flow in a stream or river varies continuously, the extent depending principally on the amount and rate of the runoff. Other factors such as seasonal effects, with their consequent influence on vegetation, or differing land usage throughout a year have a lesser influence. Runoff will vary throughout the year, being generally greater in the winter months, whereby the reduced losses in the catchment noted in the previous section, will tend to produce the largest average flows in the winter. Rain falling onto frozen ground or runoff from melting snow may also lead to a substantial increase in flow. The majority of the most severe floods in the past, however, have resulted from the intense localised summer thunderstorms which occur in July through to September; the well-known flood at Lynmouth in 1952 was of this type.

Thus flow in a stream or river can increase dramatically after very heavy rainfall and often to a substantially greater amount than thought likely or seen on previous occasions. Ultimate flows of up to several hundred times the average flow could be possible.

The estimation of flood flows is an inexact science and is particularly so for the smaller catchments generally associated with reservoirs and which fall within the scope of this guide. The characteristics affecting runoff, as discussed in Section 6, will all influence the amount and speed of the runoff and hence the maximum flood flow to some extent. The main factors that should be considered for flood assessment are:

- flood producing rainfall
- catchment area, surface water areas and areas of rapid runoff
- stream and catchment slopes
- soil and geology
- vegetation and the extent of afforestation.

The smaller the catchment, the steeper the slope, the thinner the soil and vegetation cover and the lower the surface permeability, the greater will be the magnitude of the flood per hectare of the catchment for a given rainfall. Thus steep rocky catchments in high rainfall regions are likely to produce high storm flows per hectare; those with a flat permeable catchment in low average rainfall regions will produce lower storm flows per hectare. Urban development within the catchment will tend to increase the flood by producing rapid and near total runoff from impermeable concrete and similar surfaces, whilst surface water areas, including the reservoir, will affect the flood flow from the reservoir. The effects of vegetation on runoff are discussed in Section 6 and can vary.

If detailed runoff observations are available, together with the appropriate rainfall records, these can be considered to help the assessment of extreme floods. Such work is beyond the scope of this guide and specialist advice should be sought if this is required. This data is not normally available, and thus recourse has to be made to other statistical methods to estimate flood flows based on the catchment and climatic characteristics and parameters.

It is customary to consider flood flows in terms of the frequency of occurrence or return period. Given adequate records, statistical methods will show that floods of a certain size may, on average, be expected once every year, every ten years, every 50 years etc. The floods are random events and thus the 50-year flood (that is the flood that will occur on average once in 50 years) may not occur for 200 years, or could happen several times in the next 50 years, next year or tomorrow. In any given period, however short, there is a small risk of occurrence of floods beyond a certain size. This aspect is particularly important in assessing the risk of flood flows during construction. The required return period to provide for is related to conditions downstream of the reservoir, as discussed fully in Section 7.2.1.

7.1.2 Flood producing rainfall

The amount of recent rainfall that has fallen has a significant effect on whether subsequent rainfall infiltrates into the ground or runs off. A period of five days is taken to provide a recent history for detailed forecasting, but a sufficiently accurate assessment of the flood-producing rainfall can be gained from an index denoted as RSMD. Typical values of RSMD are given in Figure 7.1. This is defined as the one-day rainfall of a five-year return period less the effective mean soil moisture deficit (see Box 6.2). This index parameter can vary by up to 100 mm in the mountainous areas in the north and west of the British Isles, but is typically between 25 and 40 mm over most of the central, southern and eastern areas.

The procedure to follow for calculating RSMD and the peak flood flow, Q_m, is given in Figure 7.2. The flow, expressed in terms of flow per unit area of the catchment, (Q_m/A_C), can be determined from Figure 7.3.

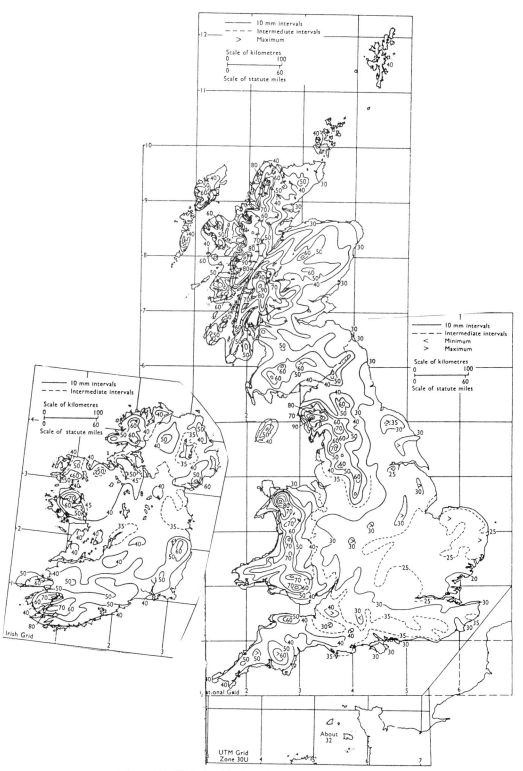

Figure 7.1 RSMD (mm) in Great Britain (Ref 54)

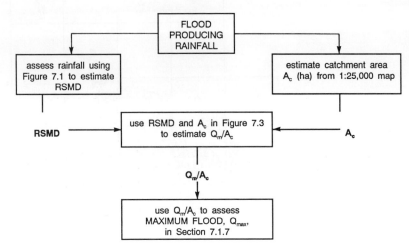

Figure 7.2. Assessment of peak flood flow

7.1.3 Catchment area, water surface areas and areas of rapid runoff

The relative difference between the catchment area, the area of surface water (including the proposed reservoir) and the area of rapid runoff within the catchment is very important. The catchment area A_C may have to be amended to take account of these relative differences. The amended catchment area is expressed in terms of a catchment area factor A.

Catchment area, expressed in hectares, can be best assessed from a 1:25 000 map. The extent of the catchment can be plotted by drawing a boundary along the topographical water divide; this must be continuous and extend around the catchment from one end of the embankment to the other. The boundary will tend to cross contours perpendicularly and run along the highest positions of a ridge, often along the route of a crest track, with the ground surface dropping away on both sides. A typical example is shown in Figure 7.4. The boundary thus encloses all of the area draining into the watercourse and runs along the local drainage divide. It is possible for areas outside the catchment to either contribute or drain water as a result of the underlying geology, but the amount is normally small for flood assessment and ignored. If highly porous or permeable strata are known to dip towards or away from the catchment as also shown in Figure 7.4, their effect on flood runoff may need to be considered. A cave or pothole system may also allow substantial quantities of water to leave or enter the catchment. Similarly any man-made features which, either deliberately or accidentally, allow water to enter or leave the catchment should be included. The effect of these natural or man-made features will depend on several aspects and specialist advice should be sought.

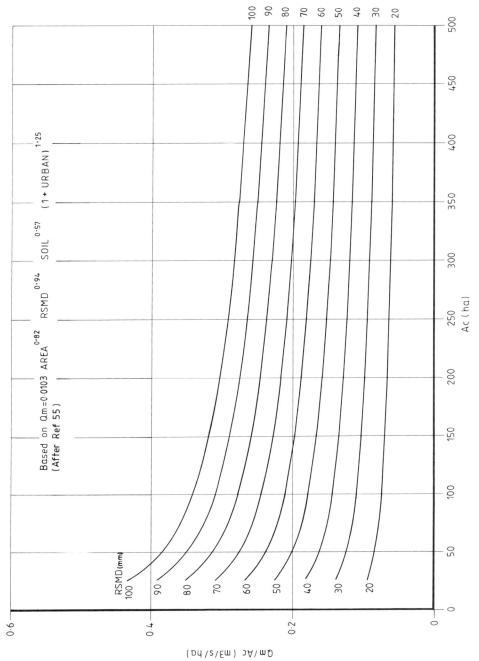

Figure 7.3 Flood peak intensity for small reservoir catchments (After Ref 54)

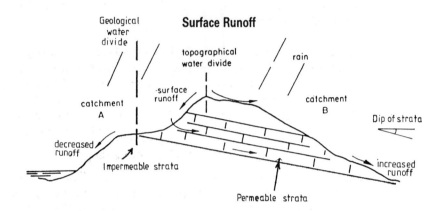

Influence of Permeable Strata

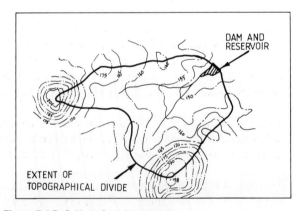

Figure 7.4 Definition of catchment area

Reservoirs falling within the scope of this guide are likely to have a catchment of a few tens to a few hundred hectares. Any urban development, whether existing or planned, or other feature which could allow rapid runoff, as discussed in Section 6.2.1, should also be estimated from the map as an equivalent area in hectares. A conservative estimate should be taken. If the development is concentrated into one or two areas, particularly if these are near the downstream end of the catchment, the runoff may be enhanced considerably and therefore specialist advice should be sought.

The catchment area may be defined topographically or geologically, as shown below. In most cases there will be limited information to determine the geological water divide.

Both urban development and water surface areas may have changed significantly since the map was last updated. In all cases, the review of the map must be supplemented by a walkover of the catchment, so far as is possible, to ensure that these features are not significantly different from those indicated on the map. A close visual inspection will also help assist in assessing the actual extent of the catchment and may reveal other features which might affect the flood runoff from heavy storms.

The procedure to follow to calculate the catchment area A_C, the water surface area A_W, the area of rapid runoff A_R and hence the amended catchment area factor A, required for calculating the maximum flood Q_{max}, is given in Figure 7.5.

7.1.4 Stream and catchment slopes

The steepness of a catchment is expressed in terms of the gradient of the main watercourse in the catchment, defined as the S1085 value. A preliminary value can be estimated initially from Table 7.1. It is obvious from this table that an estimate of a 'Hilly' catchment may encompass a broad range of S1085 values. The value of S1085 may fall anywhere in this range, but for the smaller catchments is more likely to be in the undulating or hilly categories ranging between 10 and 40 m/km. The value of S1085 should be calculated following the procedure outlined in Figure 7.6 to improve the initial estimate.

The S1085 value can be used to assess the stream slope factor G, from Table 7.1 and to consider the implications of the steepness of the catchment on runoff. The stream factor G, can be calculated following the procedure shown in Figure 7.6.

The mean catchment slope, expressed as 1V:gH, also influences the amount and rate of runoff. The value is broadly related to the S1085 value but is also influenced by the catchment slope and contour pattern. A sufficiently accurate value can be gauged by averaging the slope perpendicularly across the contours on the 1:25 000 map from the local drainage divide to the watercourse. A sufficient number of determinations must be carried out around the catchment to produce a realistic mean value.

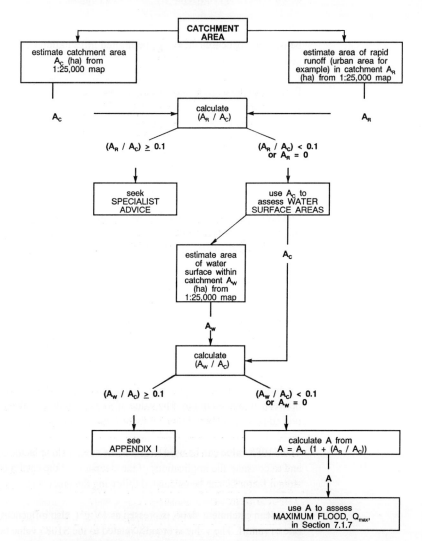

Figure 7.5 Assessment of catchment area, water surface areas and areas of rapid runoff

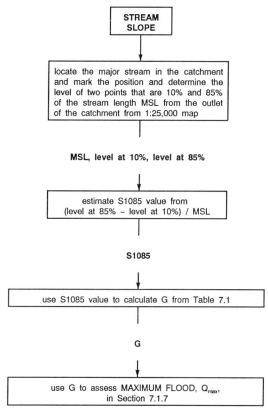

Figure 7.6 Assessment of stream slope factor

Table 7.1 Stream slope factors

TOPOGRAPHICAL CLASS	S1085 VALUE (m/km)	G
Mountainous	more than 100	1.40
Hilly	80–100	1.34
	60–80	1.28
	40–60	1.22
	20–40	1.12
Undulating	10–20	1.00
Flat	less than 10	0.94

7.1.5 Soil and geology

The runoff will be strongly dependent on the ability of the surface and subsurface soil to accept rainfall. This, in turn, is essentially related to the underlying soil or rock type, groundwater conditions and ground slope. Thus the runoff will be dependent on four properties:

- drainage ability of the underlying soil or rock and the level of the groundwater
- depth to impermeable horizon
- permeability of surface and subsurface soils
- mean slope of the catchment towards the watercourse.

The drainage ability can be readily assessed from the depth and duration of waterlogging. This should be evident from the appearance of the catchment throughout the year, but information from the site investigation (Section 8) should allow the general depth to groundwater, the drainage ability of the surface, subsurface and deeper soils and rocks, and the depth to the impermeable horizon to be found. The permeability of the surface layers should be evident from visual observations in wet weather or from the simple tests described in Section 8. The mean slope towards the watercourse can be visually assessed as discussed in Section 7.1.4. The four properties are defined in Table 7.2 and are combined in Table 7.3 where the effects of the soil types and topographical setting on runoff can be assessed. The table details five major soil classes; Soil Class 1 has the lowest runoff potential, whilst Soil Class 5 has the greatest runoff potential.

Table 7.2 Classification of soil properties – (Ref 56)

PROPERTY	CLASSES	
Drainage Group	1	Rarely waterlogged within 600 mm of ground level at any time (well and moderately well-drained)
	2	Commonly waterlogged within 600 mm during winter (imperfect and poor)
	3	Commonly waterlogged within 600 mm during winter and summer (very poorly drained)
Depth to impermeable horizon		More than 800 mm 800 to 400 mm less than 400 mm
Permeability of surface and subsurface soils		Rapid Medium Slow
Slope 1V:gH		g more than 28 (less than 2°) g between 28 to 7 (2°–8°) g less than 7 (more than 8°)

Table 7.3 Classification of soils by runoff potential – (Ref 56)

Drainage Loss	Depth to impermeable horizon (mm)	Slope Classes (g)								
		more than 28 (0 – 2°)			28 – 7 (2 – 8°)			less than 7 (more than 8°)		
		Permeability of surface and subsurface								
		Rapid	Medium	Slow	Rapid	Medium	Slow	Rapid	Medium	Slow
Well and moderately well–drained	more than 800	1	1	1	1	1	1	1	2	3
	400 – 800	1	1	1	2	2	2	3	3	4
	less than 400	-	-	-	-	-	-	-	-	-
Imperfect and poor drainage	more than 800	2	2	2	3	3	3	-	-	-
	400 – 800	2	2	2	3	3	3	4	4	4
	less than 400	3	3	3	3	3	3	4	4	4
Very poorly drained	more than 800	3	3	3	5	5	5	-	-	-
	400 – 800	3	3	3	5	5	5	-	-	-
	less than 400	3	3	3	5	5	5	-	-	-

NOTE: Run-off potential: 1 = very low; 2 = low; 3 = moderate; 4 = high; 5 = very high
Percentage areas of Classes 1 to 5 are S_1 to S_5 respectively.

High runoff potential is favoured by:

- low permeability surface and subsurface soils and rocks
- high groundwater levels
- shallow depths to impermeable horizons
- steep slopes.

Rapid movement of water likely to cause flooding during or after heavy rain, tends to occur near or over the soil surface; movements through deeper soils or permeable rock tend to have little or no effect on short term flood flows.

Table 7.4 gives a description of soils falling typically within each class and may be used as a first estimate for classifying the soils within the catchment.

Where differing soil classes are present within a catchment as a result of varying drainage abilities, permeabilities or slopes, the percentage areas of each soil class should be assessed (S_1 to S_5) and an overall soil index, denoted SOIL, calculated. Many reservoirs within the scope of this guide are likely to fall entirely within one class and thus SOIL is calculated on the basis of this class with the remaining percentage areas taken as zero. It should be noted that SOIL can only vary between 0.15 and 0.50.

The effects of soil type on runoff can be assessed in terms of a soil factor, denoted V, in Table 7.5. The procedure to follow to calculate both SOIL and V is given in Figure 7.7.

Table 7.4 Description of typical soil classes – (Ref 56)

Soil class		Typical soils
1	(i)	Well-drained, permeable sandy or loamy soils and similar shallower soils over highly permeable limestone, chalk, sandstone or related drifts.
	(ii)	Earthy peat soils drained by dikes and pumps.
	(iii)	Less permeable loamy over clayey soils on high ground adjacent to very permeable soils in valleys.
2	(i)	Very permeable soils with shallow ground-water.
	(ii)	Permeable soils over rock or frangipan, commonly on slopes in western Britain associated with smaller areas of less permeable wet soils.
	(iii)	Moderately permeable soils, some with slowly permeable subsoils.
3	(i)	Relatively impermeable soils clays, and in alluvium, especially in eastern England.
	(ii)	Permeable soils with shallow ground-water in low lying areas.
	(iii)	Mixed areas of permeable and impermeable soils, in approximately equal proportions.
4		Clayey, or loamy over clayey soils with an impermeable layer at shallow depth.
5		Soils of the wet uplands (i) with peaty or humus-rich surface horizons and impermeable layers at shallow depth, (ii) deep raw peat associated with gentle upland slopes or basin sites, (iii) bare rock cliffs and screes and (iv) shallow, permeable rocky soils on steep slopes.

NOTE: Urban areas are unclassified

Table 7.5 Soil factors

SOIL	V
0.5	1.00
0.45	0.95
0.4	0.89
0.35	0.83
0.3	0.76
0.25	0.69
0.20	0.61
0.15	0.53

Certain catchments can be more complex than first imagined; chalk, for example, has a low runoff potential (Class S1), but many chalk valleys have a less permeable layer over their lower levels and thus an enhanced runoff from these areas. Conversely, low permeability catchments may have gravels beneath the valley floor which may allow flow beneath a surface clay layer supporting a small reservoir.

The effects of a variable geology over a catchment cannot be readily assessed and may result in runoff differing significantly under various climatic conditions and from different areas of the catchment. Thus, where there is any doubt or the soil conditions are complex or uncertain, a value for SOIL of 0.5 should be used unless specialist advice is sought.

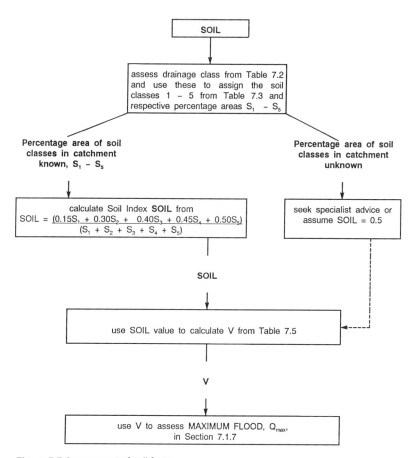

Figure 7.7 Assessment of soil factor

7.1.6 Vegetation

Vegetation in the catchment can have a significant effect on the runoff, but this is dependent on both the vegetation type and age, the style of management and the average drainage ability of the catchment. The effects of future removal of the vegetation should be considered as this would potentially lead to changes in the runoff as discussed in Box 6.2. The overflow arrangements provided might then be insufficient to pass the previous design flood and the return period of the flood that could be safely passed by the embankment would be reduced (i.e. the risk of an overtopping of the embankment would be increased).

The effects of forested areas within the catchment on runoff can be quantified in terms of a factor W, based on the forestry index F in Table 7.6. The procedure to follow to calculate W is given in Figure 7.8.

Table 7.6 Forestry index

Conditions	Soil class	F
Mature woodland with dense undergrowth and no man-made drainage	1	0.8
	3	0.7
	5	0.6
Mature woodland with dense undergrowth and drainage ditches	1	0.9
	3	0.9
	5	0.8
Immature woodland with no drainage ditches	1	1.0
	3	1.0
	5	1.0
Immature woodland with drainage ditches	1	1.1
	3	1.1
	5	1.1
Newly planted with no drainage ditches	1	1.1
	3	1.1
	5	1.2
Newly planted with drainage ditches	1	1.1
	3	1.2
	5	1.3

NOTE: The index for intermediate classes may be interpolated. If forestry conditions are not known use:

$F = 1.0$ for mature or established woodland
$F = 1.2$ for newly planted woodland

7.1.7 Maximum flood

The maximum flood, Q_{max}, can be estimated using the factors calculated in Sections 7.1.2–7.1.6. The procedure to follow is given in Figure 7.9. An example of a flood calculation in which all the factors and procedures required to calculate Q_{max} have been incorporated is shown in Appendix H.

7.1.8 Other factors

A small reservoir sited in the flood plain of a large river will be subject to disproportionately large flood flows from the extensive catchment of the river. This would require significant works to pass small floods of low return period and the embankment would tend to become merely a weir across the watercourse. A large catchment area beyond the limit of Figure 7.3 is also likely to require extensive overflow arrangements even if a small flow in the watercourse has been recorded. The size of critical catchment would reduce with increasing steepness in the catchment, i.e. increasing S1085 value

(see Section 7.1.4), and thus in hilly and mountainous areas smaller maximum catchments would be applicable of perhaps 200–300 hectares.
These catchments are beyond the scope of this guide and specialist advice should be sought.

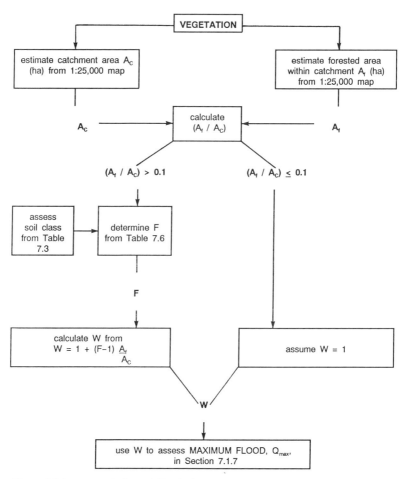

Figure 7.8 Assessment of vegetation factor

The presence of one or more upstream reservoirs or areas of surface water will require careful consideration depending on their relative sizes, types and condition of embankment, type and size of overflow arrangements etc. It is likely that the flood flows through the series of reservoirs will not be straightforward to assess, and this is particularly evident when two or more reservoirs are in sequence down a valley. This aspect is further discussed in Appendix I, but it is recommended that specialist advice be sought in all cases.

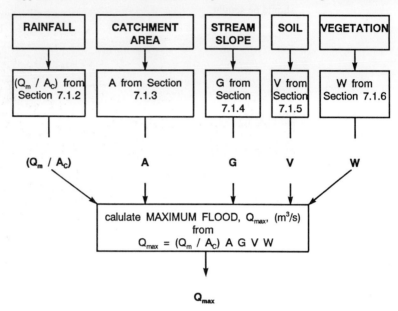

Figure 7.9 Assessment of maximum flood

7.1.9 Further advice

ADAS offer a peak runoff estimation service using a computerised system that can be used for rapid calculation of peak runoff. It may be used to check the calculations carried out following the procedures outlined in this section or as an independent service. For further information contact the ADAS Soil and Water Research Centre. Alternatively, the calculations can be carried out and advice provided by private sector consultants whose names can be obtained from ACE. The addresses are given in Appendix D.

7.2 PASSING DESIGN FLOOD

7.2.1 Overflow assessment

Flood flows entering the reservoir must be passed safely and discharged downstream of the embankment by means of a suitable overflow structure or structures. Failure to provide adequate arrangements to pass the flows will lead to discharge over the top of the dam (Figure 7.10), threatening the integrity of the embankment (Figure 7.11). Floods during construction pose a major threat to the partly completed embankment and are discussed further in Section 7.5.

Figure 7.10 High reservoir level threatening overtopping

Figure 7.11 Damage as a result of overtopping

The passage of the flood flow through the reservoir will require a head of water to discharge the flow over the overflow and thus the water level in the reservoir must rise. If this level is too great, and the rise in flood level must be considered in conjunction with any wave action as subsequently discussed, water will discharge over the embankment crest and the dam is then said to be overtopped. The risk of damage to the embankment arising from overtopping depends on several factors. These include:

- amount and duration of overtopping
- condition of the embankment
- uniformity of level along and across the top of the dam
- extent and condition of vegetation on the downstream slope
- existing features on the downstream slope
- gradient and slope length of the downstream slope
- type of fill material in the embankment
- crest surfacing and width
- time of year

Prolonged overtopping may be sufficient to cause erosion of the downstream slope; the junction of the embankment fill and original ground being particularly vulnerable. This may result in localised gulleying by flowing water, but may ultimately lead to washing out of the embankment and the release of a large uncontrolled quantity of water.

Protection standards must resolve acceptably the conflicting claims of safety and economy. Although it is now considered possible to design an overflow for the total protection of a dam against overtopping, there is a clear possibility that a smaller overflow built at less expense would survive several generations without any disaster or damage occurring downstream of the reservoir. This is particularly relevant if minor damage to the embankment after a severe flood is repaired to the previous standards. It is not, however, simply a matter of economic judgement as the design must recognise the many factors which may not be definable in monetary terms, particularly those which arise from effects on persons or communities downstream of the reservoir. Some dams, even if they are overtopped, are most unlikely to breach; increasing experience of erosion-resistance makes it possible to include this additional factor. It is necessary, therefore, to specify a design flood, in combination with an allowance for wave action, which the embankment must be capable of withstanding.

The passage of this flow through the reservoir should cause no fundamental structural damage to the embankment and the embankment must be fully capable of maintaining its intended function after such a flood. The same standards are not necessary downstream of the embankment and rare flows outside any channel or watercourse may be tolerated, provided that any damage does not affect the embankment and is repaired at the earliest opportunity.

The design flood, Q_i, is given by:

$$Q_i = M\, Q_{max}\ (m^3/s)$$

Where M = multiplier from Table 7.7
Q_{max} = maximum flood from Section 7.1.7 (m^3/s)

The class should be assessed from a detailed inspection of the area downstream of the reservoir on the 1:25 000 map, supplemented by a visual survey along the watercourse. It is envisaged that most reservoirs falling within the scope of this guide would be in Classes 3 and 4, but certain instances may require a higher multiplier. The comments in Box 7.1 should facilitate selection of the appropriate multiplier when read in conjunction with Table 7.7.

Table 7.7 Reservoir classification – (based on Ref 54)

Class	Description	Multiplier (M)	Approximate return period (yrs)	Minimum freeboard (m)
1	Reservoirs where a breach will endanger lives in a community of not less than ten persons	1.0	much more than 10 000	0.6
2	Reservoirs where a breach may endanger lives not in a community or result in extensive damage	0.5	10 000	0.6
3	Reservoirs where a breach will pose negligible risk to life or cause limited damage	0.3	1000	0.4
4	Reservoirs where no loss of life can be foreseen as the result of a breach, and very limited additional flood damage will be caused	0.2	150	0.3

A conservative approach should be adopted in all cases. Advice should be sought from a specialist adviser in case of doubt. Development downstream of the reservoir following construction may alter the class of the reservoir, an aspect which should be considered during the operation of the reservoir in subsequent years.

The multiplier, M, depends on the risk posed to life and property downstream of the reservoir. This risk is classified using four classes, 1–4; Class 1 being the highest multiplier with $M = 1.0$ to Class 4, the lowest, with a multiplier of 0.2. These classes are summarised in Table 7.7. This classification is based upon the principles recommended by the Institution of Civil Engineers in the publication: 'Floods and Reservoir Safety: An engineering guide'[54], in which reservoirs are given a category from A to D. The appropriate return period of the design flood (see Section 7.1.1) for each class can be assessed approximately. These assessments are also given in Table 7.7. These range from far in excess of 10 000 years for Class 1 to 150 years for Class 4.

Box 7.1 Selection of appropriate multiplier

Class 1
- village/housing development downstream of a reservoir
- isolated community buildings (such as schools and hospitals)
- industrial and commercial properties
- frequently used sports stadia and viewing areas.

Class 2
- railway stations
- permanent and intermittently used camping or caravan sites
- isolated houses
- well-used road, rail or other transport facilities
- extensive damage to agricultural land
- extensive damage to transport facilities
- rarely used sports stadia and viewing areas.

Class 3
- isolated individuals working in or crossing the area downstream of the reservoir
- loss of livestock
- loss of crops
- playing fields, golf courses, equestrian facilities with no viewing facilities.

Class 4
- guideline taken is that the amount stored would add not more than ten percent to the volume of the flood in the event of a breach which is assumed to be at the position at which damage could occur to a feature of value downstream (such as a mill or road bridge). If $1500 A_c/V_r$ is greater than ten, adopt Class 4.

Where A_c = catchment area (ha)
 V_r = reservoir volume (m^3)

7.2.2 Flood attenuation

The presence of a reservoir along the line of a watercourse affects the flood flows passing downstream. The large surface water area relative to the water course allows a volume of water to be contained temporarily within the reservoir and thus the flood passing into the reservoir is greater than that passing over the spillway i.e. the flood is reduced or attenuated. This is achieved by the water level increasing to accommodate the additional volume of water. Thus the quantity of water passing down the watercourse is not affected, but the maximum flood flow is reduced as the time of discharge is increased. This is in addition to any storage below the overflow level which will further help to reduce the overflow. This additional storage is clearly dependent on the operation of the reservoir and can only be available where a means of drawoff from the reservoir is present. As its availability cannot be relied upon, it should not be considered in any attenuation of the flood flow.

This process of flood flow reduction through the reservoir is known as 'flood routing' and can be readily and simply assessed for a reservoir from knowledge of the maximum flood rise that can be accommodated, the area of the water surface at the mid-height of this flood rise, and the maximum flood inflow into the reservoir, a factor dependent on the slope class of the catchment and the catchment area.

If the reservoir area is less than ten percent of the catchment area, it is unlikely that a significant reduction in outflow could be achieved and so this routing procedure is not worthwhile. Where the area is larger, some worthwhile reduction in the flood rise and hence also of the overflow size, may be possible, depending on the relative size of the reservoir and the permissible flood rise. This aspect is also discussed in Appendix I and an example of a flood calculation given.

Other reservoirs and water bodies upstream of the reservoir will similarly affect the flood flow in the watercourse, depending on their relative sizes within their catchments. This aspect is discussed further in Appendix I.

Where the reservoir area forms a very large part of the catchment, much of the rainfall falling onto the catchment will enter the reservoir directly. Some attenuation will still occur, however, as the water level rises, and thus the peak outflood from the reservoir will be reduced.

7.2.3 Overflow arrangements

The overflow arrangements must be designed specifically for a particular reservoir site and catchment area. Provision can be made quite simply and cheaply in some cases; in others more complex arrangements may be required which may prove uneconomic. Photographs of various options and types are given in Figures 7.12 to 7.19 and discussed in Sections 7.2.4 and 7.2.5. The basic requirement is that the overflows from the reservoir must be of adequate capacity to pass the design flood, Q_i, with sufficient freeboard (allowance for wave action) to avoid overtopping of the embankment. The overflows must also be able to withstand erosion for such periods as the flows are likely to occur.

Once an impounding reservoir is full, a continuous outflow from the reservoir will occur unless the inflow dries up or losses from the reservoir exceed this inflow. Thus overflow facilities must be provided which can take a continuous flow. A single overflow, if used must have adequate capacity to pass all flows up to the design flood Q_i, and, as it is in continuous use, must be constructed of material with high resistance to long-term erosion. This type of overflow is likely to be suitable only for very small catchments; separate overflow arrangements for low and high flood flows are likely to be more economic in the majority of cases.

Figure 7.12 Concrete weir chamber (with provision for adjustable weir boards)

Thus low flows should be taken by a main overflow constructed of concrete or other material with a high resistance to long-term erosion. Greater flows which will occur on a much less frequent basis should be taken by an auxiliary overflow which is set at a higher level than the main overflow. The latter will only be used on very infrequent occasions for short durations and is normally an earth channel of shallow depth, protected by a dense cover of grass or reinforced grass. The basic arrangements for the main and auxiliary overflow are shown in Figure 7.20, assuming the auxiliary overflow is located on the dam. Other arrangements and positions for the auxiliary overflow are possible depending on topographical and other constraints.

Assessment of flows is in any event an inexact process and thus a conservative approach must be adopted in assessing the design flow for the main overflow. A minimum value of ten percent of the maximum flood Q_{max}, entering the reservoir, as calculated in Section 7.1.7, should be used as the design flood for the main overflow. The remainder of the design flood, as shown in

Section 7.2.1 should be passed over the auxiliary overflow, with a maximum head over the overflow of 300 mm plus the required freeboard to crest level.

Figure 7.13 Vertical pipe with bellmouth (operating at top water level)

The total discharge capacity of the two overflows should be considered jointly when assessing the sizing of the auxiliary overflow, as an increased discharge head will be present on the main overflow. This will result in a flow in excess of the design flow for this structure at the lower reservoir level.

Where a multiplier has been applied to reduce the maximum flood flow Q_{max}, to obtain the design flood flow Q_i, in Section 7.2.1, it should be noted that additional discharge capacity is available within the freeboard if a flood in excess of the design flood occurs. Thus substantially increased flows will actually be capable of being passed before the embankment is overtopped other than by wave splash.

The values of discharge heads of water required to pass the various flows may be attenuated to take account of surface water areas, where relevant, in accordance with Appendix I.

The discharge ability of the overflow arrangement must never be compromised by the installation of fish screens or other methods to retain fish in the reservoir nor works to raise the top water level, even for a temporary period. These will prevent the overflow arrangements passing the required flow and may lead to overtopping and possible failure of the dam.

Figure 7.14 Simple concrete main overflow

Figure 7.15 Stone-faced overflow with fishpass

Figure 7.16 Blockwork overflow with dropweir

Figure 7.17 Side entry overflow

Figure 7.18 Grassed auxiliary overflow

Figure 7.19 Gabion-lined auxiliary overflow

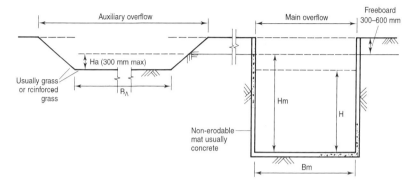

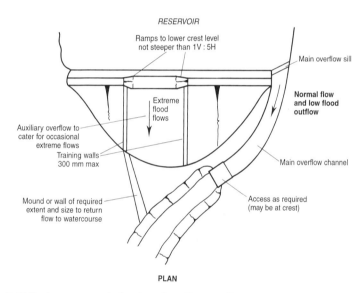

Figure 7.20 Basic arrangement of main and auxiliary overflows

7.2.4 Main overflow

The main weir can take two basic arrangements depending on the size of the flow to be taken by the main overflow and other operational, maintenance or environmental requirements. Small catchments may allow the use of a concrete or blockwork weir chamber or vertical pipe sited in undisturbed ground at the edge of the reservoir which then functions as the controlling level for the reservoir. The length of weir (i.e. the top of the chamber or the perimeter of the pipe) must be sufficient to pass the design outflow with a maximum head over the weir of 300 mm to maintain the hydraulic efficiency of the weir chamber or pipe. Table 7.8 gives the length of overflow weir for a given capacity, based on discharge heads up to this maximum value, whilst Section 9.10.2 and Figure 9.22 give information on other aspects of design and the required size of outlet pipe from the vertical pipe or chamber. Where a vertical pipe is to be used, the top of the pipe may be widened to form a bellmouth of the required perimeter given by Table 7.8. Subject to the other constraints on size in Section 9.10.2, the main section of vertical pipe must have a diameter of not less than twice the maximum head over the top of the pipe and ideally not less than 600 mm. If the flow is in excess of about 1.5 m³/s, the weir chamber or vertical pipe arrangement is not satisfactory and alternative overflow arrangements, as subsequently discussed, must be used.

Table 7.8 Length of weir to pass design discharge

Head over weir (mm)	Length (B_m – m) required for discharge capacity (m³/s)								
	0.1	0.2	0.3	0.4	0.5	0.6	0.8	1.0	1.2	1.5
100	1.9	3.7	5.6	7.5	–	–	–	–	–	–
200	0.7	1.3	2.0	2.7	3.3	4.0	5.3	6.6	7.9	–
300	0.4	0.8	1.1	1.5	1.8	2.2	2.9	3.6	4.3	5.4

NOTES:
1. Based on discharge capacity = $1.7 B_m H_D^{3/2}$ (H_D and B_m measured in metres)
2. Max head to be 300 mm, with maximum discharge limited to 1.5 m³/s
3. Weir chamber unlikely to be appropriate if length of overflow exceeds 8m.

Whilst the outflow from small catchments may be passed by a weir chamber or vertical pipe, the majority of cases are likely to produce a design flow which is too great to enable this option to be used. Flow must then be passed by means of a non-erodible weir and spillway located on undisturbed ground at or just beyond the end of the embankment. The top of the weir then functions as the controlling level, or sill, for the reservoir and should be of sufficient length such that the required flow can be passed. A greater discharge head to the level of the auxiliary overflow is possible with this arrangement than for a weir chamber or vertical pipe. Various combinations of discharge head and length for the main overflow should be considered. The main overflow should be designed to pass a flow of 10% of the maximum flood Q_{max} entering the reservoir, as described in Section 7.2.3 from:

$$B_m H_D^{3/2} = \frac{0.1 Q_{max}}{1.7}$$

Where Q_{max} = maximum flood flow (m³/s)
B_m = length of main overflow weir (m)
H_D = discharge head over main overflow (m)

The controlling level may be fixed by the elevation of a relatively narrow sill, as Figures 7.16 and 7.17, or a broad sill as Figures 7.14 and 7.15. In the latter instance, the broad sill should slope in a downstream direction at a slope of not less than 1V:100H from its highest level.

Values of Q_{max}, B_m and H_D can also be determined from Figure 7.21 derived from the formula given. The final choice of the length and discharge head is related to many factors, primarily the economics of construction and environmental constraints. Normally the length is unlikely to exceed five metres, and probably somewhat less if access is required across the overflow or environmental constraints limit the size of the structure. The line of the overflow may be curved or form a flat V-shape in plan, which will allow a longer length of overflow along the line of the embankment. If this approach is adopted, each individual limb should be not less than $5H_D$, with the required length of weir, B_m, measured along the downstream side of the weir. For hydraulic efficiency, the angle between the limbs should not be less than 90° nor the curvature exceed this value. A maximum depth of about 600 mm for the discharge head at the design flow is recommended, although the length of the weir should be not less than $5H_D$. A weir that is very long will result in a very shallow depth of water under normal flow conditions.

A longer weir may be obtained by use of a side entry spillway as shown in Figure 7.17. This provides a length of weir set essentially at right angles to the dam leading into an open channel or culvert located on original ground at or just beyond the end of the embankment. This approach potentially allows the length of weir to be increased with a consequent reduction in the discharge head. This may be particularly useful where the abutments are relatively steep and insufficient length is available along the line of embankment. The design is complex, however, and should only be undertaken by an experienced specialist adviser.

The design of the weir and channel are discussed in Section 9, but for hydraulic efficiency, the structure on the upstream side should funnel the water towards the weir, whilst the channel on the downstream side should maintain a width not less than the length of the weir. Changes in the slope or direction of the channel should be minimised and any necessary bends should be curved at the maximum possible radius.

7.2.5 Auxiliary overflow

The level of the auxiliary overflow should be set so that it begins to operate when the main overflow design flow is reached. The design for the auxiliary overflow will vary substantially depending on the downstream situation, as discussed in Section 7.2.1. The combined capacity of both overflows may vary from twice that for the main overflow to several times that amount, depending on the multiplier applied to the maximum peak flood. Thus the length of the auxiliary overflow may vary significantly and can form a considerable proportion of the length of the embankment in some instances. The length of the auxiliary overflow weir and the discharge head can be calculated from the standard formula for weir discharge by taking account of both overflows as follows:

$$Q_i = 1.7\, B_m H_m^{3/2} + 1.7\, B_A H_A^{3/2}$$

Where Q_i = design flood outflow (m³/s)
B_m = length of main overflow weir (m)
H_m = discharge head over main overflow weir (m)

$\qquad$ = $H_D + H_A$ with H_A normally limited to not more than 300 mm,

H_D, as determined in Section 7.2.4

B_A = length of auxiliary overflow weir (m)

H_A = discharge head over auxiliary overflow weir (m)

An initial value of B_A can be calculated assuming $H_A = 300$ mm. Values of Q_i can also be used in Figure 7.21 to determine combinations of B_m, B_A, H_m and H_A. It should be noted that the discharge head for the main overflow will be greater than that used in Section 7.2.4 if this is a concrete weir and channel, due to the increased rise in reservoir level with flows over the auxiliary overflow. The increase in flow over a weir chamber is small and should be ignored. The required freeboard must be present above this level. Where the lowest multiplier of 0.2 in Section 7.2.1 is used, and the main overflow is designed with a water depth of 600 mm or less, the overall design flow can be virtually passed by the main overflow alone as a result of the increased discharge head. An auxiliary overflow then appears to be either of very short length or not required. In this instance, a nominal length of auxiliary overflow should be provided, with one metre of length for each 25 hectares of catchment.

The auxiliary overflow can be located off one end of the embankment if the topography permits suitable excavation and final slopes; flows can then be discharged over undisturbed ground clear of the embankment either down a steep slope or in a gently sloping channel. In many instances, however, this is not possible if the ground slope is relatively steep or a substantial length is required, and so the auxiliary overflow flow must be passed over the embankment. If a gently sloping channel is used, the depth of the discharge head can be increased and the value of H_A can be increased to 400 mm.

The ultimate discharge capacity of both overflows can be calculated by considering the total flow from Figure 7.21 with the reservoir at crest level and recalculating with the increased discharge heads. This allows a comparison with the estimated flood inflow to be made.

The design of the auxiliary overflow is discussed in Section 8, but the downstream slope must be fully capable of taking the discharge in this instance. This may require a reduced downstream slope or special protective measures, as described, to limit the discharge velocities and minimise the erosion risks.

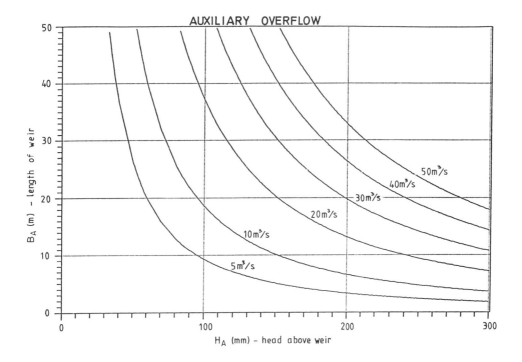
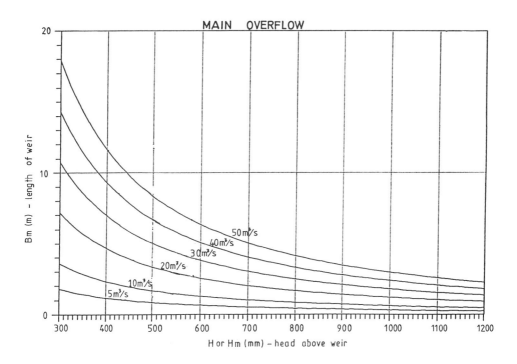

Figure 7.21 Sizing overflow structures

7.2.6 Other aspects

The size of overflows may be reduced by decreasing the amount of flood flow into the reservoir in some instances. This would require the construction of a bywash channel which diverts flood flows around the reservoir to discharge downstream of the dam. This normally requires extensive excavation and reshaping above the top water level and is normally only possible on steeper catchments. Extensive protection works are usually necessary and thus this option has been considered only on the larger reservoirs in the past to handle relatively small floods. This technique is not normally used for modern reservoir construction, and its application to small reservoirs is inappropriate in view of the imprecise nature of the flood assessment.

Non-impounding reservoirs fed by gravity flow along channels require special consideration under flood conditions. Substantial inflows may result at such times unless careful diversion facilities have been provided and the design of the inlet works requires consideration by a specialist adviser.

A reservoir which is essentially non-impounding, but which has a very small localised catchment due to its method of construction, has been referred to in Section 7.1.1 and overflow facilities are required as detailed in Section 7.2. Non-impounding reservoirs with no direct catchment will also be subject to a flood rise in the event of heavy rainfall by water falling directly onto the water surface and inside slopes of the reservoir. Facilities for passing a flow of $0.1 \text{ m}^3/\text{s}$ for each 1000 m^2 of reservoir surface area should be provided to allow for intense rainfall. Excessive pumping into the reservoir may also be possible unless measures are provided to limit this. Thus the reservoir level may rise above the intended level and so it would be necessary to provide either a small non-erodible overflow weir to allow a similar discharge flow plus the allowance for rainfall, or an overflow pipe capable of carrying the total outflow. This should be designed to pass the required flow in accordance with Table 7.8 or Figure 7.21 and must be at least the size of the inlet pipe.

7.3 WAVES AND FREEBOARD

The previous sections have referred to freeboard required above the reservoir rise resulting from the flood flow. This is necessary to provide sufficient height between the increased reservoir level and the embankment crest to cater for any wave action. It is dependent on the class of the reservoir and the minimum values that must be provided are given in Table 7.7. The required freeboard must be provided in all instances for flows up to the required design flood.

Where the reservoir is subject to certain climatic or topographic influences, the freeboard may need to be increased. Most of these are either beyond the scope of reservoirs in this guide or need only be considered in certain locations. Reservoirs in areas where the average hourly annual maximum wind speed exceeds 20 m/s should have a further 0.1 m added to the minimum freeboard. These locations can be obtained from the Meteorological Office, but essentially comprise Devon and Cornwall, most of West and North Wales and much of Scotland and Northern Ireland. Similarly increasing the size of the reservoir will increase the wave action, and an increase in freeboard of 100 mm should be made if the reservoir fetch is greater than 200 m, or 200 mm if greater than 500 m. Reservoir fetch can be assessed by measuring the largest, straight across-water line from the embankment towards the upstream end of the reservoir.

7.4 FLOOD EVENT PROCEDURES

7.4.1 Emergency procedures

A simple written procedure should be prepared detailing the course of action to be carried out in the event of a flood warning. The flood warning may be a formal alert from the local drainage authority if very heavy rainfall is expected or it may be a general warning put out on radio or television by the Meteorological Office. Anticipated very heavy rainfall after a prolonged wet period or a rapid thaw may lead to such warnings, but the typical intense thunderstorms of late summer often occur with little warning.

The emergency procedure should detail:

- any requirements for opening any low-level outlet or other controls on outflow
- a check to ensure the main and auxiliary overflows are clear of debris
- warnings to persons downstream who may be affected by flow from the auxiliary overflow if overland flow is anticipated
- movement of livestock or vehicles and plant within the anticipated flow path
- removal or lashing of boats on the reservoir.

The sequence of operations may be staged depending on the level of the warning, but the initial two items are particularly important.

All persons associated with the operation and maintenance of the reservoir should be aware of the procedure and should be given a copy. Where particular responsibilities are involved, this should be pointed out to them from the start of their involvement and any change must be notified. Where particular items are needed such as a hand wheel to open a valve, or keys to locked structures, the location of these should be included in the procedure and a named person made responsible for their maintenance and safe keeping. Contingency arrangements should be made for when the person made responsible is absent for whatever reason.

7.4.2 Procedure following substantial flow

Following a substantial flood flow over the overflow works, the embankment should be walked over and inspected and any problems identified, as discussed in detail in Section 12. Debris may have collected around the entrance to and within the overflow structures, and this should be removed, together with other debris in the reservoir which is likely to find its way ultimately to the overflows. Any erosion or other damage should be repaired at an early stage and any vegetation on or adjacent to the embankment or around the reservoir that has been affected should be dealt with as required.

Where possible, the extent of flood flows should be identified and compared with any unusual observations at the peak of the flood. The presence of debris, flattened grass, damage to vegetation etc. may help to delineate the extent of water flow. The depth of water flowing over the main and auxiliary overflows should be noted where possible, together with the depth and extent over the crest. A brief record of the event is useful, together with any appropriate photographs and this ideally should be kept with the written procedure. Following the flood, the emergency procedure should be reviewed in light of experience and amended as required.

7.4.3 Other meteorological events

Warnings of other forms of severe weather which could affect the reservoir may also be given by the Meteorological Office. These may take the form of strong wind, snowfalls or severe cold. Each can have an effect on the reservoir and embankment and may necessitate some action.

Strong winds may cause loss of branches or toppling of trees both on the embankment and around or upstream of the reservoir. Well maintained and balanced trees on the embankment should not suffer unless the winds are very extreme. Some damage was noted in the October 1987 storm in Southern England, but in most cases this appeared to be limited to dead or to trees that were badly out-of-balance. Loss of branches or toppling of trees around the reservoir, together with other debris which may end up in the reservoir, will tend to collect ultimately at the overflows and should be removed at an early stage.

Snowfalls are not likely to affect the reservoir directly other than their possible effects on vegetation. Significant snowfalls, however, may block access to the reservoir and preclude any necessary inspection or maintenance work. Problems are more likely to follow a thaw as the volume of water in the snow lying on the catchment is released as runoff. A rapid thaw of a thick covering of snow may generate significant runoff and moderately large flows over the main and auxiliary overflows. Where the thaw is associated with heavy rainfall, the flow will be substantially greater. A mixture of partially thawed ice and water can clog channels under certain conditions.

Ice on the reservoir may develop in a prolonged cold spell and this may drift with the prevailing wind building up to quite substantial thicknesses of tens of centimetres or more. Where the ice is able to build up adjacent to the embankment, damage is possible from both impact and lateral pressure. Small structures such as pipe inlets or outlets may be at risk and the ice should be removed where possible and not allowed to build up to any significant degree. Weir chambers and access works can be particularly vulnerable to ice damage.

Following any severe meteorological event, the embankment should be walked over and inspected and any problems identified, as discussed, following a substantial flood. A record of the event, and any subsequent damage or other effects, is often useful and allows a record of the reservoir behaviour to be maintained.

7.5 FLOODS DURING CONSTRUCTION

The control of floods is critical at all times for an impounding reservoir, but particularly so during the construction of the reservoir. Unanticipated floods during the construction works, particularly midway through, may breach the embankment and release water impounded behind a partially completed embankment. Provision to pass floods during construction generally takes the form of a cofferdam to prevent water flooding the construction area by temporarily increasing the storage, whilst a diversion pipe discharges the flood downstream through or around the proposed dam. The options of leaving out a section of embankment to pass flows which is infilled rapidly once the remainder of the embankment has been completed or provision of a temporary diversion channel around the dam might be applicable for small reservoirs, but should only be adopted following specialist advice.

Most reservoirs within the scope of this guide will be constructed within one earthmoving season and reasonable provision must be made to pass the likely flood flow safely. Construction will normally be during the summer time when typical flows are least, but the risk of a large flood during the construction period will be always present and it is only realistic to provide a certain capacity for diversion flows and accept that these may be exceeded on rare occasions. A minimum pipe diameter of 400 mm should be provided for catchments of up to 150 ha and 600 mm for larger catchments. A cofferdam should be provided to a height above the pipe invert of at least four times the pipe diameter plus a further 600 mm as a safety margin and to allow for any wave action. Where the catchment exceeds the scope of Section 7.1.8 specialist advice should be sought in sizing the pipe diameter and height of cofferdam and assessing the influence of construction flows on the works.

It is essential that any temporary diversion arrangements are adequately sealed, where required, when reservoir construction is completed to facilitate impounding. Specialist advice should be sought to ensure this is carried out adequately.

Flood damage to downstream areas as a result of the uncontrolled release of water during construction must be adequately covered in pre-contractual arrangements (Section 11), and adequate insurance cover should be provided (Section 3).

8 Site investigation

8.1 INTRODUCTION

8.1.1 General

Whatever the scale of works, it is essential to carry out a thorough investigation of the intended site. The site investigation for both non-impounding and impounding reservoir sites should address the following points:

- watertightness — establish the ability of a site to hold water
- foundation conditions — confirm the ability of the ground to support an embankment and prevent excessive seepage from occurring through the foundation beneath the base of the embankment
- construction materials — assess the permeability of the potential fill materials and establish the suitability and amount of locally available material for embankment construction
- stability requirements — ensure land within the reservoir area and adjacent land affected by reservoir is currently stable and will remain stable after partial or total filling of the reservoir
- access requirements — ensure suitable access for construction plant and imported materials as well as access requirements dependent on the eventual use of the reservoir.

Guidance on good practise for site investigation work can be found in BS 5930: Code of Practice for Site Investigations, 1981[57] and CIRIA Special Publication 25 'Site Investigation Manual'[58]. Although these cover more complex investigation techniques than are generally required for a small reservoir, they do provide simple, concise advice.

If the investigation work is to be carried out by a third party, it is important that a simple but adequately comprehensive contract is set up with the party carrying out the works, as described in Section 12. If the feasibility stage of the project suggests that specialist advice is required, it is important that this is brought in at a sufficiently early stage to allow the site investigation to be designed to suit any specific requirements of the adviser.

8.1.2 Site investigation approach

The following phased approach should be adopted to obtain sufficient information to progress beyond the feasibility stage for a scheme:

- information study
  - collect and assess data about the ground conditions of the site and adjacent areas from existing records
  - seek information on buried and other services
  - establish land ownership if in doubt and confirm rights of way
- site survey
  - walkover of the site to record the topography, vegetation and general ground conditions
  - carry out levelling survey by specialist surveyor
- ground investigation
  - establish the soil profile and groundwater conditions beneath the most suitable potential reservoir and dam site using exploratory holes
  - carry out laboratory testing if required
- assessment
  - determine whether the site is suitable
  - prove sufficient and suitable construction materials
  - establish that there is sufficient information for the design of the dam.

This site investigation procedure is shown diagrammatically in Figure 8.1. It is important that the site investigation is well structured and planned so that information obtained from each stage can be assessed and used to modify the programme as new information is acquired.

In many cases there may only be one site available to the undertaker. The site investigation is then a relatively simple procedure in which information as to the nature of the ground is gradually built up from the information study, site survey and ground investigation. A full assessment of the site can then be made to enable the undertaker to prepare a design of the dam and reservoir and be aware of associated potential problems.

If the undertaker is fortunate to have a number of site options, these should be investigated in the information study and site survey before the most suitable site is selected. A levelling survey and a ground investigation should then be carried out to enable the undertaker to make a full assessment of the selected site.

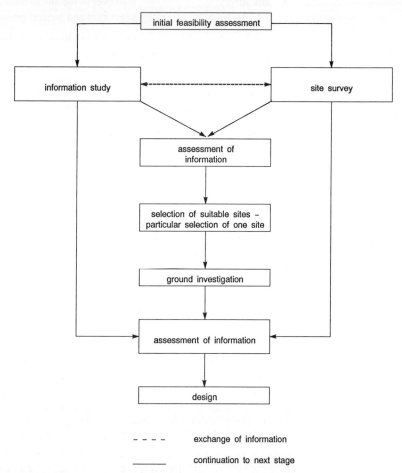

Figure 8.1 Site investigation procedure

8.2 INFORMATION STUDY

8.2.1 Approach

The information study involves the collection and appraisal of relevant available information for the site and the adjacent area. The information should be used to make a preliminary assessment of the ground conditions and site suitability. The information study should be carried out as the initial stage in the site investigation programme.

8.2.2 Sources of information

The information sources that should be reviewed as part of the information study may include the following:

- topographical maps and plans
- geological maps and published information
- soil survey maps and published information
- mining and mineral extraction records
- aerial photographs
- previous ground investigations in the area
- local utilities
- local building and civil engineering works
- local planning authority (LPA)
- relevant licensing authority (RLA)
- local clubs and societies
- local educational establishments.

A summary of typical information which can be obtained from these sources is provided in Appendix J. The British Map Library, in particular, should hold copies of all past published maps of the area and should allow previous land use to be assessed. Appendix D gives addresses of the relevant publishers and suppliers of this information.

Local enquiries may also yield relevant information although unsubstantiated verbal comments should be treated with caution.

8.2.3 Buried and other services

The presence of services which may be privately owned or belong to statutory bodies and public authorities within and adjacent to the site of the proposed reservoir must be established. The main organisation, including those for water, gas, electricity and telephone, should be informed of the proposed works and a large-scale plan marked up with the likely extent of the reservoir. The national grid reference of the site should also be provided. A copy of the plan would normally be returned with the approximate location of any services in the vicinity.

Other organisations may also have services in the vicinity which might include pipelines, telecommunications drainage facilities, etc. The information study should reveal their potential presence and contact should be made to confirm their position within or adjacent to the site.

8.3 SITE SURVEY

8.3.1 Approach

A walkover of the site should be made at an early stage during the course of the information study to identify and record features of geological and topographical interest. Information on groundwater levels, water features and drainage (hydrological information) for the site should also be acquired. Data collected from a site survey can then be used to supplement and clarify information collected in the information study.

8.3.2 Levelling survey

A levelling survey of the site must be carried out by a specialist surveyor with the results presented ideally on a 1:200 or larger scale plan. The survey should extend across and several metres beyond the area of the proposed reservoir. Any areas of concern (e.g. potentially unstable slopes around the reservoir area) should also be included within the survey. This plan can then be used to assist in locating the exploratory holes for the ground investigation, and the design of the reservoir and the embankment.

8.3.3 Information to record from site walkover

Details should ideally be noted on a large scale plan at 1:200 or larger, and on a suitable field sheet. A typical hand-sketched field sheet is shown in Appendix K, but where possible the information should be added to a true scale plan.

The proposed site should be inspected carefully and methodically for conditions that might cause construction difficulties. Topography is important. Slopes greater than 1V:10H may be subject to soil creep as indicated, for example, by tilting walls, trees and fences. Abrupt changes in local topography may indicate changes in ground type and a consequent variation in soil characteristics across the site. In particular, the extent of flat areas in the bottom of a valley should be identified; these may delineate the extent of softer or weaker material infilling the valley floor.

Vegetation is an important indicator of soil types and groundwater levels. Reeds, rushes and willows indicate a high water table, whereas bracken and gorse usually denote a well-drained soil with a low water table. Abrupt changes in vegetation may indicate important variations in ground or groundwater characteristics. The extent and type of vegetation across the site should be recorded and linked, where possible, with topography and other features.

Removal of vegetation, particularly mature trees and large shrubs, may lead to groundwater level changes and ground movement. Evidence of past ground deformations due to soil moisture changes and movements may be revealed in existing buildings or structures on the site. Current and past land use, where known, should be noted. Previous land use by man may have resulted in

realigned watercourses, areas of infilling or dumped material ('made ground') or abandoned underground workings. Dumped material is likely to have been tipped without control and may contain voids, as well as combustible or hazardous material. Where household refuse or debris and residues from industrial processes are suspected, the site must be treated with care.
The material should not be disturbed unless the necessary precautions are taken as detailed in Section 9. The presence of such material is likely to make most sites unsuitable and specialist advice must be sought.

The typical features which should be identified in a site survey are listed in Box 8.1. The survey should be carried out on foot, whereby the items listed in Box 8.1 may prove useful.

Box 8.1 Features to note during a site walkover survey

Topographical Information
• access
• present land use
• evidence of former land use
• signs of made ground or the deposition of material
• gradients of slopes steeper than 1V:10H
• tilting trees, walls or other structures
• abrupt changes in slope or topography
• current and recently removed vegetation.
Geological Information
• soil type where visible or exposed by hand digging or exposures
• evidence of slipped material
• changes in vegetation
• breaks in slope.
Hydrological Information
• watercourses, whether running, damp or dry
• seepage, springs and flowing water
• areas of standing water
• waterlogged and boggy ground
• changes in vegetation
• evidence of former flooding.
Useful Items for Walkover Survey
• true scale plan or map — at least two photocopies of site plan to mark up with features and observations
• spade — to dig small holes for initial appraisal of soil type
• bags and ties — for soil samples
• compass — use with site plan to establish orientation and position on site
• tape measure — for approximate measurement of features of interest
• camera and film — for a complete record of the site or a feature when used in conjunction with the site plan and notes made during the survey.

A considerable saving can be made if a thorough information study and site survey is carried out. These should identify a potentially suitable site for the embankment and reservoir and locate where exploratory holes should be sunk, to allow a detailed assessment of the soil profile to be made.

8.4 GROUND INVESTIGATION

8.4.1 Approach

The ground investigation comprises the work carried out to gain detailed information below the ground surface and involves the excavation of exploratory holes possibly followed by laboratory testing to assess the soil parameters. This work may be arranged and carried out by the undertaker without the need to involve a specialist ground investigation contractor but a specialist may be required if boreholes or laboratory testing are carried out. The ground investigation work is generally carried out by excavating trial holes using a mechanical excavator. A backhoe tractor is ideal for this purpose. A soil auger may also be used if available, whilst boreholes will normally be necessary if information is required below 4 m to 5 m depth or if groundwater conditions prevent the excavation of open trial holes.

Exploratory holes should be located across the floor of the proposed reservoir and over any area which would form the foundation for the dam. In addition, exploratory holes should also be sunk in the location of potential borrow pit areas to assess the suitability and amount of material for use as embankment fill, and elsewhere where unstable slopes might be present.

The number of exploratory holes required will depend on the size of the site and the variability of materials across the site. For even the smallest site there should be a minimum of three initial exploratory holes positioned in a triangular arrangement at the corners of the reservoir site, with one at each end of the embankment. These initial exploratory holes should indicate the general nature and variability of the ground across the site. Additional holes should then be sunk in the centre of the proposed reservoir and along the embankment line to confirm or investigate any further areas as required.

The exploratory holes should extend through the superficial material and not less than 1.5 m into the underlying sound insitu ground. The superficial materials comprise the generally weaker and softer material in the valley bottom and, to a lesser extent, on the hillside slopes. It also includes alluvial clay and gravel, boulder clay and granular deposits, material moved downslope as a result of natural processes, slipped material, highly weathered ground and any made ground. On the embankment line, the exploratory holes should generally be taken to a depth of not less than the embankment height, and a minimum depth of 4 m.

Representative soil and groundwater samples should be obtained from the exploratory holes as detailed in Box 8.2 to allow the soils to be classified and their properties found. Where laboratory testing is required, the minimum sample sizes given in Table 8.1 should be obtained. Specialist groundwater testing is discussed in Section 12.4.

Box 8.2 Sampling methods and frequency

> **Trial holes (unshored)**
> Up to 1.2 m depth, take representative sample from base or lower sides of trial hole. Beyond 1.2m, take representative sample from bucket of excavator.
> **Trial holes (shored)**
> Take representative sample from base or lower sides of trial hole.
> **Hand auger**
> Take sample from returns in auger.
> **Boreholes**
> Take samples from returns from shell. Obtain open-hole, undisturbed samples if required for specialist laboratory testing or detailed examination of insitu soil profile.
> **Frequency of sampling**
> At each change of soil type or consistency and at one metre spacings where soil type remains constant. Samples should be placed in strong plastic bags/sacks, securely tied and labelled with details of the exploratory hole number and depth. Samples for moisture content should be placed in screw-top jars with the jars fully filled to minimise drying out.
> **Groundwater**
> Sample each time groundwater is found, in screw-top jar after washing out in groundwater.
> **Size of sampling**
> See Table 8.1.

Table 8.1 Mass of soil sample required for laboratory tests

Laboratory test	Soil type	Minimum mass of soil sample required (kg)
• liquid and plastic limits (on clay and silt samples only)	clay, silt, sand	1
• particle size distribution	fine and medium gravel	5
• moisture content	coarse gravel	30
• chemical analysis		

The exploratory techniques described are generally sufficient to enable the reservoir and embankment to be designed from the data collected. Where unsuitable or questionable ground conditions are thought to be present from the findings of the investigation, a more thorough investigation may be required. Specialist advice should then be sought to assist in the planning and the location of the additional exploratory holes and the extent of the investigation. The cost of a more detailed ground investigation, however, may exceed the feasibility of the project. If there are several site options, it may be prudent to abandon the site for one which is geologically more suitable rather than spend large amounts on an extensive ground investigation with compensatory design modifications to the scheme.

8.4.2 Trial holes

Trial holes allow a good visual assessment to be made of the soil profile and groundwater seepages. Soil and water samples should be taken for examination and possible laboratory testing. Their depths should be noted. The holes will need to be supported in weak ground if they are to be entered and any hole greater than 1.2 m depth should not be entered unless the sides have been shored up. The sides of trial holes and other excavations may be liable to collapse if unsupported or inadequately supported (Figure 8.2) and a careful approach must be adopted. Further information on shoring methods is given in CIRIA Report 97 'Trenching Practice'[59]. Colour photographs of the excavated faces should be taken wherever possible with a prominent scale and method of identifying the photographed face. The position of the trial holes should be marked on the site plan used for the site survey. Trial holes can be excavated to a maximum depth of up to 5 m, depending on the ground conditions, the position of the water table and the size of excavator used. On completion, trial holes should be backfilled carefully with the excavated material. This is particularly important to ensure that the ground is returned to as near to its original condition as possible. The procedure to adopt when excavating trial holes is detailed in Box 8.3.

Figure 8.2 Inadequate support to trial holes

8.4.3 Soil augers

Augers are used in cohesive soils (clays) and are unlikely to be successful in granular soils (sands and gravels). Advancing the auger may prove difficult in stony or very stiff soil, when the water table is reached, and in frozen ground. Detailed soil descriptions should be made from material removed from the auger, with soil samples taken for closer examination and laboratory testing. The augured material should be laid out in sequence on the ground and photographed to give the soil profile. The depth of water seepages may possibly be estimated, but little information on groundwater levels is usually available with this method of investigation. Auger holes can be taken to a maximum depth of up to six metres, depending on the ground conditions, and the position of the water table. A selection of soil augers is shown in Figure 8.3.

Box 8.3 Excavation of trial holes

> **Excavation and sampling**
> - strip turves and place to one side
> - strip topsoil and place to one side (preferably on plastic sheeting)
> - excavate materials and place separately (preferably on plastic sheeting)
> - describe, sample and photograph soil profile (sampling to be as Table 7.3).
>
> **Backfilling**
> - replace materials in the order they were excavated (those excavated last should be backfilled first)
> - avoid mixing excavated materials as much as possible (this may change the properties of the respective materials)
> - place backfilled materials in thin layers (approximately 150 mm) and compact with the bucket of the excavator (this is particularly important when backfilling cohesive materials)
> - replace stripped topsoil on completion and replace turves or scatter grass seed where required
> - leave surface 'humped' to allow for settlement.

8.4.4 Boreholes

Boreholes allow the extent of exploration to be continued to a substantial depth if required and enable an exploratory hole to be sunk through very soft, loose or waterlogged ground which cannot be investigated by other types of exploratory holes. The boreholes are normally sunk by percussive means using a tripod rig, as shown in Figure 8.4. Where necessary, they can also be sunk through water, if required, by means of a platform above the water level. Improved sampling and seepage observations are generally obtained, but boreholes require specialist expertise to sink and interpret the hole. Consequently, boreholes are more expensive than the more simple exploratory holes, but may be required by site conditions.

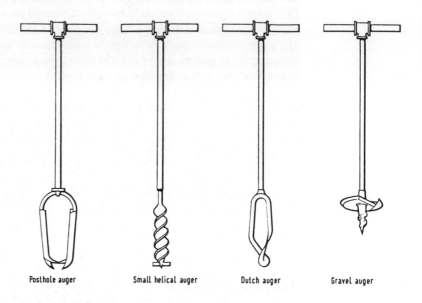
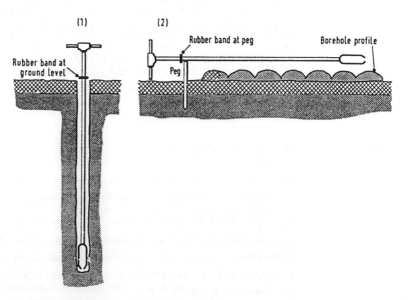

Figure 8.3 Selection of soil augers (after Ref 58)

Figure 8.4 Cable percussive boring rig

8.4.5 Information Required

Information recorded from the exploratory holes should include:

- thickness of each layer encountered (including topsoil and subsoil)
- type, classification and variability of material present in each layer
- compactness/strength and permeability of the material present in each layer
- position of the water table (if present) and any seepage, giving details of the estimated rate of flow.

A description of the material in each layer should be made on the basis of visual observations and the response to simple field tests. Tables 8.2 and 8.3, based on Table 6 of BS 5930[57] enable the basic soil types to be assessed and the materials to be classified from the relative proportions of the different sized constituents and the consistency/ strength. The basic colour and any variations should also be recorded. Reference 60 offers guidance for a preliminary assessment of the soil texture and Figure 8.5 is reproduced from this guidance note. The simple jar test detailed in Figure 8.6 also allows the proportions of the constituent materials to be assessed. The percentage amount of clay can thus be estimated from these tests. An accurate assessment of the variously sized materials, however, requires a number of particle size distribution tests to be carried out in a laboratory, but these simple methods allow an initial estimate to be made at an early stage. Particular attention should be paid to the compactness/strength of the soils encountered and the permeability (defined as the rate at which water flows through the soil). These can be determined from simple field tests and observations given in Tables 8.4 and 8.5 and Box 8.4.

In clayey soils the presence of any planar or undulating polished surfaces, identified from trial holes, should be clearly noted. These may be of a differing colour and a lower strength than the adjacent ground. A typical example is shown in Figure 8.7. The implications of these are discussed in Sections 8.5.1, 9.3.2 and 9.3.8.

If imported fill is to be used (Section 8.5.2), samples should be taken from the source, with appropriate and sufficient testing and visual assessment carried out. This should be sufficient to assess the type, classification and variability of the material and the compactness/strength and permeability when used as fill.

Table 8.2 Field identification and description of cohesive and organic soils

Basic soil type	Particle size (mm)	Visual identification	Plasticity	Composite soil types			Compactness/strength		Colour
				Scale of secondary constitutents			Term	Field test	
Silt	–006	Only coarse silt is barely visible to the naked eye	Non plastic or low plasticity	Term		% of sand or gravel	Soft or loose	Easily moulded or crushed in the fingers	Red Pink Yellow
	Coarse	It exhibits little plasticity and marked dilatancy		Sandy Gravelly	Clay or silt	35 to 65	Firm or dense	Can be moulded or crushed by strong pressure in the finger	Brown Olive Green
	–002	Slightly granular or silky to the touch Disintegrates in water		Clay		under 35			Blue White Grey Black
	Medium –0006	Lumps dry quickly, but can be powdered easily between fingers and dusted off hands when dry							
	Fine –0002								
Clay	less than 0002	Dry lumps can be broken but not powdered between fingers	Plasticity				Very soft	Exudes between fingers when squeezed	Supplement as necessary with:
		Clays disintegrate at a slower rate than silt in water					Soft	Moulded by light finger pressure	Light Dark mottled
		Exhibits plasticity, but no dilatancy					Firm	Can be moulded by strong finger pressure	Pinkish Brownish
		Sticks to fingers and drys slowly Feels smooth					Stiff	Cannot be moulded by fingers Can be indented by thumb	Reddish Yellowish
		Shrinks appreciable in drying, usually showing cracks					Very stiff	Can be indented by thumb nail	
Organic	Variable	Contains a substantial amount of vegetable matter					Firm	Fibres already compressed together	Black
Peats	Variable	Predominantly composed of plant remains Often has distinctive smell Fibrous					Spongy structure	Very compressible and open	Dark Brown
							Plastic smears	Can be moulded in hand and fingers	

Table 8.3 Field identification and description of granular soils

Basic soil type	Particle size (mm)	Visual identification	Composite soil types (mixtures of basic soil types)			Compactness/strength		Colour
			Scale of secondary constituent			Term	Field test	
			Term		% of clay or silt			
Boulders Cobbles	200	Only seen complete in final pits or exposures. Tennis ball to bowling ball size, difficult to recover from auger holes				Loose Dense	By inspection of voids and particle packing	Red Pink Yellow
	–60–		Slightly clayey	gravel or sand	Under 5	Loose		Brown Olive Green
			Clayey Silty	gravel or sand	5 to 10	Dense		Blue White Grey Black
	Coarse	Easily visible to the naked eye	Very clayey Very silty	gravel or sand	15 to 35	Slightly Cemented	Can be excavated with a spade 50 mm wooden peg can be easily driven	Supplemented as necessary with:
	–20 Medium	Particle shape: angular subangular subrounded rounded flat elongate					Requires pick for excavation 50mm wooden peg hard to drive	Light Dark Pinkish Brownish Reddish Yellowish, etc
	–6	Well graded: wide range of grain sizes well distributed					Visual examination: Pick removes soil in lumps which can be abraded	
	Fine –2–	Poorly graded: grains of uniform size	Sandy GRAVEL Gravelly SAND	sand or gravel and important second constituent of the course fraction				
Sand	Coarse –06 Medium –02 Fine –06–	Visible to the naked eye. Sand is friable when dry, and easy to break up. Fines and may exhibit dilatancy						

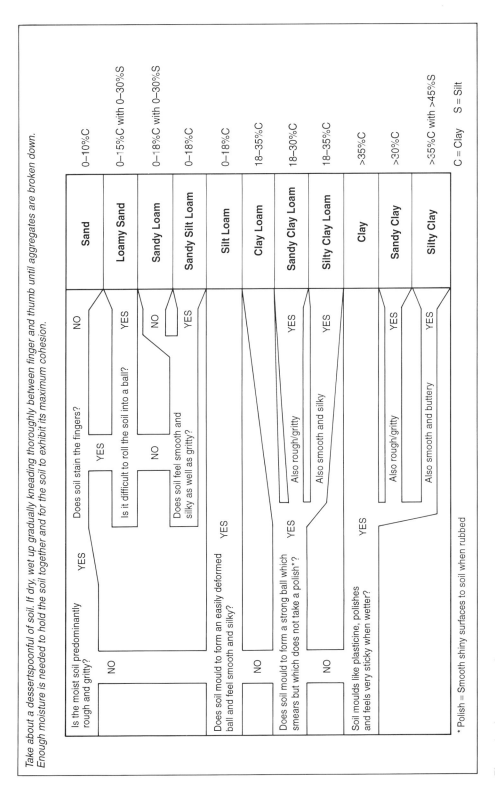

Figure 8.5 Assessment of clay content by hand texturing (Ref 60)

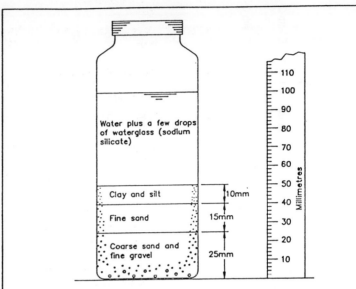

The proportions of the various component materials that form a particular soil may be estimated with reasonable accuracy by making the following test. Half fill a narrow parallel-sided glass bottle and shake well to mix the soil and water thoroughly. If waterglass (sodium silicate) is available add 2 or 3 drops to water and stir the soil and water mixture vigorously. Stand the bottle on a firm surface and allow the soil to settle for 24 hours. Coarse sand will settle out immediately, fine sand within a few minutes and the silt and clay last. These will stratify into clearly visible layers from which the approximate solid proportions can be estimated by measuring the depth of each layer.

For example:

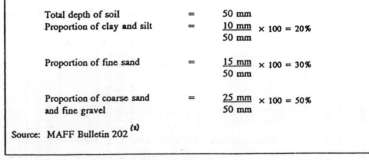

Source: MAFF Bulletin 202 [2]

Figure 8.6 Initial assessment of particle size from settlement test (Ref 2)

Table 8.4 Estimation of soil compactness/strength from simple field tests and observations

Soil type	Term	Field test	Trial hole behaviour
Cohesive			
Clay	Very soft	Extrudes between fingers when squeezed	Unsupported trial hole sides bulging and collapsing
	Soft	Moulded by light finger pressure	
	Firm	Moulded by strong finger pressure	Unsupported trial hole sides likely to remain vertical and stable
	Stiff	Can be moulded by finger pressure or be indented by thumb nail	Unsupported trial holes sides remain vertical and stable
	Very stiff	Cannot be indented by thumb nail	
Silt	Soft or Loose	Easily moulded or crushed in fingers	Unsupported trial hole sides bulging and collapsing
	Firm or Dense	Moulded or crushed by strong finger pressure	Unsupported trial hole likely to remain vertical and stable
Granular			
Sands Gravels	Loose	Can be excavated with a spade. 50 mm wooden peg can be easily driven	Unsupported trial hole side collapsing
	Dense	Requires pick for excavation. 50 mm wooden peg hard to drive	Unsupported trial holes unlikely to remain vertical unless high clay and silt content
	Slightly Cemented	Visual examination. Pick removes soil in lumps which can be abraded	
Cobbles Boulders	Loose Dense	By inspection of voids and particle packing	
Organic	Firm	Fibres already compressed together	Variable, but likely to collapse
Peats	Spongy	Very compressible and open structure	
	Plastic	Can be moulded in hand and smears fingers	

NOTE: Trial hole behaviour for material above the water table. Granular materials below the water table will generally flow immediately upon excavation into the layer, leading to collapse of the trial hole sides.

Clay or silt underlain by granular material will tend to be undercut by loss of granular material, leading to probable collapse of trial hole sides.

Loading from excavated material or plant adjacent to trial hole will increase likelihood of collapse.

Table 8.5 Soil permeability

Soil type	Approximate permeability (m/sec)	Permeability/drainage characteristics
Clay	less than 10^{-8}	practically impermeable
Silt	$10^{-8} - 10^{-6}$	low permeability
		poor drainage
Sand	$10^{-6} - 10^{-3}$	medium/high permeability moderately free draining
Gravel/cobbles	greater than 10^{-3}	high permeability free draining
Organic/peat	variable	variable

Box 8.4 Estimation of insitu permeability by simple infiltration test

Methods
- a length of pipe, approximately 150 mm diameter and about 500 mm long, and some soft clay for sealing the pipe are required
- excavate hole at required depth in trial hole, leaving a flat bottom to hole
- stand pipe vertical in hole and seal the bottom with hand compacted soft clay to form seal to prevent upward flow of water
- fill with water to top of pipe and allow to stand, topping up as required
- if the drop in water level is minimal, measure time for a drop of 50 mm and calculate average inflow from:

    pipe area x drop in level/time

- if the drop in water level is rapid, adjust the inflow (from a hose or similar) so that the water level is maintained just below the top of the pipe. The average inflow can then be calculated by timing the filling of a container of known volume
- calculate the permeability (k) from:

    $k = Q/5.5\, rH$ (m/s)
    where Q = average inflow (m^3/s)
    r = pipe radius (m)
    H = length of pipe (m)

- repeat to give three tests and take average rate
- compare results of the tests across the reservoir area
- where different materials are present across the site, carry out three tests in each different type and average the values for a representative estimate of the mass permeability.

Results
- rapid water loss indicates a high permeability and shows the material is predominantly granular. Large flows will also occur in desiccated or fissured cohesive soils where water loss occurs along the cracks and fissures
- a very slow water loss indicates the material is predominantly cohesive

NOTE: The test results are valid only if the soil beneath the test area is of one general material type. The presence of a thin layer of sand or gravel in an otherwise cohesive material will invalidate the results.

A very approximate assessment of the permeability can be found from the rate of drop in water level in a borehole or augerhole. Provided the drop in level is not more than about 20% of the depth of the hole from an initially near full level, the permeability can be approximately assessed as the rate of drop/50.

Figure 8.7 Example of planar polished surfaces

8.4.6 Laboratory testing

Soils can be classified in the field according to their basic type and particle size. However, for a more accurate assessment, particularly where there is doubt in the description or properties of the soils, a selected number of samples should be sent to a materials testing laboratory. The local ADAS adviser should be able to recommend a suitable laboratory. Only basic inexpensive classification tests are normally required and these should include the following, referred to BS 1377: Methods of test for soils for civil engineering purposes[61]:

- cohesive soils
  - moisture content determinations
    Part 2; Sub-clause 3.2
  - liquid limit
    Part 2; Sub-clause 4.3
  - plastic limit
    Part 2; Sub-clause 5.3
  - particle size distribution
    Part 2: Sub-clauses 9.2, 9.4 and 9.5
  - density/moisture content relationship
    Part 4; Sub-clauses 3.3 and 3.4
- granular soil
  - moisture content determinations
    Part 2; Sub-clause 3.2
  - particle size distribution
    Part 2; Sub-clause 9.2
- soil and water samples
  - chemical analysis
    Part 3; Sub-clauses 5.5 and 9.3.

The significance of these tests is discussed later in this section and in Section 10. In some instances, additional or more complex testing may be required to measure the strength and permeability of the foundation and/or potential fill materials (References 62 and 63).

Further information on these tests can be found in CIRIA Special Publication 25[58] and BS 5930[57].

8.5 GEOTECHNICAL ASSESSMENT

8.5.1 Ground conditions

The data collected by the information study and site survey should be used in conjunction with the results of the ground investigation to assess the ground conditions at the site. Subsurface profiles of the ground should be prepared from the information and thickness of the various layers in the exploratory holes, with a minimum of one section along the line of the embankment and a further section approximately at right angles. Additional subsurface profiles will be necessary for more complex sites, where a number of embankments are required, or where a containing embankment for a non-impounding reservoir is planned. Sections across the proposed borrow area will also be required to assess the availability of materials and constraints on the area. These will allow the variation of the ground across the site to be assessed and, together with the properties of each soil layer, will provide the basic information to assess the suitability of the site for reservoir construction.

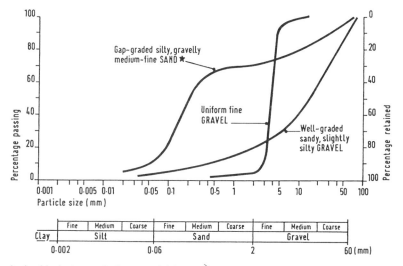

* In this instance, lacks material between approximately 0.5 and 5mm in size

Figure 8.8 Grading curves for coarse-grained soils

In a fine grained or cohesive soil (predominantly silt and clay), the engineering behaviour is best denoted by a description which takes into account the marked influence of the silt and clay sized fractions. Box 8.5 summarises the differing effects of these fractions and the methods to measure and assess these.

In a coarse grained soil (predominantly sand and gravel), the engineering behaviour is best denoted by the particle size distribution, the shape of the particles (whether these are mainly rounded, angular, etc.) and the compactness of the material. Box 8.6 summarises the main features affecting the engineering behaviour and the methods to identify these.

Box 8.7 summarises the properties of the main soil types and their suitability to provide a watertight reservoir and to support the embankment. In general, sites which are comprised of cohesive materials to a depth well below the reservoir floor are ideal for reservoir and embankment construction. Granular soils with a high permeability will be more problematical and may require substantial works to limit the loss of water. In some instances, the necessary works would be uneconomic and may rule out the proposed reservoir site, whilst others may be used with careful attention to the design and the incorporation of appropriate features.

Box 8.5 Engineering behaviour of fine (cohesive) soils

> In a fine soil, the engineering behaviour is best denoted by a description which takes into account the marked influence of the silt and clay sized fractions. A relatively small amount of clay sized material can result in cohesive behaviour and can be sufficient to warrant description as a 'clay'. A distinction between the clay and silt constituents is required as their behaviour is markedly different and this is given by the 'plasticity' of the soil. It is the characteristic property of clay soils to take and retain a new shape when compressed or moulded. The plasticity varies markedly with the moisture content – which is the ratio of water mass to soil mass expressed as a percentage value – and the size and nature of the actual clay mineral particles. The upper limit, denoted as the liquid limit, is the moisture content at which the soil begins to flow, whilst the lower limited, denoted as the plastic limit is defined by the moisture content where a 3 mm diameter thread of soil can just be rolled between the palms.
>
> These limits, also known as Atterberg Limits, have a standard laboratory test procedure to ensure comparability of the results and are determined on samples which have all material in excess of 425 microns (just under half a millimetre) removed. The difference between these values is defined as the plasticity index and is the range of moisture contents throughout which the soil exhibits plastic behaviour. A silt will have a very low plasticity, whilst a clay will have a higher value. The presence of organic material in a soil will influence these values and tend to give a reduced plasticity. A knowledge of these values allows the material to be classified according to its position on a plasticity chart. This should normally be undertaken by a specialist adviser, but Reference 58 offers some simple concise advice on the interpretation of these test results.
>
> Soils with a substantial silt content will exhibit 'dilatancy' whereas clays will not. A pat of soil which is moulded in the hand will develop a film of water on the surface; when squeezed this will recede if the material is predominantly silt. This is a simple test which can readily be used to identify soil types during the excavation of the exploratory holes.

Certain material types are unsuitable if present in the dam foundation and should be avoided, particularly if significant amounts of these materials occur. Box 8.8 summarises these materials and outlines the main problems that are likely. The past history of the site may also have resulted in features which are problematical and these are also detailed briefly in Box 8.8. In particular, existing polished surfaces in clays should be regarded with caution, whilst the presence of made ground is likely to result in an unsuitable location for an embankment and reservoir. If difficult ground conditions are encountered, and sites containing unsuitable materials cannot be avoided or they cannot be removed by a reasonable depth of excavation, specialist advice should be taken. Ground improvement schemes may be possible, but it will often be more straightforward and economic to consider an alternative site.

Box 8.6 Engineering behaviour of coarse (granular) soils

The soils are classified in terms of the main constituents which are typically described as:

Material	Composition (by weight)
Slightly sandy gravel	up to 5% sand
Sandy gravel	5 to 20% sand
Very sandy gravel	over 20% sand
Gravel/sand	about equal proportions of gravel and sand
Very gravelly sand	over 20% gravel
Gravelly sand	5 to 20% gravel
Slightly gravelly sand	up to 5% gravel

Minor constituents present are described by suitable qualifying terms (e.g. 'with occasional gravel' or 'with little silt and clay'). In general terms, higher strengths are given by the more angular fragments, a wider range of constituent sizes (well graded) and an increased compactness of the material. The fragment shape can be assessed visually whilst the simple means of estimating the compactness are included in Table 8.4. Typical particle size distributions are given in Figure 8.8 and show some naturally occurring coarse soils. In general, gap graded soils (which lack material of certain sizes) or uniform materials (which have a substantial amount of material in a narrow size range) are likely to be less compact and more permeable than a well graded soil.

8.5.2 Construction materials

The potential construction materials must be reviewed and their adequacy and availability confirmed for the embankment construction. This should include the following:

- suitability as embankment fill material
- impermeability assessment
- available quantity
- ease of winning, transporting and placing
- distance from dam.

The locally available material should normally be used for embankment construction to minimise the length of haulage and avoid the importation of expensive fill from off site. In many cases, suitable designs can be prepared to incorporate otherwise unsatisfactory fill materials by a range of technical solutions including flattening slopes, providing drainage and other measures. Such designs are outside the scope of this guide and require specialist advice at an early stage. Despite the additional costs of the alternative construction techniques, possibly greater fill qualities and the advice, this approach is still likely to prove more economic than importing better quality fill.

Box 8.7 Properties of main soil types

Gravels
- will be of sufficient strength to support embankment
- will encourage seepage beneath dam with risk of erosion of fine material and ultimately undermining of embankment. Excavate and replace with low permeability material if of shallow depth, or seek advice if depth in excess of three to four metres
- will lead to seepage from reservoir with possible loss of fine material and problems at discharge position. Provide liner and control groundwater below liner
- mix with sand and clay to form stable construction material for shoulders of embankment
- clean gravels may be suitable for drainage and slope and crest protection purposes.

Sands
- will normally be of sufficient strength to support embankment, unless material is disturbed or loose tipped. Remove unsuitable material
- will encourage seepage beneath the dam and from reservoir as gravels, but possible lower flows depending on grading and compactness of material
- possibly prone to erosion by flowing water and wave action requiring adequate protective measures
- mix with clay to form stable construction material for shoulders of embankment.

Silts
- unsuitable for foundation material or construction purposes.

Clays – very soft to soft
- will be of adequately low permeability beneath the embankment and reservoir
- will have insufficient strength beneath embankment and should be removed
- may possibly lead to instability around reservoir perimeter, but unlikely to be significant

- moisture content will tend to be too high for construction material and will need careful selection and placement. May need to be limited to central portion of embankment
- high plasticity material will need to be excluded from the shoulders to avoid cracking in dry summers.

Clays – firm to hard
- will be of adequate strength to support embankment if not subject to past landslipping or deformation
- hard clays may be liable to swelling on saturation from reservoir but unlikely to be significant
- will form suitably low permeability foundation, provided clays are not fissured. Rework surface or provide low permeability seal of softer clay if fissured
- will be of adequately low permeability beneath the reservoir
- may need to be reworked and watered or careful selection to provide suitable construction material for central portion of dam, but adequate for shoulders
- high plasticity material will need to be excluded from the shoulders to avoid cracking in dry summers.

Organic clays/peat
- unsuitable for foundation material or construction purposes
- possible source of construction material for landscaping purposes or surface soiling.

Made ground
- unsuitable for foundation and may lead to seepage from the reservoir
- possible risk of contamination problems with reservoir water
- possible health and safety problems
- possible source of fill material if investigated sufficiently and close construction control maintained, but not likely to be economic.

Box 8.8 Unsuitable embankment foundation

Very soft to soft ground
- highly compressible
- low strength
- settlement/stability problems
- construction difficulties
- probably slow to drain

Predominantly silts
- inherent instability when wet
- lack of cohesion and poor structure
- settlement/stability problem
- rapidly affected by rainfall during construction
- slow to drain when wet
- readily eroded by flowing water

Organic clays/peat
- highly compressible
- low strength
- settlement/stability problems
- probably slow to drain

Made ground
- likely to have very variable properties
- probable low density
- risk of voids and loosely placed material
- likelihood of degradation and settlement
- settlement problems
- may contain very soft/soft or organic materials
- possible contaminated ground/hazardous materials
- possible gas or leachate problems
- possible chemical problems

Clays with polished surfaces – may be very smooth planar or undulating, probably stained or differing colour on surface.
- probably indicative of past shallow landslipping
- reduced strength
- risk of further movement during construction
- risk of further movement when groundwater levels rise with filling of reservoir

Normally, the cost and availability of suitable imported materials need only be considered if there is a shortage of potential fill materials on site. The constraints on access, both on and off the site, may also need to be taken into account if the importation of fill is considered. This option may dramatically influence the design of the embankment if the use of potentially expensive imported fill is to be minimised. The required changes in design may also necessitate the testing, assessment and consideration of different geotechnical properties for the fill materials.

The unsuitable material types included in Box 8.7 should not be used in embankment construction unless compensatory design and construction control is carried out by a specialist adviser. They may provide a suitable source of landscaping fill, however, or be used to form islands or other features around the reservoir.

8.5.3 Permeability assessment

Assessment of the permeability characteristics of a site or a particular soil with any degree of accuracy and confidence is not straightforward and should only be entrusted to a suitably experienced civil engineer or soils expert. Two methods for making an initial appraisal for the potential fill material, prior to laboratory testing or employing a specialist, are described in Figures 8.5 and 8.6. Soil from at least three locations should be tested from exploratory holes spaced evenly about the potential borrow area, with two samples at varying depths selected at each location. The former method assesses the clay content by hand texturing based on an approach by ADAS and the Soil Survey and Land Research Centre[60]. The latter method, which can give misleading results if not carried out correctly, allows the proportions of gravel, sand, silt and clay to be estimated directly. Whenever possible, the final assessment of the permeability should be based on permeability tests on both insitu foundation materials and laboratory tests on fill materials. In some instances, the permeability may be able to be assessed from representative particle size distribution tests by a specialist adviser if permeability tests are unavailable or not consistent with visual observations.

The recommended acceptance criteria for use of the soils as fill for the impermeable barrier or core, or for the foundation, are discussed in Section 9.2.1. The main criteria are a permeability of not more than 10^{-9} m/s (i.e. 10^{-10}, 10^{-11} m/s, etc.) and a minimum clay content of 10% although a value of 20% to 30% is desirable. Table 8.5 shows the ranges of permeabilities for various soil types and gives an indication of the meaning of 10^{-9} m/s in practice.

Where clay material is to be imported to the site to form the impermeable barrier, the material must be fully considered and sufficient tests carried out to assess the clay content and the permeability when recompacted as fill. The amount of testing will be dependent on the quality control arrangement for the source, but a minimum of five sampling locations should be selected.

8.5.4 Embankment fill appraisal

A suitable soil to use for the construction of the core or a homogeneous embankment should contain generally not less than 20% nor more than 30% of clay, the remainder being well-graded sand and gravel. Such a soil is likely to be stable even when subject to significant changes in moisture content. Embankments made from soil with a clay content much less than 20% are likely to be subject to moderate water losses and depending on the purpose of the reservoir and availability of water, may require an impermeable liner. On the other hand, soil with a clay content higher than 30% is likely to shrink and crack on drying and such material should ideally be limited to the centre of an embankment to form the impermeable zone. If the tests suggest that the clay content is outside the recommended range of 20% to 30%, a homogeneous form of construction is unlikely to be suitable and a more complex form of embankment will be necessary. Alternatively, a specialist adviser could be consulted to assess the materials in greater detail and more advanced laboratory testing, possibly with tests for strength and other properties, may be required.

Where the clay content is in excess of 30%, the liquid limit is in excess of 90% or the plasticity index (equal to the difference between the liquid and plastic limits – see Box 8.5) is in excess of 25%, the soil should not be used in the outer zones of the embankment unless specialist advice is taken. Where the liquid limit is in excess of 90% or the plasticity index is more than 65%, the soil should not be used as embankment fill material.

8.5.5 Design aspects

The findings of the information study, site survey and ground investigation must be considered together to assess fully the effects of constructing the reservoir at the chosen location. This must include assessment of all factors which might affect the reservoir, particularly those aspects which might influence reservoir safety. The effects on the surrounding area must also be considered during the construction and future operation of the reservoir.

The potential aspects that should be considered are fully detailed throughout the guide and it is important that these should be adequately addressed. It is prudent for all these aspects to be reviewed by a specialist adviser unless the site is known to be straightforward and no possible difficulties have been indicated by the information study, site survey and ground investigation. Many sites which appear at first sight suitable for a reservoir have hidden problems which make it unlikely they will hold water. A cautious, methodical approach to the site investigation will help both to assimilate the various aspects and to provide sufficient information to allow the design to proceed.

9 Design aspects

9.1 RESERVOIR DESIGN

9.1.1 General

All the intended uses for the reservoir should be established during the design phase. The design should also ideally provide for future alternative reservoir use, where these may be required. Excavation within the reservoir area, islands, shoreline slopes, access points, depth variations across the reservoir and long-term planting arrangements must also be considered as part of the design process.

Where buried or above ground services, access routes, rights of way or other features are presently within or across the reservoir, their diversion or replacement must be considered at the design stage. This will require liaison with the appropriate owners and agreement on the route and details or the diversion or replacement.

9.1.2 Slope stability

Slopes adjacent to the reservoir must be adequately stable under the influence of the reservoir. Section 9.3.8 discusses suitable long term slopes for borrow areas. Similar comments apply to man-made and natural slopes around the reservoir. Drainage may help stabilise unstable areas of the reservoir perimeter if caused by groundwater.

9.1.3 Reservoir perimeter

Existing reservoir slopes may need to be flattened for various reasons other than being potentially unstable. Wildlife may require localised steeply dipping or shallow areas or platforms near water level. Except for very localised excavations for hollows, small cliffs for environmental reasons etc., slopes should not be steeper than 1V:4 or 5H.

Areas around or adjacent to the reservoir which have special planning controls and land use management policies or have archaeological or environmental constraints may require special measures which will need to be agreed with the relevant body or LPA. Such works will have normally been discussed at the feasibility stage and the requirements should be incorporated at the design stage.

9.1.4 Erosion

Erosion protection measures around the reservoir, other than by careful planting and management of vegetation, are generally not needed. In some instances, however, it may be necessary to provide localised measures as discussed for the dam itself. Generally this is best left to the construction stage and the need assessed at that time.

Areas that may need protection include:
- around structures, such as inlet structures, where their presence results in rapid flow or discharge over the existing ground
- where drains, drainage channels or small streams run into the reservoir
- where vegetation intended for reservoir protection has to develop and short term measures are needed
- where public access or fishermen result in the need for some form of protection against surface abrasion and/or wave erosion
- south west facing areas subject to the most frequent wave action. Local features, topography etc., may result in other directions being prone to more frequent wave action.

Figures 9.1 to 9.4 show alternative forms of protection for the reservoir perimeter.

9.1.5 Islands

Islands can be designed to use unsuitable material arising from the embankment construction. They should be formed at a maximum slope of 1V:4H but the properties of the material or environmental requirements may necessitate flatter slopes.

9.1.6 Sedimentation

The reservoir may silt up with material transported down from further upstream by the watercourse. Measures should be provided to remove the silt or stop it from entering the reservoir. These are discussed in greater detail in Section 9.8.4.

9.2 EMBANKMENT DESIGN

9.2.1 General

The basic principle of design is to produce a functional structure at minimum cost. Consideration must be given to maintenance requirements to ensure that economies achieved in design will not result in excessive maintenance costs.

Figure 9.1 Slope protection of boards and stakes to reservoir perimeter

Figure 9.2 Slope protection of logs and stakes to reservoir perimeter

Figure 9.3 Slope protection of stone near top water level around reservoir perimeter

Figure 9.4 Slope protection of vegetation near top water level around reservoir perimeter

The design and construction of the embankment is, by necessity, site-specific and must take account of all relevant aspects at that particular site. It involves naturally occurring soils with all the variability which that implies, rather than manufactured materials with known and predictable properties. The design of the embankment should be carried out by a suitably qualified and experienced civil engineer. The guidance in this document is necessarily of a general and reasonably conservative nature. On some sites it is likely that a well engineered design, taking account of site conditions and measured soil properties, may result in a more economic embankment.

Although this guide discusses impermeable soils for seepage barriers or zones of the embankment, complete impermeability is impossible to achieve. Normally, a soil should be considered to form an adequately impermeable barrier if it has a permeability of less than 10^{-9} m/s (i.e. 10^{-10}, 10^{-11} m/s etc.), and a minimum thickness of 500 mm . Table 8.5 shows the ranges of permeabilities for various soil types and gives an indication of the meaning of a permeability of 10^{-9} m/s in practice. Methods of assessing the insitu permeability of the foundation soil at a site are given in Sections 8.4.5 and 8.5.3, whilst the means to ensure the soil is placed and compacted to the required permeability are discussed in Section 10.3.11.

The design phase must not be considered to end once the proposals are completed and drawings prepared. Additional information from construction must be considered and the design modified. The generally conservative approach adopted in this guide will help to minimise changes but the design must be reviewed regularly throughout the creation of the reservoir.

9.2.2 Design requirements

The principal requirement for the design of a small water retaining embankment dam is to ensure that it is safe and stable during all phases of its construction and operation. The following criteria must be met:

- the dam must not impose excessive stresses upon the foundation
- the embankment slopes must be stable under all conditions, including rapid drawdown of the reservoir
- seepage through and beneath the dam and abutments must be controlled so as to not affect the stability of the dam nor the function of the reservoir
- the dam must be safe against overtopping by provision of adequate spillway and outlet works capacity
- the upstream slope must be protected against wave action and the crest and downstream slope must be protected against erosion or damage by wind and precipitation
- the embankment must be protected against damage by foot or vehicular traffic, stock, wild animals and vandalism.

The stability of a small embankment depends upon the type of fill, the type of soil underlying the embankment, the past geological history and land use of the area, the slope of the upstream and downstream faces, the overall foundation width, and the measures taken to prevent surface and internal erosion. The stability also depends greatly upon the care with which an embankment is constructed and the extent to which the design provisions are put into effect. The stability of excavated slopes also needs careful consideration.

Instability may be initiated within a dam or the natural slopes if the water level of the reservoir is lowered rapidly. Permeable/semi-permeable shoulder fill materials help to reduce the risk of this problem.

Seepage through the foundation and the embankment is important, not only because of the water lost, but also because seepage causes internal erosion and collapse of the embankment. Erosion proceeds slowly at first, the minute streams of water carrying away the finer particles of the soil, but at a certain stage, the rate of erosion increases rapidly and is followed by the sudden collapse of a section of the embankment. This process of internal erosion is sometimes referred to as 'piping'.

It is not possible to construct an earth embankment which is completely watertight. Seepage, however insignificant, will always occur. Thus, to ensure stability, the embankment and foundation interface should be designed and constructed so that the rate of seepage is reduced to a minimum and seepage paths are adequately controlled by suitable drainage and filtering.

To be an effective and durable water-retaining structure, the embankment and its foundation must be designed and considered as one unit.

The fill material must be adequately compacted to maximise the density and reduce the amount of air and/or water-filled voids left in the soil. These are micro voids between the individual fragments of fill, not visible to the eye. Compaction will also increase the strength, decrease the permeability and minimise post construction settlements.

9.2.3 Design options

Stable embankments may be constructed on a suitable site if the following conditions are met:

- an acceptably level, sound foundation is present
- the soil types are as stated
- the fill is compacted properly
- the required cutoff and seepage control measures are provided
- drainage measures are included to control seepage
- the height is not in excess of 5 m, measured from the crest to the excavation level
- the upstream slope should not be steeper than 1V:3H
- the downstream slope should not be steeper than 1V:3H
- the crest width of the embankment should be a minimum of three metres.

These slopes may be adopted where a site investigation has found no adverse features. Specialist advice must be obtained if unfavourable features are indicated or where some aspect is intended to be varied from the conditions listed above or described in Section 9.3.3.

Small embankments constructed of high permeability soils, e.g. sands and gravels, with an impermeable membrane on the upstream slope without soil cover, may be built with slightly steeper slopes not exceeding 1V in 2.5H due to a general increase in fill and foundation strength associated with the dry conditions. Similarly steeper slopes may be possible with such materials for the downstream slope of a zoned dam.

If adequate information is available from a site investigation, slope angles may possibly be amended following specialist advice on embankment stability.

Four embankment design types are applicable to small embankment-type reservoirs. These are influenced by the type of soil available as fill and as a foundation. These are summarised as:

- homogeneous design – comprising impermeable soil throughout
- zoned design – comprising more permeable soil with a core of impermeable soil
- diaphragm design – comprising permeable soil with a thin core or diaphragm of impermeable material
- lined design – comprising permeable soil with an impermeable lining or blanket on the upstream slope and reservoir floor.

Whichever design is adopted, the stability and safety of the embankment must be ensured and a conservative approach must be adopted at all stages of the design.

The latter two options are likely to be more expensive and complex.

9.2.4 Homogeneous design

Where suitable soil is available and the height of the dam is small (less than three metres) this design is likely to be adopted. The soil should be well graded and should contain at least 20% to 30% clay. Soils with a very high clay content will lead to the risk of excessive shrinkage and maintenance problems unless flat slopes are used for the embankment design. Where the height is greater than 3 m, a zoned construction can offer some advantages in construction and operation.

Homogenous designs result in seepage at high levels through the embankment, which could result in seepage appearing on the downstream face unless a toe drain is provided to draw down the seepage below the downstream face.

Typical designs of homogeneous dams are shown in Figures 9.5a and 9.5b. In Figure 9.5a, the embankment is built directly of an impermeable foundation soil; the easiest and cheapest form of construction. Figure 9.5b shows how a cutoff trench can be incorporated where up to three metres of permeable soil overlays the impermeable soil thereby avoiding extensive deep and difficult excavation. For depths of less than 1m, it would be preferable to excavate down to the impermeable soil over the full embankment width.

9.2.5 Zoned design

A zoned design should be adopted if a sufficient variety of soils is available. An impermeable soil is used to form a seepage barrier or core within an embankment otherwise constructed of more permeable but stable soil. If the supply of impermeable soil is sufficient, the cheapest and easiest solution is to construct as much of the interior of the dam as possible with this soil to form the core, and to use the more permeable soil in the outer portions of the dam. The core should be as wide as possible if it is to offer an adequate barrier to seepage, with a recommended width at the base of the dam of twice the height.

Where the fill material is of low permeability and reasonably homogeneous, it is often desirable to form a zoned dam, with a central core of the selected more impermeable material and the more varied material in the outer portions of the dam.

Typical designs of zoned dams are shown in Figures 9.6a and 9.6b.

9.2.6 Diaphragm design

Where the supply of soil with an adequate clay content is insufficient to form a core of the required width, a central barrier or diaphragm of impermeable material can be incorporated. This is normally installed from the crest of the completed homogeneous embankment by specialist plant and comprises a zone of bentonite cement or a bentonitic clay material. The technique is normally used only on larger embankments and is not likely to be cost-effective for embankments within the scope of this guide. Means of installing shallow diaphragms down to five metres depth have been developed using adapted standard excavation plant and self hardening bentonite cement. This approach may prove viable where cost is not a major concern.

If this design approach is considered feasible, specialist advice will be necessary through the design and construction stages.

A typical design of a dam incorporating a diaphragm is shown in Figure 9.7.

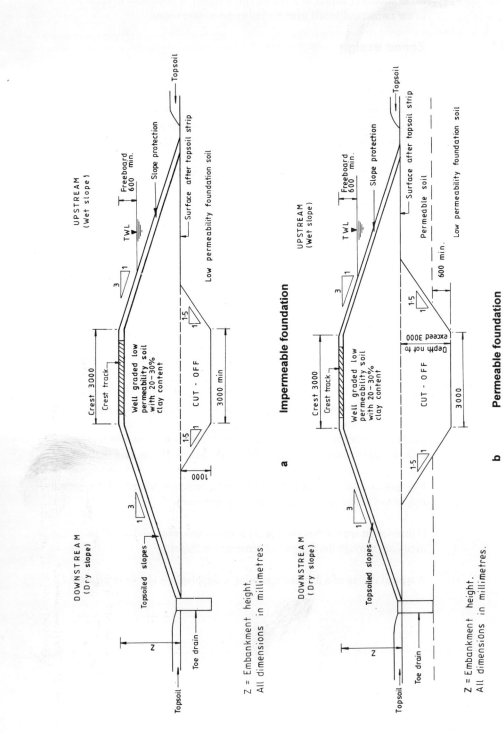

Figure 9.5 Typical designs for a homogeneous dam (a) on impermeable foundation (b) shallow depth of permeable foundation

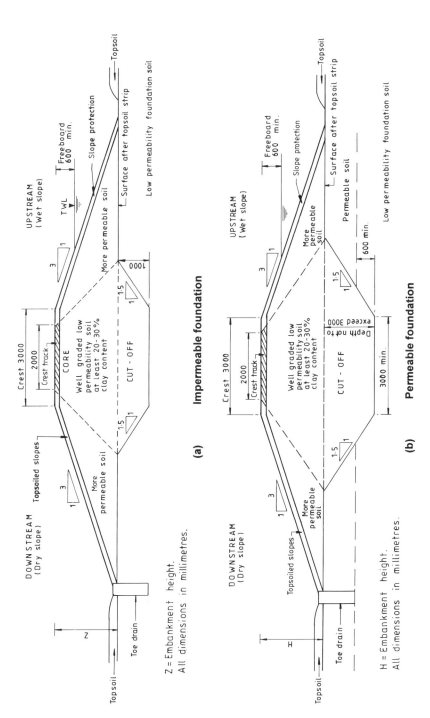

Figure 9.6 Typical design for a zoned dam (a) on impermeable foundation (b) on shallow depth of permeable foundation

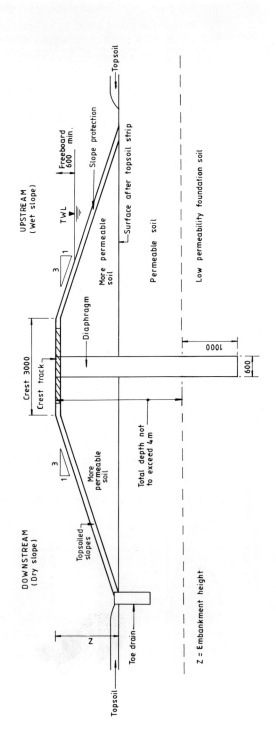

Note: All dimensions in millimetres.

Figure 9.7 Typical design for a dam incorporating a central diaphragm

9.2.7 Lined design

Where the foundation soil for a proposed reservoir is permeable and extends for a depth in excess of 4 m, a lined reservoir may be considered. This should be seen as a costly option, although one that can be very effective provided good working practice is followed from the lining selection stage through construction and filling to the maintenance stage. Reservoir linings may be constructed using geomembranes, clay or bentonite materials. These are discussed in detail in Section 9.4. This option is likely to require more maintenance work than the others and is also likely to prove more expensive if remedial works are required at a later stage.

Typical designs for a dam incorporating a liner are shown in Figures 9.8 and 9.8b.

9.2.8 Vegetation on the embankment

Vegetation must be considered as an integral part of the design, operation and maintenance of the dam. Vegetation should be sited with care and the following aspects considered carefully at the design stage:

- reservoir purpose
- purpose of vegetation
- size of branch spread and root spread
- maintenance requirements
- life expectancy.

Vegetation may add an extra margin of confidence to stability and reduce the likelihood of maintenance or remedial work. It may also help guard against long-term changes or local variations in embankment conditions.

The principal benefits of established vegetation on embankment dams are:

- improved slope stability
- protection against surface erosion
- control against flowing water or abrasion effects
- wave erosion control on the upstream slope in some instances.

Other lesser benefits, in some instances, may include:

- noise attenuation
- wind shelter
- barriers to unauthorised access.

The vegetation may also act as an indicator by way of colour changes, more lush growth or changes in inclination which may indicate variations in the conditions.

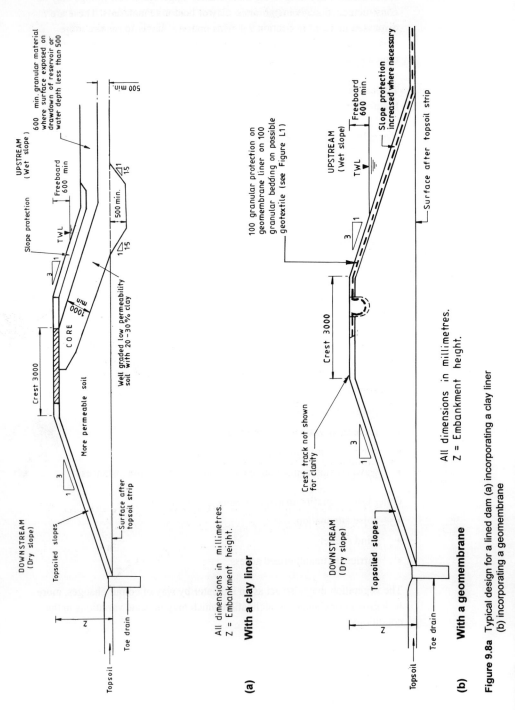

Figure 9.8a Typical design for a lined dam (a) incorporating a clay liner (b) incorporating a geomembrane

The root system is of particular importance. Both grass and herbs have a shallow rooting system largely within the uppermost 50 mm of the topsoil. Most roots from other herbaceous vegetation or shrubs are found within a few hundred millimetres of the surface but may extend typically to a depth of 3 m for mature deciduous trees. Conifers typically have very shallow root systems which spread laterally.

Certain species have a high water demand and a very extensive root system. Experience has shown that such species should normally be excluded on newly constructed dams at all times. These include:

- willow
- poplar
- alder.

Other species have a moderate water demand and may develop a relatively extensive root system. These include:

- oak
- rowan/service tree
- cherry/plum
- horse chestnut
- hawthorn
- sycamore/maple (less usually).

Use of this latter group, together with most other tree species, is not detrimental to dams provided they are excluded from certain parts of the dam including:

- the upstream slope
- the crest
- the uppermost two metres of the downstream slope
- within and beyond the auxiliary overflow
- within three metres of drains or ditches
- within three metres of structures and their associated systems.

especially if the reservoir can operate at lower levels for extended period

A CIRIA publication entitled 'Uses of Vegetation in Civil Engineering'[64] offers worthwhile advice concerning vegetation use.

9.3 EARTHWORKS

9.3.1 General

The design of earthworks using naturally occurring soils is discussed briefly in Section 9.2.1. The properties of the in-situ, undisturbed soil are likely to vary both laterally and vertically. Groundwater, vegetation and previous land usage, erosive, meteorological and weathering processes, excavation, transportation, placement and compaction of the soil as a fill material will further modify the properties.

Selection of the appropriate design techniques and soil parameters is often not straightforward and specialist advice may be required from an experienced geotechnical or dam engineer. Earthworks design is partially dependent on past experience, but a solution for one site should not be used elsewhere without due consideration of all the relevant aspects. Some useful general guidance on earthworks design is contained in BS 6031: Code of practice for earthworks[65]. Normally, it is more straightforward and economical to use the locally available materials in preference to importing better fill materials from off site. Suitable designs can often be prepared using apparently unsuitable materials by seeking specialist advice at an early stage. This approach is discussed in greater detail in Section 8.5.2.

Guidance on the design of structural foundations is given in References 66 and 67 whilst advice on earth retaining structures, including sheet piled walls, is included in References 68 and 69.

9.3.2 Embankment foundations

An embankment foundation must have adequate strength to support the loading of the embankment under any condition without unacceptable settlement or displacement. It must also provide adequate resistance to sliding. Most soils have sufficient strength to carry the weight of small embankment dams which fall within the scope of this guide, but unacceptable materials that should be avoided are detailed in Box 8.8. Disposal of these materials is discussed in Section 10.3.6, but much of the material may be suitable for use as low-grade fill material for landscaping purposes or to create islands or other features in or around the reservoir.

Where vegetation is established on an existing dam that is to be altered or enlarged, the effects, both of its presence and removal, should be considered at the design stage (see Sections 12.5 and 14.5).

Problems with sliding of the dam on, or at a shallow depth within its foundation, are unlikely if unsuitable material is removed. However, polished surfaces at shallow depths in the foundation, as detailed in Box 8.8, and particularly on the sloping ground forming the abutments, may be indicative of past ground instability. Excavation, even of limited extent and depth, fill placing or stockpiling of materials may be sufficient to regenerate instability.

These surfaces may also threaten the safe functioning of the completed embankment with raised groundwater levels following the filling of the reservoir. If such surfaces are suspected or discovered, specialist advice should be sought. If their locations are reliably known, excavation extending below these depths may be feasible, with a careful assessment of the implications on the proposed works. Alternatively, they may be left in-situ, but this will affect other design aspects, and specialist advice must be again sought.

The slope of the foundation across the dam will also affect its stability. No problems are likely if the ground slope is less than five degrees (approximately 1V:10H), but steeper slopes should be excavated to give a foundation slope not steeper than this value. This aspect is particularly important for non-impounding reservoirs, which may be sited on sidelong ground, often with steeper ground slopes than along a watercourse.

Recently cleared areas of woodland or orchard will have soil with a depleted water content as a result of the trees extracting water over a prolonged period. In clay soils, this depletion may take many months to dissipate, whilst in sandy soils dissipation will be quicker. Where a dam is to be founded on a wooded site which is on clay, the trees should be cleared as early as possible so that the soil can adjust to the new conditions. During this time the ground will swell and thus it is important that construction is not started until the soil has achieved its new equilibrium moisture content. This is particularly relevant where structures and pipelines are to be constructed.

Existing trees outside the embankment area and the reservoir itself should normally be retained. Removal of trees that are close to a new structure may be potentially more harmful, due to ground swelling in certain clay soils, than leaving them in place.

9.3.3 Seepage control measures

Seepage through permeable foundation soils which overlie an impermeable layer can be controlled by constructing a cutoff along the line of the embankment. This is normally a trench of sufficient width excavated down and into the impermeable layer and filled with soil which has a suitable clay content. The method of filling the cutoff is the same as that described for the construction of the embankment. Bentonite mixed with a permeable soil, such as some sands and gravels, may produce an impermeable material which is suitable for cutoff filling.

An earth embankment founded on rock will be a relatively costly structure arising from the need to excavate a cutoff trench into the rock to control seepage along the fissures in the rock surface. In many rocks only a shallow trench may be needed to extend into the unfissured rock, but stratified rock will require greater depths owing to the existence of many deep-seated, water-bearing fissures and may be a very costly operation. In most circumstances it is unlikely that an embankment can be constructed on rock within the economic limits of small reservoir construction.

In all cases where a cutoff is required, it must intercept all the permeable soils below the dam and extend along the full length of the dam. Depending on the thickness and permeability of the permeable layer, it may be necessary to extend the cutoff beyond the limit of the dam. In many instances, if the layer is sufficiently permeable, more extensive measures may be required and specialist advice should be sought.

Alternatively, seepage flows and uplift forces can be reduced by increasing the length of the seepage path through the foundation. This can be achieved by placing a clay blanket on the upstream part of the foundation and under the upstream shoulder, either connected to the core or to the fill of a homogeneous dam. Blankets of this form have been adopted when a cutoff to an impermeable strata beneath the embankment is not practicable because of excessive depth and the upper levels of the foundation are only of moderate permeability (i.e. fine or silty sands). A suitable design guide to the blanket thickness is 10% of the maximum reservoir depth with a minimum thickness of 500 mm. The length of blanket depends on the permeability of the foundation, the width of dam and the depth of water. The blanket length beyond the upstream toe should be not less than five times the maximum reservoir depth, although the final design should always be assessed by a specialist adviser.

Many sites for reservoirs will have had land drains installed in the past. These will provide a ready path for seepage running in unfavourable directions and thus measures must be included to intercept or divert these as part of the dam construction. Whilst they may not be encountered in trial pits or no evidence of installation might be available, it is prudent to assume they are present and to allow for their interception and cutoff along the full length of the dam. Most land drains are within 1 m of the ground surface and thus the cutoff trench should extend for not less than 1 m below the general excavation level for the dam or to a depth of not less than 300 mm below the drains. Any land drains that are encountered should be sealed with concrete on both sides of the trench to stop any future seepage. A cutoff trench of this form will also intercept any previous shallow trench, e.g. for services, earlier excavations or infilled watercourses, which may otherwise provide a possible seepage path beneath the dam.

Any pipe, culvert, trench or excavation passing through or beneath an embankment below top water level must be backfilled with a cohesive material to prevent seepage along the trench. A pipe or culvert should also have anti-seepage collars to prevent the internal erosion of soil by seepage flow along the periphery. These collars, which increase the length of the seepage path, should be made of concrete and be at least 150 mm thick along the line of the pipe. The number and dimensions of these collars should be determined using Figure 9.9 with a normal minimum spacing of 6 m. It is important that the collars are constructed to the required projection around the entire perimeter of the pipe and extend into the ground beyond the pipeline trench by a minimum of 500 mm. Further comments on their construction are included in Section 10.4.3.

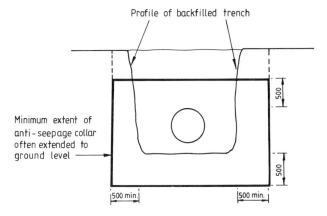

Section Through Anti-Seepage Collar
(NOTE: Dimensions (mm) are minimum and should be increased to suit)

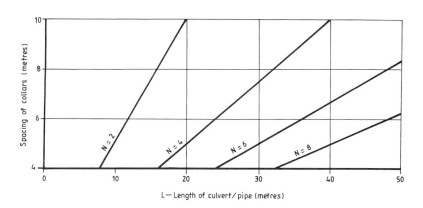

Assessment of Number of Collars (N) Along Pipe or Culvert

Figure 9.9 Typical design of anti-seepage collars

9.3.4 Slope and crest design

The recommended slopes and crest width are given in Section 9.2.3. These should only be used if a site investigation has been carried out and any specialist advice has been considered. Steeper slopes are not recommended and must not be used without agreement from the specialist adviser.

Flatter slopes and a wider crest width may be used for landscaping purposes as required. An embankment with flatter slopes and wider crest width requires larger quantities of fill material, however, and occupies a larger area; the borrow area will also be correspondingly larger. Regular vehicular access is not generally recommended across embankments unless a track is provided, but the recommended width allows occasional access for maintenance, remedial work or emergency access.

Where embanking is to be carried out adjacent to excavated slopes to provide extra depth, the embankments should preferably be set back from the top edge of the excavation by not less than ten metres. This will increase the stability of the combined slopes and facilitates construction by providing a working space and access for construction plant.

9.3.5 Slope protection and freeboard

Embankment and excavation slopes may need to be protected from erosion caused by wind-generated wave action. The factors influencing erosion are:

- size and fetch of the reservoir
- upstream slope angle
- upstream fill material
- aspect of embankment (i.e. orientation to prevailing winds)
- vegetation
- location and elevation of the reservoir
- whether the reservoir is to operate at top water level or at variable levels (i.e. abstraction reservoir).

Freeboard is provided to contain the flood surcharge plus wave splash without overtopping the embankment. The required freeboard depends on the situation of the reservoir and should be assessed from Table 7.7.

The severity of wave action will dictate the method of protection to be used. Other reservoirs in the area should be visited and the effects of wave erosion and protective measures assessed. A flexible approach is needed, however, and measures must be amended or introduced as required to control potential erosion. Various forms of erosion control are discussed in Box 9.1.

The intersection of an impounding dam with the natural ground is a position of potential severe erosion and special measures may be necessary. Slope protection on the dam is normally continued for a short distance onto natural ground beyond the embankment to reduce the risk of erosion at these vulnerable locations.

Box 9.1 Erosion control

Hard protection

Stone pitching or other hard materials such as broken stone, hardcore, etc. provide the most durable protection for the upstream or inner slope, possibly retained with a wire basket to form a gabion. A cheap source of material may be available, but this form of protection may not be an economic option. A thickness of not less than 200 mm and a fragment size of not less than 100 mm should be used. Proprietary geotextile or composite geotextile/blockwork systems are also available, but are likely to be expensive. Cement/sand filled bags may prove a viable and economic, but harsh alternative. Their use with vegetation to give a more sympathetic solution may be feasible.. Where the reservoir is consistently at top water level, the extent of the protection can be restricted to extend for not less than one metre vertically below top water level and to a similar distance above. If the crest is at a higher elevation, the remainder of the slope can normally be topsoiled and grassed.

Soft protection

This includes the use of live vegetation as a means of erosion control either alone or in conjunction with mainly natural materials such as timber, brushwood etc. Geotextiles have been used increasing in association with these materials in recent years. Marginal and emergent plants are frequently used to form a protective margin along the waterline which alleviates wave action whilst the surface root structures restrain and reinforce the soil. These plants will have a much reduced effect if the water level is subject to variation. The principal protective action of vegetation below water level is reinforcement by the roots of trees and shrubs, although the roots of smaller vegetation, often in the form of rhizomes, can also help to restrain, reinforce and buttress the soil. The major species of vegetation used for soft protection are reeds, grasses and trees such as willow, poplar and alder, which are water loving and thrive in wet conditions. The use of these trees on the embankment, however, should be considered carefully (see Section 9.2.8). Selection of the appropriate vegetation should include consideration of the problems of establishing an artificially introduced range of plants involving a limited number of mature plants. It should also consider any potential management problems likely to arise in maintaining the range and preventing the uncontrolled spread of a dominant species.

Depending on the exposure of any particular site, including the direction and fetch of waves onto the embankment, vegetation alone may be insufficient. This may often be the case before any planting has become established and provision for short-term erosion control may be necessary. This can be achieved by the use of the appropriate forms of enhanced protection shown in Figures 9.10a and 9.10b; once vegetation has become established, these measures may be less important or they may, themselves, become the dominant protection (e.g. willow poles and osiers may shoot and become self-generating). Some of the more commonly used methods have included tree trunks or bundles of faggots (fascines) anchored at the water-line to break the wave energy and prevent damage to the banks until such time as vegetation is properly established, but their anchorage details need careful consideration. Boards set vertically and pegged have been used in the past, but frequently this approach has been found to be unsuccessful and prone to damage or vandalism.

Protection for varying water levels

Particular problems are present with reservoirs which are subject to a wide range of operating water levels as protection is ideally needed over the full extent of the upstream slope. Vegetation protection is not suitable other than for the uppermost levels and no suitable vegetation can normally develop on a slope which varies from dry to being under several metres of water. Hard protection is normally uneconomic at the level required, particularly if the reservoir is of the non-impounding type with a relatively extensive length of embankment. Irrigation dams have suffered extensive erosion in the past with erosion rates which have approached tens of centimetres per year following first filling. A possible solution which may prove economic, is to construct a sacrificial layer of fill on the upstream slope which can then be allowed to erode. A minimum width of 2 m should be provided over the full embankment height and this should be sufficient in most instances for an adequate period, particularly if vegetation is encouraged to develop around top water level. A typical example of this approach is also shown on Figure 9.10b.

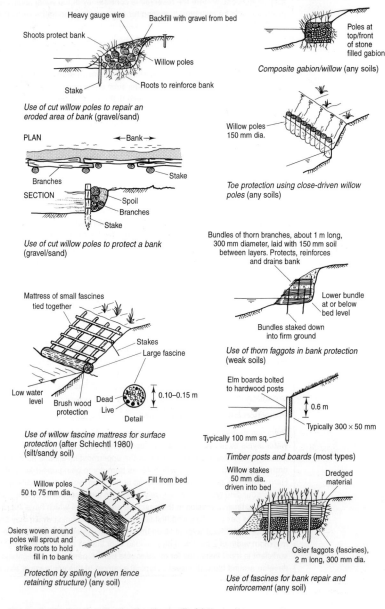

Figure 9.10a Bank protection methods (Ref 64)

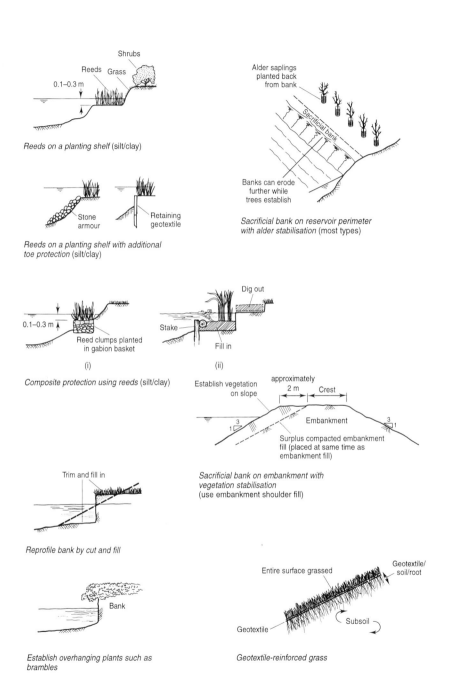

Figure 9.10b Bank protection methods (Ref 64)

CIRIA Report 161

Protection on the downstream or outer slope normally comprises grass and other vegetation. Shrubs and trees help to reinforce the slope and offer other benefits as discussed in Section 9.2.8, but are particularly beneficial in reducing erosion from rainfall runoff and from traffic or trampling effects. The intersection of the downstream or outer slope with the sides of the valley is prone to more concentrated runoff and damage and the contours should be rounded to minimise the effects on the junction of the fill and insitu material. In certain instances, for example if the runoff from the valley sides is excessive, drains may be required along the intersections.

9.3.6 Crest protection

The shoulder fill should continue above the core to provide some protection, but the exact details of the crest will be dependent on access requirements. The crest can be topsoiled, but it is preferable for a surface layer of crushed stone or well graded gravel to be used as a surface material, particularly if the crest is to be used by the public. Vehicular access to the dam should be prevented if possible, as it increases the likelihood of erosion along the crest and shoulders and may result in rutting or other damage. It will also provide a temporary surcharge load that may cause instability. Storage of any materials on the dam crest or shoulders should be avoided as it will prevent the establishment of vegetation and may result in instability of the dam.

A camber should be provided along the dam, with the maximum value at the deepest section to allow for settlement on completion of construction. This should be reckoned as 5% of the embankment height, i.e. 50 mm for each metre of embankment height.

9.3.7 Drainage

Unless the dam is founded on a permeable soil, a toe drain should be provided to control both surface runoff from the downstream slope and seepage through the dam or upper levels of the foundation. Typical details are shown in Figure 9.11. Particular attention should be paid to the detail of the outlet of the drain: headwalls should be constructed or prefabricated headwalls used as described in Section 10.5.5. In some instances, if runoff is likely to be concentrated at the junction of the downstream shoulder and original ground, drains may be required along part or all of this junction.

In theory, drains can function at a slope of as little as 1V:1000H. In practice, the slope should be 1V:250H or steeper, to allow for construction tolerances. Gradients should be uniform across a system. On sloping ground there is more scope for selecting the gradient and 1V:50H may be reasonably adopted. In general, drains should be laid at depths of 0.75 m in low permeability soils to between 1.25 and 1.5 m where required in more permeable soils. Extra depth may be necessary if heavy traffic will pass over the drain.

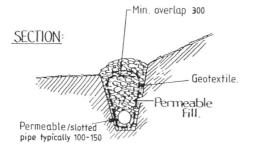

Moderate interception.
Susceptible to overgrowth.
Susceptible to ploughing.
Use as toe drain

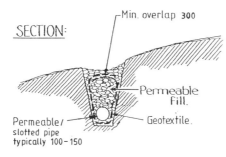

Susceptible to overgrowth.
When overgrown, interception
capacity for downslope
flow reduced.

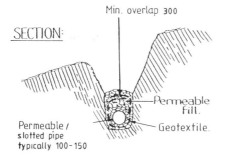

Interception very good.
Cannot be ploughed.
Risk of erosion of slopes

Figure 9.11 Drainage details

Where the proposed foundation is soft or wet and an alternative site cannot be used, a drainage blanket, normally 200 mm thick may be placed directly on the excavated surface and connected to a drainage outlet. Under no circumstance should fill be placed beneath the blanket which would inhibit flow and dissipation of water pressure. A drainage blanket will tend to be expensive in terms of materials and should normally only be incorporated in a design following specialist advice on its suitability and use.

Drainage measures to structures are also likely to be required as described in Section 10.5.4.

Virtually all drainage materials are likely to be imported and this aspect is further discussed in Sections 9.5 to 9.7.

9.3.8 Borrow areas

A borrow area is an area of land from which materials are extracted for embankment construction. It should be sited, if possible, within the reservoir area so that excavation results in increased reservoir storage without the need to increase embankment height. An estimate of the earthfill quantities required for the proposed embankment with an allowance of 30% waste will give the approximate extent of the borrow area. Any constraints on the borrow area or effects of the area itself on existing features should be considered fully.

Slopes of borrow pits should not exceed 1V:2.5H and should be as flat as possible. Where polished surfaces occur in the ground, much flatter slopes may be necessary and specialist advice should be sought (see Section 9.3.2).

There should be no excavation immediately adjacent to the dam. Borrow areas should be sited at a minimum of 10 m distance from any part of the proposed dam. In addition, borrow areas should not be planned adjacent to roads, structures or other features, including trees which are to be retained.

Where the borrow area is to be outside the reservoir or forms part of the reservoir perimeter, the comments in Section 9.14 on habitat creation should not be forgotten.

9.3.9 Core materials

The minimum and ideal criteria for the use of a soil as a core material are given in Section 8.5.3. Even where a homogenous dam is planned, it is good practice to place the more cohesive material in the centre and the less cohesive, more granular, materials in the outer portions of the dam. Works necessary to alter or increase the moisture content of the core material (see Section 10.3.11) should be identified at the design stage. This will help at the construction stage and minimise costs.

9.3.10 Shoulder fill materials

The shoulders support and protect the clay core in zoned- or diaphragm-type embankments. Satisfactory slopes are dependent on the type of fill material and the foundation. The more granular materials should be selected for the outer portions of the shoulders, thus assisting both the stability and drainage of the fill. Materials defined as unacceptable in Box 8.8 or Section 8.5.4 should be excluded from the shoulder fill, together with any material which has a surplus of individual hard fragments or consists of single-sized material which cannot be readily compacted.

9.4 RESERVOIR LINING MATERIALS

9.4.1 General

Lining materials comprise:
- geomembranes
- clay
- bentonite based materials.

All can be used to form an impermeable barrier on a permeable site, but the choice of lining should be made in conjunction with a specialist adviser. Appendix L gives information on the selection and design of the lining and outlines the more commonly available products.

Reservoir slopes flatter than 1V:3H may need to be adopted to ensure long-term stability. The underlying bedding material will need to be graded carefully to provide a reasonably uniform surface without undue irregularities, changes in profile or localised voids or protrusions. Measures to avoid the build-up of water or gas under the lining material may be necessary in some circumstances.

Particular care will be needed where the lining is connected to concrete and other structures or pipework and appropriate design and careful construction techniques will ensure that a leakage path is not created. Subsequent works to stop leakage tend to be particularly difficult and time consuming.

9.4.2 Geomembranes

A geomembrane is a thin, man-made material that is relatively flexible and elastic. The geomembrane can thus take up irregularities in the underlying material to some extent and can accept limited deformation after placement. Large irregularities and subsequent deformations cannot be tolerated, however, and the membrane is prone to mechanical damage and piercing if not well-placed and protected adequately. Geomembranes will normally need to be covered to protect against damage, erosion or other detrimental effects. Anchorage and jointing must also be carried out carefully if the membrane is to function satisfactorily in the long-term.

9.4.3 Clay

Where clay is available at a reservoir site, it has been used to provide low permeability linings to small reservoirs with varying results. The effectiveness of a clay lining depends on a number of factors. These are:
- availability of clay material within or adjacent to the reservoir area
- nature of available clay material (clay content, plasticity and strength)
- method of excavation and placing
- method of compaction
- thickness of material placed

- post-placement deformation (fluctuating water level, groundwater influences, cracking, settlement)
- reservoir purpose.

The nature of the available clay material is of primary importance where reliance is placed on a relatively thin layer. The clay content should be sufficiently high to ensure low permeability thus reducing seepage, and the material should normally comply with the requirements for core materials in Section 9.3.9. The clay content should not be too high, however, as this leads to shrinkage and cracking on drying when the reservoir level is lowered. The risk of cracking may be reduced by covering the compacted clay with a layer of granular material at least 600 mm thick where it is susceptible to exposure. This includes all parts of the embankment exposed on the maximum drawdown plus all those covered with 300 mm or less of water.

An adequate thickness of clay material should be placed and compacted to ensure a low permeability seal to the reservoir. This layer should be at least 500 mm thick when compacted, and should be thicker where the nature of the clay material in the borrow area is considered to be variable (particularly silty or granular).

9.4.4 Bentonite

Bentonite has been used on limited occasions for reservoir lining purposes in two main applications; either as a loose material (powdered or granular) or within a composite sheet (sandwiched between geotextiles).

Loose bentonite may be used in powdered or granular form to reduce the permeability of the reservoir bed. It can be either mixed with insitu soil (normally by rotovating) and then compacted to form a layer of lower permeability than the original reservoir bed, or placed as a compacted layer of pure bentonite. In both applications the layer should be covered with a layer of selected compacted material.

A composite sheet comprises a layer of bentonite sandwiched between geotextile materials to provide a lower permeability barrier, similar in principal to the use of pure loose bentonite. The sheet should be covered with a layer of selected compacted material.

9.5 GEOTEXTILES

9.5.1 General

The use of geotextiles in the design of small embankments has arisen from relatively recent developments within the civil engineering industry. Geotextiles should be considered as permeable materials used to perform separation and filtration functions for drainage and erosion control.

9.5.2 Geotextiles for drainage applications

The geotextile ensures that the soil to be drained is separated from the material within the drain, but is sufficiently permeable to cause negligible resistance to flow. The geotextile prevents localised loss of material if the flow is sufficient to carry the material away and minimises clogging of the drain. Figure 9.11 illustrates the use of geotextiles for drainage applications.

The geotextile allows greater confidence in the combined drain than in a simple stone-filled drain. The geotextile can be designed to suit the adjacent soil to be drained based on the particle size distribution (i.e. the grading) of the soil. Selection of a suitable geotextile for a drainage application is described in Appendix M. The particle size distribution is established by carrying out tests on soil samples from the area to be drained as part of the ground investigation as described in Section 8.4.6. Different criteria apply, depending on whether the soil is predominantly granular or cohesive. Worked examples covering the basic principles of geotextile design for drainage are also given in Appendix M.

In practice for cohesive soils, a geotextile with a pore size, as defined in Appendix M, of about 0.1 mm is used in many applications. The geotextile should normally be non-woven to avoid problems with the weave of the material opening up during installation or under load.

9.5.3 Geotextiles for erosion control

Geotextiles are frequently used to give enhanced protection against erosion as part of a grass reinforcement application. A geotextile should be installed in accordance with the manufacturer's instructions and adequately lapped and staked to prevent uplift by the flowing water.

Further information on the use of geotextiles for the erosion protection of overflow structures is given in Section 9.10.3 and CIRIA Report 116[70].

9.6 PIPEWORK

9.6.1 General

Reservoir pipework can be placed in two categories, i.e. that passing over or around the dam and that passing through or beneath the embankment. The former needs to be installed at a sufficient depth to avoid damage and must be sufficiently flexible either at the joints or in the pipe itself to accommodate movement of the supporting ground. Pipes passing beneath the embankment must not only be flexible, but also sufficiently strong to withstand the high loads produced by differential settlement as a result of the embankment loading. It is prudent in the latter case to adopt the most suitable type of pipe, normally spun or ductile iron or concrete. The high cost of replacing or repairing a pipe that is experiencing problems will far outweigh any initial supply costs.

Where pipework enters or leaves a structure, allowance for settlement and differential movement should be included by the provision of short lengths of pipe adjacent to the structure as shown in Figure 9.14 and Figure 10.11.

9.6.2 Types of pipe

The types of pipe commonly used are described in Box 9.2.

Whatever type of pipe is used, the pipe and joints must be capable of withstanding the internal pressures in the pipeline. Thrust blocks are likely to be needed as discussed in Section 10.9.3. Guidance concerning the design and loading on rigid pipework is given in Reference 71 and in Reference 72 for thin-walled pipes, although these flexible pipes should only be used cautiously. Manufacturers' catalogues or publications from their Associations should also give information on the use of various types of pipe. Whatever type of pipe is adopted, the quality and care of construction are critical to the long-term functioning of the pipework as discussed in Section 10.9.

Table 9.1 Minimum trench widths for pipes (Ref 71)

Nominal pipe size (mm)	Pipe outside diameter (m)	Overall trench width (m)
100	0.13	0.55
150	0.19	0.60
225	0.28	0.70
300	0.38	0.85
375	0.50	1.05
450	0.62	1.15
525	0.67	1.20
600	0.79	1.35
675	0.88	1.45
750	0.95	1.50
825	1.04	1.60
900	1.12	1.90
975	1.20	2.00
1050	1.30	2.10

NOTE: The trench widths are intended only as a guide. The width necessary in practice depends on such factors as the actual pipe outside diameter, soil stability, excavation depth and the method of trench support used.

The choice of pipe material should include consideration of the construction stage and a conservative approach should be adopted. Where the pipework is to be buried, the depth and width of trench will effect the loadings, together with other factors, and these should be minimised where possible. Table 9.1 gives the minimum trench widths for the usual range of pipe sizes, but greater width may be required for ease of construction, methods of support, greater depth or other constraints. The presence of the pipework and backfilled trench will affect any embankment or other structure constructed above and this aspect must also be considered at the design stage.

Box 9.2 Types of pipe

Pressure pipework

- asbestos cement pipes — These pipes are rigid but a degree of flexibility is given to a pipeline as the joints incorporate rubber rings. They are not as strong as ductile iron or steel pipes and tend to be brittle and liable to fracture on impact. Water or soil containing excessive sulphates will attack these pipes and their use should be restricted to areas where access can be readily gained for maintenance/replacement purposes.

- ductile iron pipes — These pipes are rigid, but flexible joints are available and should be used. Their inherent strength and good resistance to deterioration make them suitable for use in all conditions, but their high cost usually limits their use to situations where the earth loading is high, e.g. the low level outlet to an impounding reservoir. Protection against corrosion may be necessary in some aggressive soils, particularly around certain types of flexible joints.

- steel pipes — The strength of steel and a degree of flexibility make them very suitable for use in all conditions, but their life is limited by corrosion. Flexible joints are seldom necessary. All pipes should be protected from corrosion by galvanising, bitumen coating or wrapping with inert material or other recognised methods. Their use may be limited if the pipes are buried as future replacement/remedial works may prove difficult.

- polyethylene pipes — These pipes are less strong than ductile iron or steel pipes, but stronger than unplasticised PVC pipes. Their flexibility at all temperatures makes them a good choice for pumping lines to offstream reservoirs. They have a high resistance to deterioration, but their flexibility and reduced strength requires appropriate bedding and backfill.

- unplasticised PVC pipes — These pipes are less flexible than polyethylene pipes at normal temperatures. As the temperature decreases the flexibility reduces until the pipe becomes brittle and it is inadvisable, therefore, to use uPVC pipes in exposed positions. They are lightweight and have a very high resistance to deterioration, but their low strength and lack of flexibility require a high standard of bedding and backfill. They are also less suitable for high pressures and sudden surges in the flow within the pipe.

- spun concrete pipes — These pipes are rigid and therefore flexible joints must be used. Their strength makes them suitable for use in any situation, except where the water or soil has an excessive sulphate content. It is essential to use the correct class of pipe for a particular loading.

Drainage pipework

- perforated or slotted plastic drains — Perforated or slotted plastic drains are generally manufactured using uPVC which has the advantage of being lightweight. They are fabricated as flexible coils or more rigid pipes, and are normally joined with push-on uPVC sleeve fittings. The pipes typically range in diameter from 60 to 350 mm. They can be used for surface drainage, for interceptor drains and in underdrainage systems where the loading on the pipework is small. Where the pipe is more than 2 m below the finished fill level or the ground surface, they may not have adequate strength and special advice on their use should be sought.

- porous concrete pipes — Porous concrete pipes are generally manufactured using Ordinary Portland Cement (OPC) and should conform to BS 5911 Part 14.[73] They may be manufactured using sulphate resistant cement where aggressive ground conditions are present.

9.6.3 Pipe bedding and surround

The appropriate bedding and surround to pipework must be provided to ensure long-term reliable function and durability of the pipework and no adverse effects on other aspects of the embankment and reservoir. The bedding must be suitable to allow the required strength of the pipe to be developed and this may often mean that the excavated material cannot be used as bedding and surround. A granular bedding is required in many instances and such materials should satisfy the criteria of Box 9.3. The material must also be compatible with the type of pipe and the intended use of the reservoir. In certain instances, a concrete bedding or surround may be required. Guidance on the choice of bedding and surround materials is also given in References 71 and 72.

Where pipework passes from upstream to downstream, the bedding must comprise a cohesive or clay backfill with anti-seepage collars, as detailed in Section 9.3.3, as seepage along a granular bedding cannot be tolerated. Alternatively, a concrete surround could be used, but this would be expensive and must include provision for settlement and differential movement by the inclusion of flexible discontinuities in the concrete surround at the pipework joints.

Granular bedding or backfill to a pipe situated other than as described in the previous paragraph may also allow seepage of groundwater along the line of the pipe and, if this is suspected at the design stage, or becomes evident during construction, a series of clay or concrete cutoffs should be provided across the trench to stop flow.

9.7 DRAINAGE MATERIALS

Drainage materials for an efficient drainage system must be durable and fully compatible. Drains must retain the ability to discharge sufficient quantities of flow but must be constructed such that material does not clog the drain and reduce its efficiency. A combination of permeable fill, geotextiles and pipes is shown in Figure 9.11. Geotextiles are discussed in Section 9.5 and details of the various types of drainage pipework are included in Box 9.2. Permeable fill is generally used in conjunction with porous pipes, and more recently geotextiles, to assist the flow of water into the drain pipes. It also functions as the granular bedding for the drainage pipework. Washed gravel and crushed stone are generally used as permeable fill materials. Other materials that may be available locally include blast furnace slag, clinker and pelleted materials derived from pulverised fuel ash, although use of these materials may be limited by the intended use of the reservoir. Permeable fill should also satisfy the criteria in Box 9.3 to ensure suitability as a drainage media and be compatible with the type of pipe.

Box 9.3 Pipe bedding and surround criteria

•	size	–	wthin the range 5 mm to 50 mm, clean and free of fines (often described as 'washed')
•	toxicity	–	free of unacceptable toxic pollutants, e.g. heavy metals, fuel oils, etc.
•	solubility	–	evaporite materials such as gypsum or halite (rock salt) should be avoided. Chalk or limestone should not be used if the water is acidic in any respect
•	durability	–	the material should not breakdown during handling nor deteriorate after installation
•	strength	–	the material must have sufficient strength to withstand the effects of transportation, handling, stockpiling and placing.

9.8 INLET STRUCTURES

9.8.1 Gravity supply

A non-impounding reservoir can be filled or partly filled by gravity from an adjacent watercourse. A silt trap may be incorporated at the inlet to the reservoir to reduce sedimentation build up. A typical layout is shown in Figure 9.12.

This type of inlet needs to be designed to ensure a safe arrangement that can handle the entire range of flows to be expected in the watercourse yet provide the required inflow into the reservoir and any compensation flow required by the RLA. Specialist advice will normally be required if this approach is used.

The intake may take the form of a simple structure, formed by excavation in the bank of the watercourse, directing flow into a pipe or channel as shown in Figure 9.14. A more complex structure which contains one or more control weirs to maintain sufficient depth for abstraction purposes and/or controls the abstraction at times of low flow in the watercourse may be required.

9.8.2 Feeder channels or pipes

The flow to a non-impounding reservoir may be conveyed by a channel or by a pipeline. Maximum channel velocities and slopes for various types of soils are given in Table 9.2. If the velocity in a feeder channel could exceed the maximum for a particular soil type, then a feeder pipe or lined channel should be used.

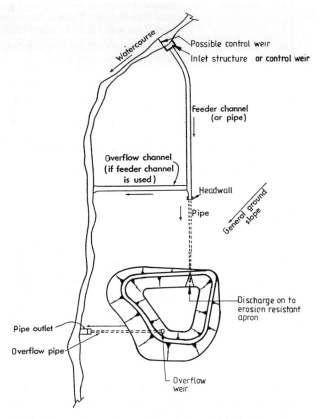

NOTE: Reservoir top water level is <u>below</u> water level at control weir

Figure 9.12 Typical arrangement for gravity filling (After Ref 2)

The channel should terminate with a headwall before reaching the reservoir, and the flow should be taken into the reservoir in a pipe of sufficient size to pass the maximum inflow. The end of the pipe should discharge onto an apron of hard material or concrete extending down to the reservoir bottom, or the pipe itself should be extended downwards ending at a small apron to prevent bank erosion occurring whenever the level of the water in the reservoir is below the inlet pipe. Alternatively, the pipe can discharge onto an area of large stones to dissipate the energy. These should be retained within a series of wire baskets or gabions, as shown in Figure 9.13, to prevent movement by flowing water or loss by vandalism. Similar precautions are necessary in a lined reservoir to prevent damage to the lining.

Table 9.2 Maximum channel velocities and slopes for feeder channels (Ref 2)

Soil type	Maximum velocity (m/s)	Maximum slope	Maximum duration of fow (hours)
Sandy soil	0.6	1V: 600H	up to 20
	0.5	1V: 900H	20 – 40
	0.4	1V:1460H	above 40
Clayey sandy silt	0.9	1V:300H	up to 20
	0.7	1V:460H	20 – 40
	0.5	1V:900H	above 40
Clay	1.2	1V:150H	up to 20
	1.0	1V:220H	20 – 40
	0.8	1V:340H	above 40

NOTE: Based on poor/scant grass cover in feeder channel

9.8.3 Watercourse control weirs

Where the abstraction rate from a watercourse is restricted, the inflow to the reservoir or pumping sump can be controlled by a weir across the entrance to the intake pipe or channel. This will ensure that water is only abstracted when flow in the watercourse is above a predetermined level. In small watercourses, a weir may also be needed to raise the water level sufficiently for the entrance weir to be effective. The relative levels and widths of the two weirs, in the latter case, will determine the proportion of the flow in each of the channels. The height of the weir in the main watercourse should be kept to a minimum.

A low weir is more easily constructed, is less costly and has less effect on the upstream water levels. Removable weir boards are an advantage as they can be used to adjust water levels or removed to leave an unobstructed channel. A hole in the weir wall below the crest level is a useful method for maintaining a reasonably constant downstream flow, as the flow through the hole is less sensitive to changes in water level. The agreement of the RLA must be obtained for any proposals for works in the watercourse and it would be prudent for the initial proposals to be discussed with the RLA at the earliest opportunity. In general, weirs across the watercourse, however, should be avoided.

Figure 9.13 Alternative protection at inlet using gabions

9.8.4 Silt traps

Any reservoir supplied by flowing water will receive an influx of silt, in addition to water. If left to accumulate, it may result ultimately in abandonment of the reservoir. A silt trap may be constructed by excavation, in concrete or masonry, or utilise natural materials as described below:

- Excavated silt traps

    A silt trap is often formed by simply deepening and widening the watercourse just before it enters the reservoir; a widening to not less than eight times the width of the watercourse and a minimum length of three times the widening is recommended. Some localised damming may also be necessary and this is often achieved by a small concrete weir. If access around the reservoir is required, this can be combined with the concrete weir to form a dished concrete slab which will allow access by farm or off-road vehicles. The silt trap may be designed as a small pond and set at a slightly higher level than the reservoir, as shown in Figure 9.15. Such a pond can be landscaped and provide a possibly differing habitat to the main reservoir.

    The silt trap must be emptied regularly to remain effective (which may be on a several years basis if of sufficient size). The design of the trap must allow suitable access for excavation and loading of the silt. It should be remembered that the excavated silt will be essentially fluid and thus access routes away from the silt trap must be sufficiently flat and straight to minimise transportation problems.

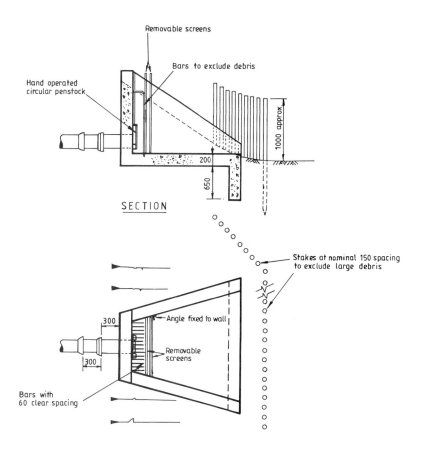

NOTE: Dimensions (mm) shown are typical and should be varied to suit

Figure 9.14 Typical arrangement of inlet structure

Figure 9.15 Excavated silt trap

- Chamber silt traps

    It may be possible to construct a chamber silt trap at the inlet to a reservoir if the watercourse is small. The chamber should be several times the width of the watercourse and thus larger watercourses would require a disproportionably large chamber. Thus the use is frequently restricted to reservoirs fed by ditches or drains. A typical arrangement is shown in Figure 9.16 and comprises an inlet, chamber and an outlet to the reservoir. This form of silt trap is far less efficient than either of the other two options and will only trap material of medium sand size and greater. The finer material will pass into the reservoir.

    It is essential that regular inspection and emptying of this form of silt trap is carried out and that the inlet and outlet facilities are regularly cleared. Normally this type of silt trap will need more frequent emptying which may arise several times annually.

- Natural silt traps

    Natural silt traps can be created by establishing emergent plants at the inlet where minor watercourses (such as ditches) flow into reservoirs. These encourage silt deposition upstream of the inlet in the minor watercourse. Provided access to the minor watercourse is available, removal of the silt should then be a simple operation. This form of silt trap has also been used for a major watercourse entering a reservoir, where a wide area of very shallow water has been established by introducing a bed of emergent plants. Regular inspections are required to ensure that silt is removed regularly and that flooding of the watercourse does not occur. Details of emergent plants that can be used for this purpose are given in Section 4.4.3.

All silt traps must be designed and constructed so as to allow the passage of flood flows along the watercourse. An excavated or natural silt trap will pose no significant resistance to flow although large flows may be sufficient to wash some of the trapped silt into the reservoir. Concrete silt traps, however, may lead to ponding upstream of the 'chamber' with localised flooding and out of channel flow. In extreme flows this may lead to severe erosion and possibly damage or destruction to the chamber unless arrangements are included to allow the safe passage of floods past the silt trap by means of an excavated channel at a high level.

9.8.5 Pumped supply

Filling by pumping gives a much wider scope for finding a good reservoir site. Pumped reservoirs can be located away from the source of supply and, in particular, on land which has little agricultural or other value. Offsetting these advantages is the cost of pumping. A typical layout is shown in Figure 9.17. The maximum rate at which water can be pumped from a stream, limited either by the amount of water available or by restrictions imposed by the abstraction licence, together with the total head required, will determine the pump duty. A practical basis for winter storage is to assume that the reservoir can be filled in 1000 hours pumping time. (i.e. for a reservoir of 18 000 m^3 this would require a pump capable of delivering 18 m^3 per hour or 5 litres per second).

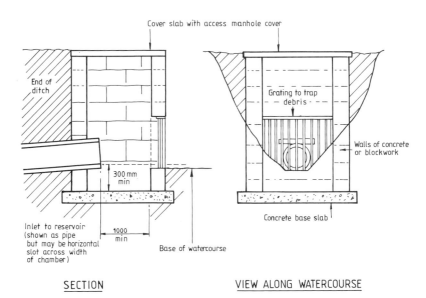

Figure 9.16 Typical chamber silt trap details

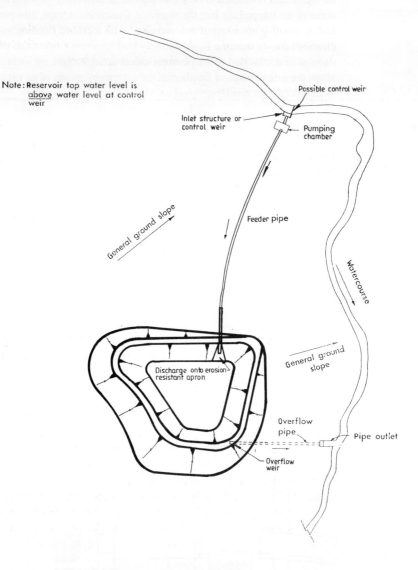

Figure 9.17 Typical pump layout (after Ref 2)

The pump selected must be capable of lifting the required amount of water against a head at least equal to the sum of the following plus an additional safety margin:

- the vertical lift, measured in metres, from the minimum water level at the pump intake to the highest point reached by the delivery pipe
- an amount, expressed in metres head, to offset the friction losses in the suction and delivery pipes

- an amount, expressed in metres head, to offset the losses due to pipe fittings, particularly valves, bends and junctions. This should be assumed to be 25% of (ii) for short pipe runs.

Table 9.3 indicates the friction losses expressed in metres per 100 m of smooth pipe bore for a range of pipe diameters and sizes. The losses will increase substantially with increasing roughness of the pipe bore.

Table 9.3 Friction losses in pumped suction and delivery pipes (Ref 2)

Pipe dia. (mm)	Pipe flow litres per second															
	1	2	3	4	5	6	5	8	9	10	12	14	16	18	20	25	30
50	0.01	2.2															
75		0.3	0.7	1.1	1.7	2.4											
100			0.3	0.4	0.6	0.8	0.9	1.2	1.4	2.0							
125						0.3	0.4	0.5	0.7	1.0	1.2	1.6	1.8				
150										0.3	0.4	0.5	0.6	0.8	1.1	1.4	
175														0.3	0.3	0.5	0.7
200																0.3	0.3

NOTE: Losses expressed as metres per 100 m of smooth pipe.

Unless the stream has sufficient depth, it will be necessary to construct a pumping sump, either by means of a weir built across the stream to increase the depth of water or by digging out the bed of the stream. This arrangement is unlikely to be acceptable to the RLA and a separate pumping bay formed by excavation into the bank of the watercourse is normally used. This will aid desilting and maintenance work and minimise the effects of pumping on flow in the watercourse. Alternatively, a separate chamber adjacent to the watercourse may be provided, fed by gravity flow from the watercourse as shown in Figure 9.18. This may tend to act as a silt trap chamber and thus the pump should be removable for cleaning out the chamber. Wherever the pump is located, the suction must be kept clear of the sump or chamber bottom and some form of screening against weeds and debris will be necessary. Figure 9.19 shows a suitable arrangement of screening and lifting facilities for a small pump.

Wherever possible the pipe should be routed through the natural ground into the reservoir rather than through the embankment fill. Within the reservoir, the discharge pipe should terminate above the maximum level of the reservoir to avoid the pipe acting as a syphon when pumping is stopped, thereby drawing water back through the pump. Water discharging from the delivery pipe into the reservoir must not cause erosion and appropriate protection measures should be provided as in the case of gravity supply.

9.9 DRAWOFF WORKS

9.9.1 General

Water may be abstracted from a reservoir by means of a drawoff pipe or by pumping. The former option is usually provided when regular or frequent abstraction is required; the latter option is usual if the water is needed at irregular or rare intervals. In all instances, a valve or penstock must be provided on the drawoff to shut off the flow and this must be located on the upstream side of the dam; under no circumstance should water at the full reservoir head, without means of isolation, be present beneath the dam. Access to the valve or penstock will normally be provided via a footbridge and an extension spindle between the valve or penstock and the controlling mechanism. Alternatively, a valve shaft of reinforced concrete or other material, which is of sufficient size and safe design to allow access, should be provided. A flow control valve should be installed at the downstream end of the drawoff pipe but, as stated previously, there must be an isolating valve or penstock at the upstream end of the pipe.

9.9.2 Drawoffs for gravity discharge

Drawoff facilities for gravity discharge are normally constructed as part of the permanent works and may pass beneath the dam or possibly be located clear of the dam. They should be installed at sufficient depth in the original ground such that construction plant or other loading, or the weight of the embankment itself, cannot cause any adverse loading on the pipe. Pipes beneath the dam must be installed in a clay backfilled trench with anti-seepage collars as discussed in Section 9.6.3. Suitable screening should be provided to the inlet to prevent the ingress of debris and Figure 9.20 shows the arrangement adopted for a recent scheme.

Alternatively, the drawoff may take the form of a channel with a suitable penstock as shown in Figure 9.21, which allows a controlled discharge from the reservoir when this is at or near top water level.

9.9.3 Low-level drawoffs

Low-level drawoff pipes should normally be provided for impounding reservoirs to allow the reservoir to be emptied for maintenance purposes. The pipe should be laid beneath the embankment in natural ground, not in fill. It must be adequately protected against failure as the embankment settles, i.e. it must be flexible and must be so designed that water cannot pass through the embankment fill, with all the inherent risks, by seeping along the outside of the pipe.

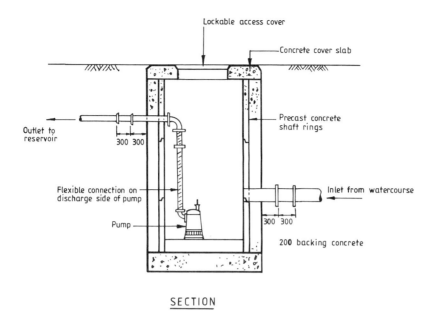

NOTE: Means of access and control not shown. Dimensions (mm) are typical and should be varied to suit

Figure 9.18 Typical arrangement of a pumping chamber

Figure 9.19 Typical pump and screen

Figure 9.20 Screened gravity drawoff

The inlet to the pipe should be sited in a pit beyond the upstream toe and a small concrete structure provided as a headwall. This will help to direct flows into the pipe and minimise the deposition of silt or localised collapse of the pit sides. No screen should be provided, other than a very coarse mesh or bars which should be not closer together than about 100 mm.

The downstream end of the outlet pipe should be well beyond the toe of the dam and allowed to discharge into the watercourse. The outlet pipe velocity is likely to be high when the reservoir is at a high level so that watercourse erosion protection is frequently required.

The low-level drawoff may be used to maintain a compensation flow downstream of the reservoir if the reservoir level drops below the overflow level and may possibly be utilised during the construction period as the diversion pipe to pass flows to the watercourse.

9.9.4 Other drawoff pipes

Water from a reservoir may be abstracted by means of the low-level outlet and diverted downstream to its intended use. Alternatively other drawoff pipes may be provided to allow the water to be abstracted directly to a distribution system. These should be designed in a similar manner to the low-level drawoff and should be located in natural ground.

Figure 9.21 Gravity drawoff with penstock control

9.9.5 Pumping arrangements

Water may be abstracted by pumping, but this is not likely to be suitable for other than occasional irrigation use in view of the costs of providing and operating the pumping facilities. The pump may be permanently sited and operated as required or may be installed temporarily for occasional operation. In the latter instance, an area of concrete hardstanding with adequate access should be provided in a suitable position at the edge of the reservoir. This should take account of any variation in operating level of the reservoir.

9.10 OVERFLOW WORKS

9.10.1 General

The assessment of the design flood to be passed over the dam is fully detailed in Section 7 and means of passing the flood flows over main and auxiliary overflows are described. The assessment of the length and flood rise for each overflow is also discussed. The dangers of inadequate overflow provision cannot be over-emphasised, particularly as the variation between normal and flood flows can be very substantial.

Photographs of various forms of overflow are included here as Figures 7.12 to 7.19.

The overflow arrangements have to be specifically designed for a particular catchment area and reservoir site. In some cases ample provision can be made quite simply and cheaply, in others it may be difficult and costly or possibly uneconomic. The basic requirements are that the overflows must be of adequate capacity to pass the design flood with sufficient freeboard against overtopping of the dam and must be able to withstand erosion for such periods as the flows are likely to occur.

Normally the most cost-effective solution is that the low flows are passed by the main overflow, constructed of a material such as concrete with a high resistance to long-term erosion. For greater flows, a separate auxiliary overflow is often provided which will be used only rarely; this is normally an earth channel protected by a dense cover of grass, or occasionally, reinforced grass.

Methods to raise the top water level, albeit temporarily, or the provision of fish screens or other means to retain fish in the reservoir must not be permitted. These will reduce the flood discharge capacity of the overflow arrangements and increase in the risk and frequency of overtopping the dam.

9.10.2 Main overflow – impounding reservoirs

Where the catchment, and hence the design flows are small, the most simple form of main overflow consists of a brick or concrete drop inlet weir chamber founded on natural ground, as shown in Figures 9.22, 7.12 and 7.16. Pre-cast concrete rings can also be used to form the chamber. The top of the chamber functions as a weir and controls the maximum level and capacity of the reservoir when it is not overflowing. The flow into the weir chamber is taken away by a pipe, laid to an uniform slope in undisturbed ground to the watercourse downstream of the dam. Table 9.4 gives the minimum pipe diameter and fall (see Figure 9.22) to pass a given flow. The pipe entry invert (i.e. the bottom of the pipe inlet) must be set at least 1 m below the weir to enable the necessary velocity to be developed to maintain the flow in the pipe and the pipe outlet should be designed so as to remain submerged. The length of pipework used in the construction of the drop inlet weir and low level outlet can be reduced if these are combined as also shown in Figure 9.22. As the chamber is close to the valley side, the latter should be protected against erosion by extending the back wall of the chamber upwards. A suitable substantial screen should be provided to prevent the entry of logs and other floating debris.

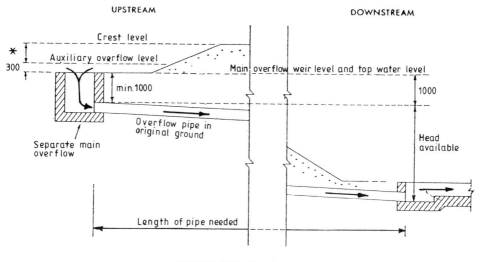

SEPARATE MAIN OVERFLOW

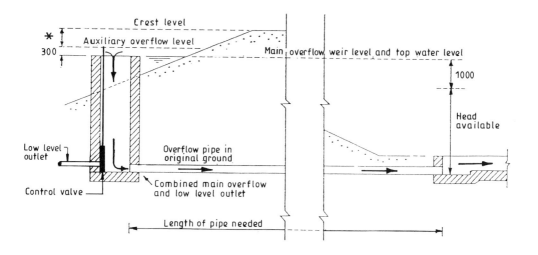

COMBINED MAIN OVERFLOW AND LOW LEVEL OUTLET

NOTES

✻ Dependant on reservoir classification and depth of flow over auxiliary overflow.

Anti seepage collars and base of fill not shown.

Access to control valve and benching to base of chamber not shown.

Figure 9.22 Typical weir chamber arrangement (after Ref 2)

Table 9.4 Weir chamber – minimum pipe diameter and fall (Ref 2)

Capacity of overflow m³/s	Length of pipe needed/head see figure 9.22	Minimum diameter of overflow pipe mm
0.1	5	175
	10	200
	20	225
	40	250
0.2	5	200
	10	250
	20	275
	40	325
0.3	5	250
	10	300
	20	325
	40	375
0.6	5	325
	10	375
	20	425
	40	475
0.9	5	375
	10	450
	20	500
	40	550
1.2	5	425
	10	500
	20	550
	40	625
1.5	5	450
	10	525
	20	600
	40	675

Note: 1. Capacity of weir chamber to be based on Table 7.8, pipe sizes and falls are adequate to carry given capacity.

2. Where combined main overflow and low-level drawoff pipe is provided, minimum pipe size and fall must be provided to pass required flow given in Section 7.2.4.

The base of the chamber should be provided with a concrete slab at least 150 mm thick to withstand the impact of the flow and to protect the underlying soil from erosion. Benching will also help to direct the flow into the pipe and should be constructed of 1:3 cement:sand (see Figure 10.11 for benching details).

An alternative arrangement for the main overflow for a small catchment is a vertical pipe, with the upper end fitted with a flared section of pipe to form a bellmouth as shown in operation in Figure 7.13. This functions as a circular weir, with the lower end connected to a pipeline passing at a slight fall below the embankment. This pipeline may also serve as a low-level outlet from the reservoir, as described for the drop inlet weir chamber. This type of overflow may be more difficult and expensive to construct than the chamber inlet type previously described. It also has the disadvantage of requiring a large diameter pipe under the dam instead of a smaller pipe which would be needed solely as a low-level outlet. Its only advantage is to enable the vertical overflow pipe to be used as a valve shaft for the valve on the upstream end of the low-level outlet pipe. Screening is particularly important with a vertical overflow pipe, otherwise debris could accumulate at the bottom of the vertical pipe which would be difficult to remove. Figure 7.13 shows how debris can be trapped around the perimeter and pass readily into the pipe if such screening is not provided.

Access to the top of the overflow chamber or pipe for valve operation or maintenance will be required if to function as part of a low-level drawoff. This can be provided by means of a light footbridge and the diameter of the chamber or pipe should be sufficient to permit access by means of a ladder or step irons for maintenance purposes. The valves can be operated at the bottom of the chamber or pipe, but generally the control spindle is extended to the top for ease of access and safe operation.

Most catchments will require a concrete weir or sill and a channel to pass the design flood. A typical plan is shown in Figure 7.20, whilst a typical arrangement is given in Figure 9.23. This is likely to be a reinforced concrete structure, possibly faced with stone (Figure 7.15) or similar to give it a more pleasing appearance, but alternatively blockwork, brick or masonry walls with a concrete slab may be feasible. The structure should be located on natural ground, probably at or towards one end of the dam.

Training walls should direct flow towards the weir and ideally these should have an inlet width of not less than 1.5 times the length of the weir. The position of the sill controlling the water level of the reservoir may be varied; normally this is on the upstream side of the crest, but its position may be altered to extend the water surface to a point below the crest. The maximum downstream position should be no further than the centre of the crest to minimise the risk of seepage around the structure.

A concrete cutoff beneath the structure should be provided to minimise any seepage and this should be continued up both sides to intercept seepage around the structure. Movement joints should be provided between the weir structure and the channel and, as necessary, along the channel to allow for any future deformation. These should be sealed with a flexible sealant to prevent water ingress through the joints. The upstream portions of the structure should be backed by impervious fill, whilst the portion beyond the crest of the dam should ideally have a granular backfill to allow drainage and relieve water pressures on the structure. A bridge may be required above the main overflow and this should be designed so that the underside of the bridge is not lower than the general crest level of the dam to allow free discharge beneath.

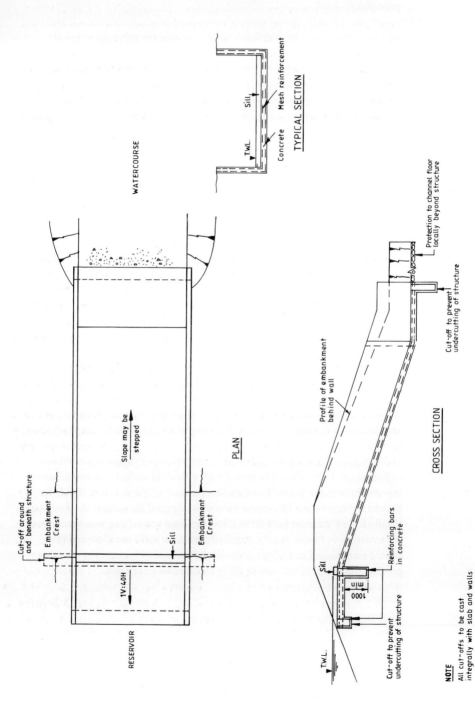

Figure 9.23 Typical concrete sill and channel arrangement

Where the valley sides are steep or a substantial length of overflow is required, a side-entry overflow structure may prove feasible as discussed in Section 7.2.4. The design is not straightforward, however, and specialist advice must be sought to ensure the design is adequate and will operate hydraulically as required.

The flow over the weir may be passed through a suitably sized culvert, rather than a channel where access is required across the crest or where the flow is required to be hidden. The provision of construction joints is particularly important and these should be formed properly and adequately sealed. The minimum dimensions of the culvert must be sufficient to allow access without difficulty for maintenance purposes. The width of a channel or culvert should be not less than the length of the weir whilst the height of a culvert should be not less than the rise over the sill when passing the design flood. This gives a conservative but safe discharge arrangement and economies may be possible if specialist advice is sought.

Backfill to the structure must, by necessity, be cohesive where the overflow structure passes through the dam or into natural ground to prevent flow along the fill/structure interface. Drainage measures may be required behind the lower portion of the channel to control surface runoff or water splash and a suitable granular backfill material is often used.

The slope of the channel or culvert will often be dictated by existing ground slopes and the arrangement of the dam. However, particularly where a structure can be constructed to take advantage of the topography, the channel or culvert may run at a relatively flat gradient. In such cases, a minimum slope of 1V:40H should be provided. The floor of the channel would normally be concrete and this may be stepped, if required, to vary the flow along the channel. Figure 9.24 shows a typical example, whilst Figure 9.25 shows a more extensive landscaped treatment to the channel. The latter, however, shows an inadequate height of channel wall which has been provided as a result of attempting to enhance the appearance of the channel as a feature. The depth of flow in a channel will decrease as speed increases along the channel, but a conservative approach should always be adopted and a wall height of not less than the design head over the weir (H_D) should always be provided. A greater height may be required if the channel floor is stepped or particularly rough as Figure 9.24.

Where the average flow is small and the normal overflow head would be very low, it may be possible to construct a short length of the overflow at a slightly lower level to concentrate the flow. The overflow level should be designed using the methods described in Section 6.2.4 and the lowered part set below this. The lowered part should be no more than 100 mm below the general sill level and extend over not more than 25% of the overflow length. Inclusion of this feature will reduce the reservoir top water level by the amount of the lowering. Alternatively, the crest level of the dam must be increased to permit an unchanged top water level.

Figure 9.24 Stepped overflow channel

Particular attention should be paid to the design of the overflow structure if facilities for varying the top water level of the reservoir are to be provided. These normally take the form of timber stoplogs which slot into grooves to allow the sill level of the weir to be raised. Figure 9.26 shows timber stoplogs in a brick overflow and channel structure with a substantial debris screen. If such an arrangement is adopted, it is possible that the RLA may impose restrictions on use of the boards at certain times of the year and require their removal at times of actual or potentially increased flows. Unless specialist advice is taken, the dam and overflow facilities should be designed to take the full flood rise with the wave freeboard discussed in Section 7. The stoplogs may need to be locked into position to avoid unauthorised tampering and any necessary maintenance requirements should be considered at the design stage. Their means and ease of removal also need to be considered at this stage assuming removal will occur at times of adverse weather and high discharges.

Figure 9.25 Landscaped stepped overflow

Debris will tend to collect at the overflow and means of preventing an excessive build-up should be considered at the design stage. Suitable screens, possibly as in Figure 9.26 may be provided, but these should have clear spaces between the bars of not less than 100 mm and preferably greater. If a screen is not provided, debris will collect at the weir itself. In either case, safe access must be provided so that debris can be cleared on a regular basis. Where a footbridge is to be installed across the overflow, this must allow a safe means of clearing a screen or sill below.

Figure 9.26 Timber stoplogs in a brickwork overflow and channel structure

9.10.3 Auxiliary overflow – impounding reservoirs

The auxiliary overflow is intended to come into use only for rare flood flows which cannot be passed by the main overflow. Normally this is located on the dam and formed by lowering a length of the crest to a constant level. A typical plan is shown in Figure 7.20 whilst a typical arrangement is given in Figure 9.27. A maximum discharge height (head) of 300 mm should be considered to pass the design flood with sufficient freeboard above as discussed in Section 7.3. The length of overflow assessed from Section 7.2.5 is the sill length; the ends of the overflow should ramp at a slope of 1V:5H or flatter if vehicle access is required. The lowered length of crest should be constructed at a uniform level with a crossfall from the upstream side of approximately 1V:40H. The gradient of the downstream slope should not be steeper than 1V:4H if the height of the dam exceeds three metres; for lesser heights the general embankment slope of 1V:3H should be adequate, but a flatter slope is preferable. Small training bunds, not less than 300 mm in height should be provided down the slope to limit the spread of discharge and direct the flow towards the watercourse downstream of the dam.

Alternatively, it may be possible to locate the auxiliary overflow on original ground beyond one end of the dam if this is compatible with the topography of the dam site. Unless the abutment is very flat, however, this is likely to result in an extensive volume and depth of excavation to produce the required length of overflow. The ground slope downstream of the overflow is also likely to require excavation and regrading works to allow a uniform depth of flow down the slope. Training walls will also be necessary and flow down the junction between the fill and the natural ground must be prevented.

More usually, an auxiliary overflow on original ground is constructed as a gently sloping channel around the dam which limits the flow velocity to a low value and avoids the need for protective works. Figure 9.28 shows a typical arrangement with a control section, which is a level, straight portion of channel extending a distance of about 8 m downstream from the auxiliary overflow crest. The function of the control section is to convert the rise in the reservoir water level above the auxiliary overflow crest into a controlled flow around the dam. The upstream portion of the auxiliary overflow should converge to guide the water towards the control section; to achieve this, the width of the auxiliary overflow entrance at the normal reservoir top water level should be 1.5 times the invert width of the channel at the control section. The convergent portion should rise towards the control section at a slope of at least 1V:40H and be such that the auxiliary overflow entrance is below the normal reservoir top water level by about 200 mm. This ensures that the funnel action comes into operation at any significant rise in reservoir water level. The downstream portion of this form of auxiliary overflow channel should fall at a slope of 1V:40H to maintain a channel velocity of approximately 1.5 m per second. This form of overflow has been used extensively in the past, but is now less common in favour of passing the flows over the dam.

In some instances, it may be possible to construct an overflow using gabion walls instead of concrete walls, particularly towards the lower end of the channel (Figure 7.19). A careful approach is necessary and subsequent repair or replacement of the gabions must be considered with the consequent effects on the dam and on the operation of the reservoir. Whilst this may appear to be an attractive, economic option in the short term, the long-term maintenance costs will be higher.

The erosion resistance of a grassed auxiliary overflow is decreased by:

- increased maximum speed of flow
- increased duration of flow
- reduced length of the grass
- irregular growth of grass or development of unsuitable species
- increased seepage flow in the soil in the direction of the slope
- reduced soil/root development
- surface irregularities or debris which cause localised drag.

A uniform slope with a well developed and maintained grass growth which is maintained at a length of 50 to 150 mm offers the best protection against erosion. The maximum velocity is dependent on the following:

- gradient of the slope
- depth of water passing over the sill
- roughness of the slope due to the grass.

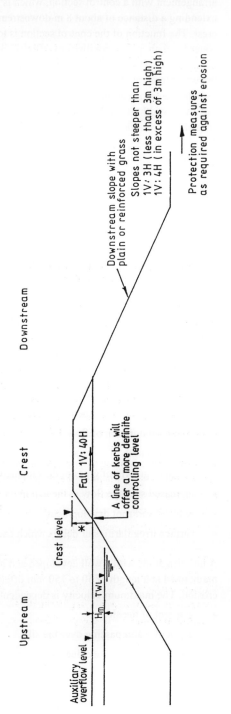

Figure 9.27 Typical auxiliary overflow arrangement for flow over a steep downstream slope

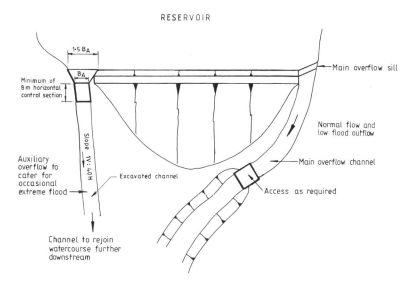

Figure 9.28 Typical auxiliary overflow arrangement for flow around the dam with a gently sloping channel

Values of the maximum velocity of flow are given in Table 9.5 whilst Figure 9.29 shows the recommended limiting velocities for erosion resistance of plain and reinforced grass. A maximum velocity for up to three hours should be considered appropriate for catchments falling within the scope of this report and the protection needs can be estimated from this figure. In most instances a good covering of grass is sufficient, providing this is properly maintained. Where higher velocities or longer durations of flow occur, grass reinforcement may be adopted. Guidelines for the design and use of grass reinforced channels are provided in CIRIA Report 116[70] or specialist advice should be sought.

If an auxiliary overflow around the dam with a gently sloping channel not steeper than 1V:40H is used, the flow velocity will be sufficiently low to allow an average cover of grass to develop and provide adequate erosion resistance. Alternatively, the channel may be protected by gabions (Figure 7.19) or other means of enhanced protection.

The small training banks should be continued beyond the downstream slope at the ends of the auxiliary overflow to channel the water into the watercourse. Adequate drainage should also be provided to prevent ponding of water at or beyond the toe or to assist in guiding the flow to the watercourse; this should be in addition to any toe drains provided for normal drainage.

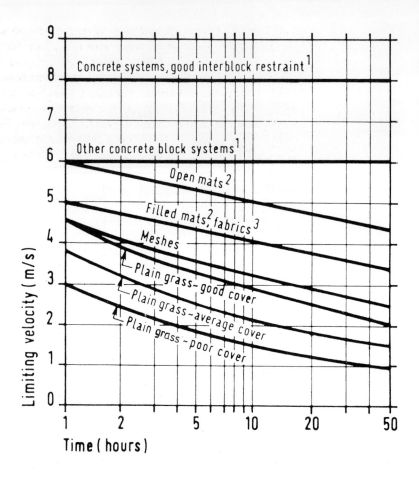

NOTES:
1. Minimum superficial mass 135 kg/m²
2. Minimum nominal thickness 20 mm.
3. Installed within 20 mm of soil surface, or in conjunction with a surface mesh.
4. See Section 9.10.3 for other criteria for geotextile reinforcement.
5. These graphs should only be used for erosion resistance to unidirectional flow. Values are based on available experience and information at date of this report.
6. All reinforced grass values assume well established, good grass cover
7. Other criteria (such as short-term protection, ease of installation and management, susceptibility to vandalism, etc.) must be considered in the choice of reinforcement.

Figure 9.29 Limiting velocities for flow over plain and reinforced grass (Ref 70)

Establishment of vegetation by laying turfs or sowing seeds should follow on the final trimming of the excavation as soon as possible. Turfing is the safest and quickest method, but if the overflow is likely to operate before the turfs have properly knitted to the underlying soil they should be staked or wired. Grass should be a low maintenance species as access onto the auxiliary overflow area should be minimised. Perennial ryegrass should be avoided and less vigorous species such as creeping red fescue, chewing fescue, smooth, stalked meadow grass and browntop should be used. These grasses are not only slow-growing, but also provide a dense cover and root formation which binds the topsoil.

Table 9.5 Maximum velocity of flow over auxiliary overflow

Depth (mm)	Flow per unit width ($m^3/s/m$)	Velocity (m/s)		
		Slope 1V:3H	Slope 1V:4H	Slope 1V:5H
50	0.02	1.50	1.33	1.13
100	0.05	2.33	2.00	1.77
150	0.1	2.97	2.54	2.20
200	0.15	3.61	2.96	2.68
250	0.21	4.13	3.87	3.33
300	0.28	4.99	4.20	3.67
400	0.43	5.37	4.23	3.99

NOTE: The velocity at slope of 1V:40H and a depth of 400 mm is about 1.5 m/s.

9.10.4 Overflow works – non-impounding reservoirs

The most suitable form of overflow from a gravity-fed non-impounding reservoir, if site conditions permit, is a channel from the feeder channel allowing surplus water to be diverted back to the watercourse before reaching the reservoir, as shown in Figure 9.12. The amount of silt entering the reservoir can also be controlled with this arrangement. If the fall of the feeder channel and the watercourse is such that the construction of an overflow channel is impossible, substantial inflow may result at times of flood unless careful diversion facilities have been provided. The design of the inlet works in this instance requires consideration by a specialist adviser. If a feeder pipe is used, then a simple overflow arrangement can be provided with an overflow pipe beneath the embankment in undisturbed ground, if the design of the reservoir permits, and discharging at some point downstream of the intake. This should be similar to the arrangement with a vertical pipe discussed in Section 9.10.2. Alternatively, a small concrete weir and channel, preferably situated on original ground, may be used to pass flood flows. The capacity of the overflow works should be greater than the capacity of the inlet channel or pipe. Allowance should also be made for the discharge of rainfall falling directly onto the reservoir and any small catchment as described in Section 7.2.6.

The overflow from a pump-fed reservoir should be similar and, as described above, discharge into a convenient water-course which can accept the maximum flow. The capacity of the overflow should be not less than the maximum pumped inflow rate, plus an allowance for rainfall.

9.11 ANCILLARY STRUCTURES

9.11.1 Bywash channels

A bywash channel is a channel constructed along a contour designed to lead floodflow around, and not into, a reservoir. It is often constructed with a bund of natural ground or fill separating it from the reservoir. The channel may be narrow and is a possible source of instability. Bywash channels also tend to be subject to erosion. Problems may occur due to blockage, and regular maintenance is recommended. These structures are generally employed at larger reservoirs and are unlikely to be suitable for reservoirs within the scope of this guide.

9.11.2 Fish passes

In certain instances, fish passes may be required to be incorporated within the main overflow structure. This may be a requirement of the RLA on certain watercourses and necessitates careful design, as well as the inclusion of means for providing attraction flows and other potentially detailed features. If a fish pass is to be included, specialist advice should be sought at an early stage and details agreed with the RLA.

9.12 CONCRETE

9.12.1 General

Most reservoir construction works involve concrete in the various structures. Much of this is exposed to flowing or standing water and the provision of impermeable, dense concrete that will resist deterioration requires careful design and construction. Poor quality design or construction will lead to long-term problems with structures resulting in expensive remedial or replacement works. The factors affecting the design and the quality of the concrete works are summarised in this section, whilst further matters to be considered during construction are discussed in Section 10.

9.12.2 Design of structures

Reinforced concrete should be designed and built to comply with BS 8110[74] and BS 8007[75]. Structures built in accordance with BS 8007, which is the recognised design code for water retaining structures, can be expected to be adequately impermeable. This is achieved by specifying a concrete mix that can be well compacted and will have a low permeability, and by specifying details

that control structural cracking. Concrete structures built to comply with
BS 8110 cannot be expected to achieve the same degree of impermeability as
those complying with BS 8007. The concrete itself may be more permeable, but
more significantly, structures to BS 8110 are not designed to control structural
cracking to the same extent. Whilst the requirements of BS 8110 are known to
most designers of concrete structures, the provisions of BS 8007 are less well
known. The design of water retaining structures is a specialised field and the
main provisions that are specific to BS 8007 are summarised in Appendix N.

The designer should specify the concrete mix, the size and thickness of the
concrete sections, the amount and positioning of the reinforcement (including
the cover), joint design and layout and all other aspects affecting the
performance of the finished structure.

9.12.3 Concrete mix specification

It is essential to ensure that the component materials (cement, aggregate, water
etc.) and mix proportions are correct to achieve a sufficiently impermeable and
durable concrete. The specification for concrete mixes is given in BS 5328[76].
The standard describes four types of concrete mix:

- designed
- prescribed
- standard
- designated.

These are described more fully in Appendix N including the means of assessing
the required materials and mix proportions. BS 5328 also gives some guidance
on typical mix applications. BS 8110 gives further general guidance on grade of
concrete (as specified in BS 5328) required for a range of exposure conditions
as summarised in Appendix N.

Most structures are constructed using Ordinary Portland Cement (OPC). Other
cements are available to overcome specific problems associated with the site,
the type of construction and the proposed method of working (e.g. early striking
of formwork may be required). The principal limitation of OPC is its low
resistance to attack by sulphates. Where sulphates are likely to be present in the
ground or groundwater, sulphate-resisting cement should be used as detailed in
Reference 77. BRE Digest 325[78] details the properties of various types of
concrete, including blended cements where OPC is replaced partially by
pozzolanic materials such as ground, granulated blast furnace slag (ggbs), fly
ash (previously known as pulverised fuel ash or pfa) or other replacements.

9.12.4 Aggregates

The size, proportion and composition of the aggregates all affect the properties
of the concrete. Aggregates should normally have a maximum size of 20 mm
for reservoir structures. A larger size of up to 40 mm may be suitable, but may
prove difficult to place and compact around reinforcement. The selection and

testing of aggregates is covered in a number of British Standards (References 79 and 80). Aggregates may consist of crushed or uncrushed naturally occurring materials or crushed concrete, although the crushed aggregates are normally restricted to the higher strength concretes. Generally, concrete containing natural, more rounded, aggregates requires less water in the mix and is easier to handle.

Aggregates must be clean, well-graded and stable. Limestone aggregates are not satisfactorily and liable to deterioration where the water is acidic for any reason.

9.12.5 Water-cement ratio

The water-cement ratio is one of the most important factors affecting concrete durability and impermeability. Unfortunately, it is also the most difficult aspect to control as site operatives can be tempted to add water to ease placing and compaction, and the sand and aggregates normally also contain water. The relationship between the concrete permeability and the water-cement ratio is illustrated in Figure 9.30 which shows the importance of minimising the water-cement ratio. As the water-cement ratio increases beyond about 0.6, the permeability of the concrete increases dramatically. The amount of water in a concrete mix should thus be just sufficient to allow the mix to be placed and compacted. Recommendations on water-cement ratio are given in BS 8110, whilst BS 8007 specifies a maximum of 0.55. The water-cement ratio can be reduced significantly by incorporating water-reducing agents into the mix without creating workability difficulties. This provides a greater tolerance against stiffening and makes it possible to control the subsequent addition of water on site.

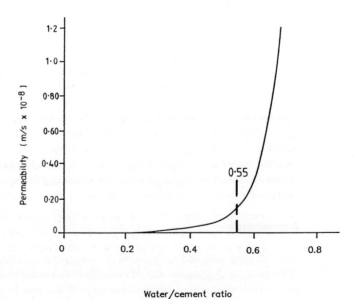

Figure 9.30 Effects of water/cement ratio on concrete permeability (Ref 37)

9.12.6 Admixtures

A wide range of admixtures are available to modify the properties of both the wet and hardened concrete. The main types are listed and described in Appendix N and are all covered by BS 5075[81]. Admixtures must be used strictly in accordance with the manufacturer's instructions and particular care must be taken if a combination is used. Most admixtures produce more than one effect and all likely effects must be considered prior to their use. The Cement Admixture Association has been set up to encourage the correct use of admixtures and provides information on availability and use.

9.12.7 Crack control and joints

Impermeable concrete requires that cracks in the finished structure are controlled and that the various elements of the structure are properly connected together. In many reservoir structures, deterioration of the concrete begins at cracks and joints and this aspect must be considered fully at the design stage. Cracks in concrete may be associated with:

- stresses due to applied loads (e.g. pressures from external backfill)
- thermal expansion or contraction
- shrinkage as the concrete hardens
- settlement of the concrete in a semi-hardened state
- poorly constructed joints
- differential settlement
- poor placement methods.

Guidance on crack limitation is given in both BS 8110 and BS 8007 which cover design and workmanship. There is evidence that water reacts with concrete to heal the surfaces of cracks up to 0.2 mm and thus crack widths should be limited to this figure. The size and location of cracks induced by applied loads can be predicted using the design methods described in the two British Standards. The width of these cracks can be reduced by appropriate structural design or the cracks can be sealed if there is a danger of leakage.

Thermal and shrinkage cracks are more difficult to predict. If the cracking process is not controlled, cracks of various widths will occur at random which will be difficult to seal effectively. Aggressive water or material entering the cracks can lead to more general and rapid deterioration of the concrete. There are two main ways of controlling the cracking which arises from contraction or expansion:

- provision of a high percentage of reinforcement, as recommended in BS 8007, to encourage a larger number of very narrow cracks which do not cause a problem. The design by this method (known as 'fully restrained') should only be undertaken by a specialist following the guidance given in BS 8007 or BS 8110
- induced or built-in controlled cracks in the form of contraction joints at predetermined positions.

Guidance on the thermal control is given in CIRIA Report R91, 'Early-Age Thermal Crack Control in Concrete'. Contraction joints relieve the stresses that would otherwise cause uncontrolled cracking. In a reinforced slab, contraction joints would normally be provided at intervals between 7 to 15 m depending upon the arrangement of the reinforcing steel. Cracks may be induced by incorporating a crack inducer in the underside of a slab or by pressing a crack former into the top surface. Provision of a preformed waterstop in the bottom face of a slab and a suitable sealant in a groove in the top face are both important if leakage is to be prevented. Figure 9.31 (top) shows a typical contraction joint detail whilst details of waterstops are shown in Figure 9.32. Dowel bars should be provided to maintain alignment between the jointed sections with the length of bar on one side of the joint fully coated with a debonding agent to ensure that there is no restraint against contraction.

Expansion joints should be provided where concrete slabs butt up to other elements of the structure. They should also be included in runs of concrete slab exceeding several tens of metres. As with contraction joints, sealants and waterstops are required to ensure watertightness. Dowel bars should be incorporated into expansion joints where appropriate to control alignment, as for contraction joints, and these should be fitted with end caps with a compressible filler. Figure 9.31 also shows a typical expansion joint detail.

Construction joints are required whenever concreting work is temporarily discontinued and between concrete elements which are constructed at different times. Construction joints are not normally specified at the design stage unless a particular arrangement or construction sequence is envisaged and the site arrangement is normally dictated by the method and speed of construction. The joints must be properly formed and the laitance (i.e. the debris, dirt and other deleterious or loose material on the previously placed concrete surface) removed before the adjacent concrete is placed. A carefully constructed construction joint may not require a waterstop, although a sealed groove at the top of the joint is good practice. If it is not possible to provide continuity of reinforcement, the joint should be designed as a contraction joint. A typical detail for a construction joint is also shown in Figure 9.31. Where slabs are cast as a series of bays, the overall construction time should be as short as possible to minimise the risk of differential shrinkage and movement.

Joint sealants must be specified carefully to ensure their satisfactory performance. They serve two main functions:
- prevention of leakage and flow through the joint
- prevention of ingress of material (e.g. dirt) that may affect the performance of the joint.

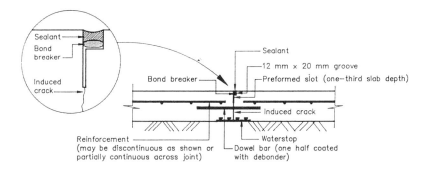

Contraction Joint Detail

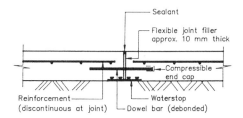

Expansion Joint Detail

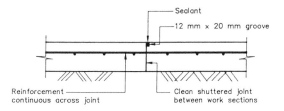

Construction Joint Detail

Figure 9.31 Concrete joint details (Ref 37)

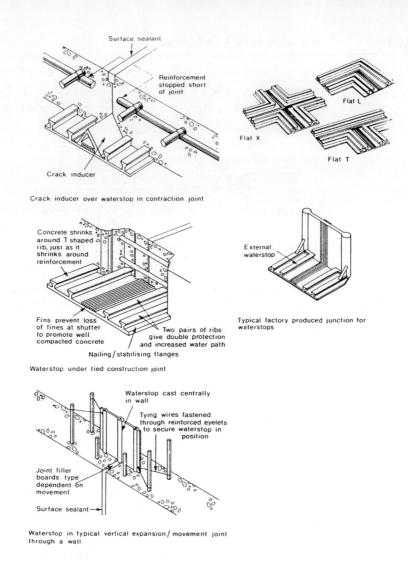

Figure 9.32 Waterstops in typical construction joints (Ref 37)

There are a large variety of sealants on the market which have been developed for particular applications. Guidance on the choice of sealant selection is given in References 82 to 85. The service life of a sealant is finite and may be limited to a few years. Once the joint sealant has broken down, the joint must be cleaned out and resealed. In practice, resealing a joint is not straightforward and the results are often unsatisfactory unless the preparatory work is carried out carefully. The use of higher quality sealants with a longer life expectancy is recommended. Rather than the cheaper bitumen-based products, these are polysulphides or polyurethane, which, when applied correctly, may last for 15 to 20 years.

9.12.8 Reinforcement and cover

The high alkalinity of concrete protects embedded steel reinforcement against corrosion. In time, the protection tends to break down as a result of:
- cracks allowing water, oxygen and other corrosive agents to reach the reinforcement
- the gradual loss of concrete alkalinity due to the ingress of carbon dioxide from the atmosphere
- the presence of chlorides in the concrete.

The degree of protection provided by the concrete is mainly related to the depth of cover. Recommendations for the minimum depth of cover are given in BS 8110[74] and detailed in Table 9.6. This is related to the concrete grade, design life and exposure conditions. In structures designed to be impermeable, the amount of cover should be restricted to the minimum necessary to satisfy durability requirements. The positioning of the reinforcement as close as possible to the surface allows better control over the surface cracking. Structures within the scope of this guide should normally be designed for a 'severe' exposure and a nominal cover of 40 mm.

Guidance on the detailing of reinforced concrete is given in Reference 86.

9.13 MASONRY

9.13.1 General

Masonry can be used to form low structures where external loadings, including water pressures, are small. Where the external pressures or the wall heights are greater, the masonry wall may require reinforcement. The structural design of masonry structures should be carried out only by suitably experienced specialist advisers in accordance with BS 5628[87]. The designer should specify the type of masonry to be used, the detailing, including fixings, mortars, the reinforcement and cover, the type of bond, joint design and layout and all other aspects affecting the finished structure.

Normally masonry construction is only used in conjunction with mass or reinforced concrete to provide the vertical elements of reservoir structures of relatively low height.

Both concrete blockwork and brickwork have been used in the past. Construction in blockwork is within the scope of many small organisations and may be more suitable in some applications than insitu reinforced concrete. The use of blockwork in recent construction has been largely at the expense of brickwork which requires more skill and takes longer to build. Where walls require reinforcement, this is also more difficult to incorporate into brickwork construction.

9.13.2 Concrete blocks

Concrete blocks should comply with BS 6073[88] and guidance on specification for various usages is given in BS 5628[87]. Concrete blocks are available in a wide variety of:

- Shape and size

Blocks may be solid, cellular or hollow, with their shape depending on the intended use. Hollow blocks must normally be used where reinforcement is needed. Block sizes range typically from approximately 390 mm × 190 mm to 590 mm × 215 mm.

- Density

Blockwork density is related to strength and durability which will both increase with greater density. Blocks with a density of 1500 kg/m^3 are not suitable for use in reservoir structures.

- Strength

Compressive strengths vary from 2.8 to 38 N/mm^2, although strengths less than 7 N/mm^2 are unlikely to be suitable. Where the blockwork is below ground or fill level and sulphates are present, advice on their suitability or special precautionary measures should be sought from the manufacturer.

9.13.3 Bricks

Bricks should comply with the following specifications:
- concrete bricks BS 6073[88]
- clay bricks BS 3921[89]

BS 5628[87] also gives recommendations on the choice of bricks for particular situations.

9.13.4 Mortars

The strength and durability of a wall is dependent not only on the properties of the masonry units but on the bond and the mortar in which they are laid. Recommendations on mortars are also given in BS 5628[87].

9.13.5 Reinforcement

Steel reinforcement should comply with the requirements of BS 5628[87] which gives various options for the choice of materials. Guidance on the cover to reinforcement is also given in this standard. Where the required cover cannot be provided, special reinforcement will be required.

Table 9.6 Concrete cover to reinforcement

Nominal cover to all reinforcement (including links) to meet durability requirements

Conditions of exposure	Nominal cover (mm)				
Mild	25	20	20	20	20
Moderate	–	35	30	25	20
Severe	–	–	40	30	25
Very severe	–	–	50	40	30
Extreme	–	–	–	60	50
Maximum free water/cement ratio	0.65	0.60	0.55	0.50	0.45
Minimum cement content (kg/m^3)	275	300	325	350	400
Lowest grade of concrete	C30	C35	C40	C45	C50

NOTE: This table relates to normal-weight aggregate of 20 mm nominal maximum size.

9.13.6 Durability and impermeability

The durability of masonry construction is dependent on the quality of the masonry units and mortar, together with the reinforcement where present, and the care taken in construction of the works. The impermeability and durability will be less than for good insitu reinforced or mass concrete, but when well-constructed and of suitable materials it should be adequate to provide a 50-year life.

9.13.7 Drainage

Masonry walls are more prone to the influences of external groundwater levels than similar concrete structures. Control of these pressures can be achieved either by suitable drainage behind walls or by providing drainage passages through the wall. In both instances, the works must be designed to ensure that the drainage capability is maintained in the long-term without clogging by use of suitable gradings and possibly geotextiles. Care must also be taken to ensure that the drainage arrangements do not permit water to pass around the outside of a water-retaining structure (e.g. an overflow).

9.14 ENVIRONMENTAL CONSIDERATIONS

9.14.1 Siting

In nature conservation terms it is preferable to locate a new water feature adjacent to existing wetland habitat, as this not only increases the size of the habitat but provides a close source of material for colonisation of the new area. The wildlife value of a site is also increased if new shrub and tree-planting links with existing woodland or hedgerow, providing extended cover. Care must be taken, however, to ensure that a more valuable habitat is not being destroyed by the construction of the reservoir.

The new reservoir should be in a quiet and secluded location to minimise disturbance to wildlife, particularly if the objective is to encourage breeding wildfowl. It is also beneficial to locate the reservoir under or close to an existing flightline used by the wildfowl leaving a clear flightpath in. If it is intended to allow the public to use the site for fishing, shooting, watersports, or informal recreation, it is important that it is easily accessible. It may also need to be located close to an area suitable for car parking.

9.14.2 Shoreline

One of the most important aspects of the design of a small reservoir is its shape. A varied shoreline which is as convoluted as possible is essential if the reservoir is to be of value to wildlife. A succession of bays and spits not only provides the variation which is visually preferable to a uniform shore, but also increases the length of shoreline. This helps to create habitat diversity and provides shelter for plants and invertebrates; both important food sources for birds and fish. Such design is also likely to increase opportunities for breeding wildfowl; small bays increase the number of nesting sites available by reducing sight lines and thus the size of an individuals nesting territory. Figure 9.33 illustrates some examples of good and bad shoreline design, whilst Figures 9.34 and 9.35 show how a varied shoreline, both in profile and vegetation type, can be obtained in practice.

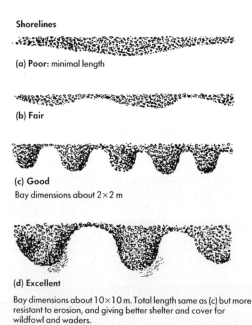

Figure 9.33 Good and bad examples of shoreline design (Ref 44)

Figure 9.34 Variable shoreline (mainly vegetated)

Figure 9.35 Variable shoreline (mainly open)

The shape of the embankment can also be varied if required by introducing flatter slopes or a wider crest width beyond the minimum requirements. In certain areas the crest level can be varied providing this does not effect any flood retention ability.

9.14.3 Bank shape

It is essential that shorelines have a gentle slope if they are to form a base for vegetation establishment and allow access to and from the water by wildlife. This is particularly important if wildfowl breeding is to be encouraged, because ducklings cannot manage steep slopes or a step which is more than 50 mm high. As a general rule shallow margins for wildlife should be graded to at least 1V:10H down to a maximum depth of 1.5 m. This can be achieved either by grading back the margins, or by depositing surplus or unsuitable fill material. If space is a limiting factor this shallow access should be provided on the northern shore, or, failing this, logs provided to help small mammals that fall into the water to climb out. If space and other constraints allow, it is desirable to create some areas of steep bank to increase habitat diversity. These should ideally be located on the eastern or western side of the reservoir to attract cliff nesting species. A gently sloping shoreline with variable soil and vegetation types is shown in Figure 9.36.

Figure 9.36 Gently sloping variable shoreline

9.14.4 Water depth

The maximum water depth will depend on the size of a water body and its purpose. If water storage is the objective, then the reservoir will need to be considerably deeper than would be required for conservation purposes. Variation in water depth, however, does maximise wildlife benefit by providing habitat variation and increasing the range of feeding areas.

Most aquatic plants thrive in the shallow water zone (less than 500 mm deep) where light penetration is good and the water warms quickly in spring. Consequently this is the zone richest in invertebrate life and of most value to other wildlife. Shallows should have a minimum depth of 50 mm grading down to a maximum of 500 mm and they should be covered with water for most of the year. These shallow ledges should grade down to deeper areas in the water body, although water deeper than 3 m becomes less valuable to wildlife. As a guide, the target depth for water surfaces over 100 m^2 designed for wildlife should be about 2.5 m and no greater than 3 m although water depth variation maximises wildlife benefit as illustrated in Figure 9.37.

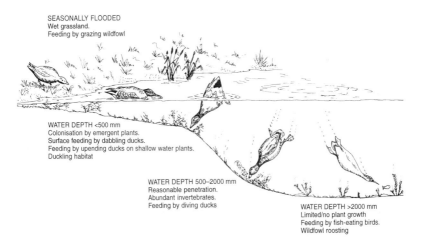

Figure 9.37 Water-depth variation to maximise wildlife benefit

Where water is to be abstracted, the reservoir level is likely to fluctuate. Such variations may be regular, seasonal or intermittent and may be slow or rapid. The ecological values of such reservoirs is potentially reduced, but careful design to provide shallows and permanent pools will help plant and wildlife. Vegetation may be more problematical if the drawdowns are substantial and frequent; few plants will thrive under such conditions, especially if the periods of drawdown are extended.

9.14.5 Islands

Islands should ideally be included in the design of a water body, although this will be feasible only at larger sites. Their form and siting are important as they provide valuable nesting habitat for wildfowl because predation and disturbance are reduced. In larger lakes they should be about 30 m away from the shore, with deep water in between. Islands suitable for breeding need only be 2 to 5 m in diameter, but they must have low gradient margins to allow access. Larger islands should ideally be cross, horseshoe or atoll shaped, providing shelter whatever the wind direction and increasing the length of available shoreline. Figure 9.38 illustrates some examples of good and bad island profiles, whilst Figures 9.39 and 9.40 show examples of island creation in practice. Figure 9.39 also shows how localised excavation can be used to create islands and retain existing vegetation.

Islands can also improve the visual quality of a reservoir, particularly if they have an irregular shape. If tree planting is envisaged on islands for landscape or conservation purposes, the trees may need to be a certain size and located at a suitable height above the water level to ensure satisfactory root conditions.

Access for maintenance works or other purposes should also be considered at the design stage and the vegetation planting planned accordingly.

Where the quantity of unsuitable or surplus fill material is limited it is often better used to create shallows and to provide rafts instead of islands. Floating rafts are also particularly useful if the water level fluctuates. There are a great variety of raft designs that can be adopted and Figures 9.41 and 9.42 illustrate two variations. Figure 9.43 shows a typical floating island soon after construction and planting.

9.14.6 Aquatic and shoreline planting

Where possible a complete planting programme should be planned that is based on the natural gradation from dry land plants, through marginals and emergents, to submerged and floating species. These zones are based on the position the plants occupy in relation to the water margin and are illustrated in Figure 4.1. Only native species should normally be introduced to maintain the ecological balance and minimise the risk of a limited number of species becoming dominant. The southern and eastern shores of the new reservoir should be left open and sunny although the limited shade cast by a few willow and alder trees, on the northern and western shores, would be of value. Comments on tree planting immediately adjacent to the reservoir are also included in Section 9.14.8.

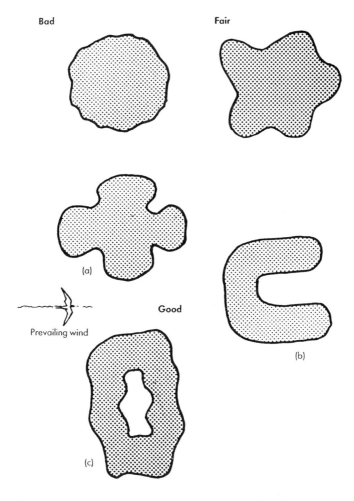

Figure 9.38 Good and bad examples of island profiles (Ref 44)

Marginal and emergent plant species grow at the water's edge down to about 1.5 m depth. They attract invertebrates and many are themselves an important source of food for wildfowl. They also have the advantage of protecting the shore from erosion, and act as a natural silt trap.

Some emergent species, such as reedmace (bulrush or common reed), are naturally competitive and invasive and can easily dominate the shallows to the detriment of other species. Emergents should only be planted in restricted areas on the margins where their spread can be managed adequately.

Open water is important for the growth of floating-leaved, rooting plants and submerged species. Free floating plants can also cover a pond if not properly managed and can be a great nuisance to fishing. They do, however, have a high value as food plants for wildfowl and some species are highly attractive when in flower.

Figure 9.39 Islands created by excavation

Figure 9.40 Islands created by filling

Lists of some common plant species that can be introduced in each zone are given in Table 9.7. These all have wildlife value, either as food plants or wildfowl or simply by providing valuable habitats for fish, invertebrates or ducklings.

Table 9.7 Common plant species of the main aquatic plant zones

Species		General comments
Marginal plants		
Water Mint	*(Mentha aquatica)*	Helps improve water quality
Soft Rush	*(Juncus effusus)*	Common in damp ground and shallows
Hard Rush	*(Juncus inflexus)*	Protects shore from erosion
Common Spike-Rush	*(Eleocharis palustris)*	Can grow in extensive and dense stands, dominating exposed shores
Common Sedge	*(Carex nigra)*	Many sedge species offer similar benefits of bank protection
		Can be invasive
Greater Willowherb	*(Epilobium hirsutum)*	Common in damp locations
Emergent plants		
Water Horsetail	*(Equisetum fluviatile)*	Erect, leafless stems
		Can take over shallow areas less than 150 mm deep
		Requires regular control
Arrowhead	*(Sagittaria sagittifolia)*	Arrow-shaped leaves with spikes of flowers growing up out of the water. Also produces floating and submerged leaves
Branched Bur-Reed	*(Sparganium erectum)*	Often grows in dense stands
		Very invasive in shallow water
		Very common
Greater Reedmace or Bulrush	*(Typha latifolia)*	Very invasive, needs careful management
		Tolerates silting and pollution
Common Reed	*(Phragmites communis)*	Very invasive and should only be introduced to large water bodies
		Good for water quality control
Yellow Flag Iris	*(Iris pseudacorus)*	Attractive flowers
		Grows in clumps in shallow water
Floating-leaved rooted plants		
White Water Lily	*(Nymphaea alba)*	Large leaves and attractive white flowers
		Can be a problem for fishery managers
Yellow Water Lily	*(Nuphar lutea)*	Very invasive especially in shallow water
		Can be a problem for fishery managers
Broad-Leaved Pondweed	*(Potamogeton natans)*	Very invasive in shallow water
Amphibious Bistort	*(Polygonum amphibium)*	Tolerates fluctuating water levels and extended drying

Table 9.7 Common plant species of the main aquatic plant zones

Species		General comments
Free-floating plants		
Frogbit	*(Hydrocharis morsus-ranae)*	Most common in south east
		Attractive lily-like leaves and white flowers
Common Duckweed	*(Lemna minor)*	Very invasive especially in shallow water
		Can be a problem for fishery managers
Submerged plants		
Water Crowfoot	*(Ranunculus spp)*	Attractive flowers
		Usually grow in dense beds and can cause problems for fishery managers
		Some plants have floating leaves
Starwort	*(Callitriche spp)*	Good oxygenator
		Can be a problem for fishery managers
Spiked Water Milfoil	*(Myriophyllum spicatum)*	Grows profusely in nutrient rich water
Mare's Tail	*(Hippuris vulgaris)*	Tolerates fluctuating water levels
		Can be a problem for fishery managers
		Also produces emergent flowering sprites
Fennel-Leaved Pondweed	*(Potamogeton pectinatus)*	Will grow in polluted or turbid water
		Can be a problem for fishery managers
Stoneworts	*(Chara and Nitella spp)*	Fast growing algae
		Good oxygenators

Where the water level will vary substantially as a result of the reservoir usage, the planting will require careful planning and consideration. Few plants will be able to flourish in the zone between the maximum and minimum water levels. Many species, however, are tolerant of, short-term changes limited and will be able to become established, despite occasional periods of flooding or lowered water levels.

9.14.7 Vegetation

The vegetation should be planned and suitable layouts drawn up at the design stage. This will enable new or existing adjacent habitats to be created or extended to enhance the environmental value of the reservoir. Native trees and shrubs should be chosen as they support a wider variety of wildlife than recently introduced species. A wide diversity of tree and shrub species will provide a more varied habitat and should result in a range of food resources becoming available at differing times.

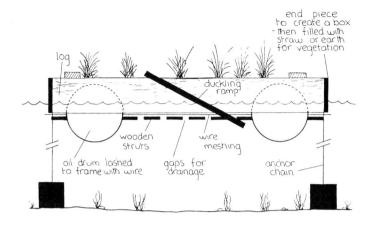

Figure 9.41 Typical design of a simple floating raft

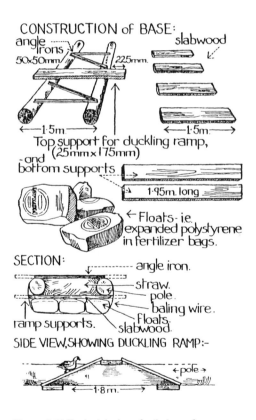

Figure 9.42 Typical design of a timber raft

Figure 9.43 Floating island showing access ramps

Where possible planting should include vertical variation as well as in plan. A three-dimension structure to areas of woodland and larger shrub plantings should be the aim; this is extremely important to wildlife as natural woodland consists of four layers:

- ground layer (low growing plants and grasses)
- field layer (tall grasses, flowers, herbs and young saplings)
- shrub layer (mature shrubs and young trees)
- canopy layer (mature trees).

Natural regeneration will also help to establish vegetation cover, but this involves a random combination of suitable seeds and growing conditions. It may be necessary to wait for colonisation and growth to occur, but the site will be attractive to the vegetation which does appear and faster growing, healthier plants are likely.

The ground and field layers can be encouraged by use of the native species of wildflower and grass seed mixes which are now widely available. Their use will enhance the natural value of the reservoir surrounds if introduced upon restoration. It is important to select a combination with appropriate species to suit site conditions; specialist advice can usually be obtained from seed suppliers. If a wildflower sward is the objective, topsoil should be stripped. This is because the higher fertility of topsoil and presence of a seedbank leads to the suppression of the desired wildflowers by more vigorous species. It is often beneficial to create some localised areas of tall, tussocky vegetation for diversity. This can easily be achieved by mounding uncompacted topsoil and allowing it to revegetate naturally with rough grasses and species such as teasels and thistles.

Whilst it is desirable to establish tall tussocky vegetation on islands for wildfowl breeding, some areas could be kept free of long vegetation for roosting and loafing (resting), or covered with shingle for certain resting birds. Trees should not be planted on islands created for wildfowl as they provide roosting areas for predators such as crows.

Patches of scrub provide food, cover and breeding sites for wildlife and should be linked to open bankside with individual bushes and tall grasses and herbs. Where possible, scrub thickets should also be linked to existing woodland. Woodland should be planned to allow ample light to reach the lower layers, otherwise they will be shaded out and lost. A woodland with only an extensive canopy will not support a significant wildlife population.

Open areas within woodland should be provided to vary the habitat and allow access. Areas of new planting should be linked whenever possible, with other areas and with existing woodland with hedges etc., to aid movement and colonisation by wildlife.

9.14.8 Tree planting

A tree-planting programme around a reservoir needs to be carefully considered to achieve maximum benefit and avoid adverse effects.

Tree species should be selected with care and mixtures of species with different growth rates and habitats should be avoided. Native species in keeping with existing woodland should be selected to create a 'natural' area in most instances and 'ornamental' species should be restricted generally to the non-rural settings. Examination of existing shrubs and trees in and adjacent to a proposed reservoir site will provide an indication of the species that are likely to establish successfully.

The location and choice of species around a water area is of particular importance. Too many trees overhanging the perimeter of reservoir will limit the penetration of sunlight to the water, unless the water area is of sufficient size, and will result in a large amount of leaf litter in the water. Both will cause dissolved oxygen deficiency which, if severe, will inhibit the ability of wildlife and vegetation to thrive, or in extreme cases, to survive. A dense surround of trees will also reduce the amount of air movement over the water, further reducing the amount of oxygen available to be absorbed. Large blocks of trees and larger shrubs, in general, should be planted at least 20 m away from the water and a clear margin left around much of the reservoir. Conifers, in particular, should never be planted in close proximity to the reservoir where their needles are able to fall or otherwise enter the water. These are highly acidic and will rapidly affect the water quality and severely restrict the ability of most vegetation and wildlife to flourish in the reservoir. Access for operational and maintenance purpose or other constraints may also require an area free of large vegetation. Some isolated trees or groups of trees may be planted at the water edge, especially when overhanging the water, to give a variety of habitat and appearance. Figures 9.4, 9.34 and 9.43 show how larger vegetation has been kept back from the waterline, whereas Figure 9.40 shows how intrusive tree growth can shade out large areas of the water surface.

Provided they are not allowed to grow out of control, trees can be very useful in providing a stabilising effect and preventing bank erosion. The shade provided by properly sited trees is also valuable for controlling aquatic weed growth, although excessive shade will seriously inhibit the growth of useful food plants. Dead and decaying leaves that drop into the water provide detritus upon which many aquatic organisms feed, although too many may accumulate and smother submerged vegetation. The likely future growth and development of the trees must not be forgotten; this is particularly important with respect to the effects of long-term shading and leaf litter when the trees are mature.

Tree planting may continue onto the dam, providing they are located in acceptable positions and are maintained in a healthy condition. Vegetation on the embankment is discussed more fully in Section 9.2.8.

9.14.9 Habitat creation for breeding wildfowl

The most important requirements when creating habitat to encourage breeding wildfowl are the provision of adequate feeding, nesting, roosting and loafing areas. Dabbling ducks, such as mallard and teal, require extensive areas of shallows less than 500 mm deep for feeding. Shoots, seeds, leaves and tubers of aquatic plants are taken as food along with freshwater invertebrates. The survival of ducklings is highly dependent on chironomids (non-biting midge larvae) which can be encouraged by aquatic planting in shallows. Diving ducks, such as pochard and the tufted duck, require water generally less than 2.5 m deep with dense beds of submerged plants. Fish such as carp, bream, roach and perch will compete with divers for food.

Nest sites should provide shelter and safety from predators and can be created by the establishment and management of tall tussocky vegetation close to the water. Islands and rafts are also excellent for this purpose. Roosting and loafing sites where the wildfowl can rest, preen, moult and sleep should also be provided by creating sheltered areas of bare ground, mudflats or short grass.

9.14.10 Habitat creation for fish

The main habitat requirements for fish are adequate oxygen in the water, plenty of food, shelter and areas for spawning. Game fish such as trout and grayling demand a much higher concentration of dissolved oxygen than coarse fish and consequently need colder water (14–16°C preferred) to survive. Game fish will survive in ponds and lakes, but need depths of at least 3 m in places to avoid the warm surface layer of summer. Ideally these deeper areas should be localised to avoid large areas of deep water which are unattractive to other wildlife.

Coarse fish in contrast are able to thrive in warm, shallow water and will tolerate relatively wide temperature fluctuations. Ideally, one third of the water body should be less than 2 m deep, one third between 2 m and 3 m deep and the remaining third should be greater than 3 m depth, providing diversity of habitat within the lake for a variety of species such as carp, tench, roach, rudd and bream.

Aquatic plants have value for fish as sources of food, shelter and spawning areas so cultivation of a diverse aquatic flora, with the associated invertebrate community, is essential. Any plants that are introduced to the system, however, should be washed carefully to avoid introducing diseases or parasites.

9.14.11 Habitat creation for other wildlife

Many of the measures described above, if adopted, will be beneficial to other forms of wetland wildlife. Dragonflies and damselflies are drawn to sheltered shallow open water with the mixture of sun and shade and the diversity of submerged and emergent vegetation has an important influence on the number of species that will be attracted. If these species are to be established successfully, a well developed grassland and scrub adjacent to the reservoir will also be necessary.

The surrounding terrestrial conditions are equally as important as the quality of the shallows for frogs and toads, who spend most of the year on land. Dense, tussocky vegetation and scrub is ideal for food and cover, and is also valuable for newts even though these species spend more time in the water.

Where a water body is being created purely for nature conservation purposes, it is important to highlight specific objectives and focus on species which it would be most beneficial to encourage. Future management of the site can then be targeted towards the conservation of certain declining or rare species rather than those already abundant.

9.14.12 Recreational and sporting use

The requirements for recreational activities such as sailing, windsurfing and rowing are detailed in Appendix C. These activities are generally restricted to larger reservoirs over several hectares in extent and are unlikely to be feasible on many of the small reservoirs falling within the scope of this guide. Angling and shooting are, however, two activities that can easily be accommodated on small reservoirs, creating a potential source of income. Any recreational or sporting use, however, is likely to be detrimental to the wildlife interests and likely to disturb bird and other wildlife.

9.14.13 Management for fishing

Stocking with fish and letting the angling is a common way of deriving an income from a reservoir, and small areas of water can be used effectively. Management of the water for fishing is relatively compatible with nature conservation management; both activities having the same objective of maintaining a rich and varied habitat with a diverse assemblage of plant and animal species. Disturbance by anglers can, however, be a problem if wildfowl are being encouraged.

It is usual to stock with a selection of species that have different feeding habitats, such as the bottom feeding carp, tench or bream with the top feeding roach or rudd to provide anglers with a choice of fishing. The newly created water body should not be stocked until the vegetation has become established and stable, and is in fact best delayed until one year after planting. Care must also be taken not to overstock the water, particularly in the early stages of development. Specialist advice on the stocking and management of fisheries can be obtained from the Institute of Fisheries Management (IFM).

Still-water trout fisheries require considerably more management than coarse fisheries, as game fish do not breed in still water and need special conditions to survive. The banks of such fisheries must be relatively clear of vegetation that would inhibit casting, and planting should be restricted to clumps, with clear casting areas provided. Whilst a still water without coarse fish may make a viable trout fishery on a small scale, it is unlikely to be feasible to plan a commercial sporting fishery on anything less than two hectares, due to the economics of management and use.

9.14.14 Management for shooting

The most common means of providing duck for shooting is the release of artificially reared mallard. It is often felt that wild duck provide greater sport and they can be encouraged to use a new water body by locating the reservoir beneath existing flightlines and providing food such as barley, frosted potatoes and acorns. This artificial feeding should principally be carried out only in harsh weather; the planting of appropriate vegetation should provide ample food under normal conditions. Wildfowl should not be encouraged until the planting is well established and can withstand being grazed.

Shooting should not be carried out too frequently to maintain the attractiveness of the site for wildfowl. Once every three weeks should be acceptable, provided the shooting stops before the last birds fly in. Hides may also have to be provided and should be positioned to prevent spent shot falling into the shallows. Specialist advice on the management of a water body for shooting can be obtained on request from the British Association for Shooting and Conservation.

9.14.15 Examples of design

The environmental aspects are site specific and must be considered on that basis. Useful information on many aspects is given in References 44 to 46 and 90 to 91 and Figures 9.44 and 9.45 show features of well designed and poorly designed reservoirs respectively. Generally good examples of design are shown in many of the figures in this section but an example of poor design is shown in

Figure 9.46. In this instance the reservoir has a straight, generally steep, shoreline with little variation and is isolated from adjacent areas of woodland by the absence of planting. It compares unfavourably with the new reservoir, shown in Figure 9.47, with its varied shoreline, perimeter reedbeds and tree planting.

9.14.16 Burrowing animals and other pest control

Earth embankments are susceptible to damage from burrowing animals. These include badgers, foxes, moles, rats, rabbits amongst others. Badgers potentially pose the worst threat as they create extensive runs of deeply penetrating burrows. Rabbits and rats dig less deeply, but may chew through lining membranes and geotextiles. Water rats and voles may burrow from the waterside and will normally form an entrance just above the waterline. These allow direct ingress of water if the level rises and may be particularly troublesome when the water level is raised after an extended period of constant, or near constant, lower level. Burrows affect the performance of an embankment in two ways:

- reduction of the distance water has to seep for leakage to occur (i.e. facilitate leakage)
- undermining of the embankment resulting in surface and internal deformation and erosion.

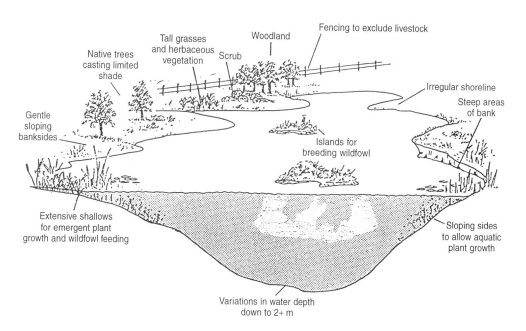

Figure 9.44 Features of a well designed reservoir for wildlife

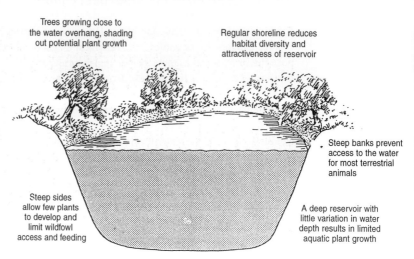

Figure 9.45 Features of a poorly designed reservoir for wildlife

Figure 9.46 Environmentally poor reservoir design

Figure 9.47 Environmentally good reservoir design

If a problem of excessive damage by burrowing animals is envisaged, advice should be sought from the local ADAS adviser on wildlife. It is important to be aware that certain animals and birds are protected by law, e.g. badgers and their setts, herons etc., and this must be clarified with ADAS before any pest control measures are included in the design. Attempts to exclude rabbits may be made by securing wire netting flush on or just below the ground, fill surface or embankment surface to prevent burrowing. The netting must extend for a sufficient distance beyond the embankment and must be adequately restrained so as to not pose an impediment to any mowing. Alternatively, rabbit-proof fencing around the embankment and set sufficiently far into the ground can be used.

If severe rat infestation is expected, reservoir liners should only be used after careful consideration and specialist advice. Placement of fine mesh wire netting may be necessary to reduce the likelihood of the rats chewing or burrowing through the impermeable membrane or layer – normally this is a problem that is only apparent after the reservoir has been in use for some while.

The effect of predatory mammals and birds at reservoirs used for particular purposes e.g. trout farming, coarse fishing, duck rearing, etc., must also be considered at the feasibility and design stage and advice on pest control sought where necessary. Before preventative measures to control pest are included for environmental or economic reasons, the following aspects should be considered:

- minimise risk by improved habitat design and management (e.g. increased cover, removal of sighting trees or branches for predators)
- improve conditions for affected species (e.g. increased feeding, better nesting options)
- removal of a predator may allow alternative, more damaging, species to establish
- make the habit less attractive for predators (e.g. slope access, islands, scaring)
- timing and extent of predatory activity may limit problem to short duration and an intensified programme for a limited period may be sufficient to minimise and control activity
- predator activity may be limited to a few or one individual within a species (e.g. an old or injured fox)
- it should be remembered that many creatures produce a large number of offspring to counteract losses and only if substantial or repetitive losses occur is action necessary. Other forms of mortality may occur if predator activity is curtailed (e.g. lack of available food, disease, etc.). In many instances, control measures are best left until the size and scope of the problem can be identified by actual experience and appropriate measures taken
- where the reservoir is to be used for fish farming, the use of a physical exclusion system, i.e. netting the entire reservoir, should be considered at the design stage.

The presence of foreign crayfish in British reservoirs is a relatively new occurrence, whether by design (i.e. crayfish farming), or by accidental introduction with other species or plants. There is evidence that their digging into the embankment may cause problems in some instances and specialist advice from ADAS or other sources should be sought at the design stage if crayfish farming is planned.

9.15 SAFETY CONSIDERATIONS

9.15.1 Public and operational safety

The design of all structures and equipment, and the construction of the reservoir, must be in accordance with the general safety provisions of the Health and Safety at Work etc. Act 1974[40] and the various associated regulations. The Act is aimed primarily at ensuring the safety of employed persons and others affected by the work activity. Consideration must also be given to the safety of the public who might have free access or might gain unauthorised access to the reservoir and adjoining land.

Owners of reservoirs or those responsible for their management should always be aware of the dangers associated with water. Whilst there is no legal requirement or necessity to fence off every reservoir, it is recommended that where access is readily available fences and warning signs should be erected. This is particularly necessary where there is an abrupt drop to the surface of the water and around inlet or control structures.

Where public access is anticipated, warning signs should be displayed where they are clearly visible. A typical example is shown in Figure 9.48. All safety signs should comply with the Safety Signs Regulations 1980[92]. Lifebelts should also be provided close to the reservoir (Figure 9.49).

Figure 9.48 Typical warning sign

The presence of chambers, voids or other restricted areas within any of the structures; manholes or other features will be hazardous in terms of access, maintenance and use. They may also collect gases or other noxious substances. Statutory constraints apply to the entry into such areas, but these should be minimised in the design whenever possible. The design should always include consideration of the construction and operation of the reservoir to minimise potential hazards wherever possible.

The inlets or outlets of pipes and other structures will prove attractive to children. Where possible, such features should be avoided by maintaining high water levels or providing locked removable screens. Other confined spaces, e.g.

manholes, valve chambers, etc., should be securely locked at all times and designed to have a free flow of air if possible.

The provision of shallow gradients around the perimeter of the reservoir where easy access can be gained will be of considerable value as a safety measure. Transitions from shallow to steeper slopes below water should be gradual to avoid a hazardous sudden change in bed level. It is also recommended that provisions are made to assist a person to climb out if steep slopes are present. A suitable inexpensive arrangement would be a 'ladder' of vehicle tyres, securely tied together and anchored to the top of the slope. Where vertical walls or structures abut deep water, hand holds and/or rungs may be necessary.

Figure 9.49 Typical lifebelt facility

Areas of soft mud or silt may be exposed if the water level drops or is lowered for any reason. These may be dangerous and may require temporary fencing and signing.

Pollution or the development of various algal growths may be a source of public danger and may require temporary fencing or signing.

9.15.2 Emergency water supply for fire fighting purposes

Any small reservoir adjacent to a structure or area that is a potential fire risk should be considered as being an emergency water supply. The main requirements for emergency water storage for fire-fighting purposes are as follows:

- minimum capacity 22 500 litres (5000 gallons)
- sited not less than 6 m and not more than 100 m from the buildings/area constituting a fire risk
- a hard road must lead to a firm pumping site which is situated not more than 1.2 m above reservoir level and not further than 5 m from the reservoir bank. The road and pumping site must be capable of supporting a fire appliance which may weigh in excess of 10 tonnes
- any gateway on the access road leading to the reservoir must be at least 3 m wide
- a deep sump should be constructed adjacent to the pumping site to facilitate water extraction from the reservoir. This sump should be constructed so as to prevent any clogging of a suction hose with debris or plant material. The area should be fenced as necessary in view of the deep water
- water sources should be clearly marked by standard indicator plates conforming with BS 3251[93].

Whenever a proposed reservoir is designed or an existing reservoir adapted to serve as a water supply for fire fighting, advice should be taken from the Fire Prevention Officer of the local Fire Service.

9.15.3 Livestock

Few small reservoirs are now used for stock watering due to health hazards. If access for watering is required, a railed or fenced drinking bay incorporated within a reservoir perimeter fence should be adopted to prevent livestock damaging the shoreline of the reservoir and to reduce the risk of livestock entering the reservoir and drowning. The drinking bay should comprise a hardstanding preferably of hardcore or concrete, to prevent excessive damage to the localised shoreline. This should extend sufficiently far into the reservoir such that the end is not exposed, even in periods of drought.

Water to be used for livestock may be pumped to an adjacent trough to prevent problems of puncturing of geomembranes or clay liners used in the embankment construction. This is a practical alternative to the provision of drinking bays, with the obvious advantage that risks of pollution due to livestock are drastically reduced.

The possible dangers to human health arising from across the water by livestock or arising from the nearby proximity of livestock should not be overlooked (Appendix E).

9.16 DESIGN RECORDS

9.16.1 General

A record of the design process should be kept for future reference. This should be drafted in simple terms with sketches where required, but should be sufficient to record the development of the scheme from initial conception, through the feasibility stage to the final design proposals. The reasons for the chosen approach and solutions to problems should be recorded together with any particular difficulties or unusual aspects.

The information is best maintained in a loose-leaf folder format and should include:

- a summary of the original site, including topography, water features, vegetation, land use etc.
- a record of the information, study, site survey, ground investigation and subsequent assessment
- notes on the reservoir yield and the flood assessment including the choice of overflow arrangements
- comments on the design, including reasons for the adoption of a feature or element of design
- appropriate photographs, titled and dated, at each stage
- copies of the drawings and the various contractual documents or construction notes

This information will prove invaluable during the construction stage and in future years for maintenance purposes. In the event of difficulties or need for remedial works, the records may prove a useful source of information. Any future owner taking over the reservoir at a later date will also require this information.

9.16.2 Subsequent records

Subsequent records of construction, maintenance, remedial or alteration works should also be kept to allow a full record of the history and the development of the reservoir to be available. A large number of titled, dated photographs is often useful.

Part 3 Construction

10 Construction aspects

10.1 GENERAL

10.1.1 Introduction

Good construction practice should follow on from the feasibility assessment, site investigation and design. It should take account of all the various requirements to produce a final structure that will retain water in a safe and satisfactory manner. It is important that a close link is maintained between the designer of the scheme and the contractor, or person constructing the reservoir if these are not the same, so that any unexpected conditions can be reassessed. These may necessitate changes to the proposed design or may require specialist advice to be sought to enable the construction work to continue. Thus the entire construction stage must be reviewed continuously and a flexible approach adopted throughout. Changes during the construction stage should be recorded, ideally as an extension of the design record.

The comments in Sections 9.2.1 and 9.3.1 regarding the variability of the soils and the consequent comments should not be forgotten.

Useful general advice on the construction aspects is contained in BS 8000 'Workmanship on Building Sites'[94], whilst construction safety is discussed in Section 10.12. It must always be remembered that the Health and Safety Executive have many statutory powers relating to construction works. They are likely to be interested in safe working practices on any construction site, particularly with regard to the safety of all personnel on the site and any other parties who might be affected. Where significant breaches of statutory requirements are evident, they may require the unsafe works to be stopped. They may also take legal action in some instances against all or any of the parties involved with the construction works.

10.1.2 Weather considerations

The effect of weather on the construction of a small reservoir is of primary importance. Earthworks in Great Britain are prone to considerable delay if commenced during the autumn and winter months. The problems of poor weather conditions are compounded with clayey soils. The principal problems arising from ground conditions due to the weather are summarised in Table 10.1 and the results of ill-considered access after wet weather are shown in Figure 10.1.

Table 10.1 Implications of weather conditions

Frost	Will affect concrete pours and necessitate extensive precautions
	Unsuitable material for filling and compaction (material behind face of the borrow pit will be frost free)Increased strength of sub-surface for haulage routes
Rainfall	If substantial, may affect concrete pours
	Rutting increases roll-resistance of the plant and reduces speeds
	Existing placed fill will be prone to waterlogging
	Fill material will have increased water content in borrow area and fill placing should be avoided after heavy rainfall
Hot weather	Will affect concrete pours due to rapid drying
	Drying and cracking of cohesive material may necessitate watering of fill
	Haulage speed increases due to reduced moisture content and lack of rutting
Wind	Tends to give same effects as hot weather
	Possible dust problems.

Figure 10.1 Effects of ill-considered access after wet weather

10.1.3 Working limits and setting out

The site working limits should be defined upon a plan of the site area. This should outline to the contractor the surrounding land use and possible requirement of the undertaker's need for access during construction. The working site may require enclosure to control access of livestock or the general public. Provision for adequate temporary fences, such as chestnut paling, and warning signs should be made.

Particular features within the working limits of the site such as specific areas of vegetation or existing structures which are to be retained may need to be fenced off and/or suitably protected.

Setting out of the embankment and structures should be carried out immediately and prior to the start of the construction, using pegs. Care should be taken not to disturb these during construction and key pegs should be related to adjacent permanent features (e.g. kerbs, trees, solid posts etc.) in case they are disturbed.

Pegs should be used to mark the following positions:
- dam centre line
- crest of embankment (both sides)
- toe position (upstream and downstream)
- core position (if the core is central and has the same width as the crest, the crest pegs can be used)
- overflow extent
- any structures and drains
- the position of any known buried services or other hidden features which must not be disturbed.

The dam centre line should be set out by the undertaker or a surveyor so that the basic position of the works can be established. Where appropriate, the contractor should set out all the other positions and these should be verified by the undertaker. A series of pegs at 25 m spacings is usually sufficient.

10.1.4 Construction access and temporary works

Any possible site access problems should be foreseen at the feasibility stage of the development. Access may be restricted in a number of ways:
- services   –   overhead
                  –   buried (consult statutory bodies see Section 10.2)
- narrow or restricted access to site from public roads
- narrow or restricted access upon undertaker's land between proposed site and public roads or between borrow pit and construction site
- weight restrictions on roads or structures
- wet, soft or otherwise poor ground conditions.

Temporary access roads (haul roads) are likely to be required to allow the construction plant to gain access into and about the site. Where the ground is satisfactory and the potential usage is small, it may be possible to form these by simply defining an access route, but in many instances the topsoil will need to be removed to minimise damage as detailed in Section 10.13.12. If the ground is soft or otherwise prone to damage, a temporary surface of crushed stone, hardcore or similar may be necessary to allow the plant to move about the site. In some instances, especially if the haul road is likely to be heavily trafficked, an underlayer of a suitable geotextile may be required, possibly together with temporary drainage measures, to prevent the road material being lost into the ground.

Following the completion of the works, the temporary haul road should be removed and the ground reinstated as described in Section 10.3.12.

It must be remembered when planning haul roads that construction plant is significantly larger and less manoeuvrable than normal vehicles, requiring gentle gradients and curves to access roads. Temporary removal of fences to allow reduced haulage lengths can be cost effective, reducing the construction time and fuel consumption.

Where a contractor is employed to construct the reservoir, haul road locations and details will be decided by him as part of his method of carrying out the construction works. Other operations which do not directly affect the finished reservoir (the permanent works) will also be decided by him as temporary works and may include, for example, method of working in the borrow area, sequence of construction, etc. Normally the contractor should be allowed to plan and carry out his works as he requires, but any constraints on access or other restrictions should be detailed clearly in the contract documentation (Section 11).

Access may be required in some instances along or across public roads and this aspect should be discussed with the local highway authority. Vehicles will need to be suitably licensed for highway use, warning signs will be required and provision for regular road cleaning should be made.

10.1.5 Construction monitoring

Monitoring during construction should seek to check and ensure that:

- construction complies with the design, specification and drawings
- the correct materials and techniques are used. This is especially necessary for fill materials whose properties may vary depending on the method of selection, excavation, transportation, placement and compaction
- completed works are not damaged or do not deteriorate in any way, whether from construction activities, inclement weather or any other cause. Placed fill materials are particularly vulnerable and appropriate precautionary measures should be taken to minimise the risk of subsequent damage or deterioration.

Construction supervision is covered in Section 11.8.

10.2 BURIED AND OTHER SERVICES

10.2.1 General

The presence of services which are privately owned or belong to statutory bodies and public authorities within a proposed reservoir site must be established and the exact position clarified prior to commencement. Safe working practise beneath or adjacent to services must be adopted and advice concerning this sought from the relevant owners, statutory bodies and public authorities. Where appropriate, warning notices and barriers should be erected.

10.2.2 Services

The statutory bodies, public authorities and companies responsible for river control and drainage, water, gas, electricity and telephone services should be informed of the proposed works. The national grid reference of the reservoir site and a large-scale plan marked up with the proposed working areas should be sent not less than two months before the construction is to start to ensure a reply. A copy of the site plan would normally be returned, marked with the positions of any services. If adequate information is not obtained prior to commencement of the construction work, the relevant statutory body, public authority or company should be requested to inspect the site and be informed of the date of commencement.

10.2.3 Inspection holes

Excavation of areas where services are thought to be present should be advanced with caution. Holes should be excavated by hand to a depth of approximately 1 m to allow inspection of the ground and to establish the presence of services prior to excavation continuing by mechanical means.

10.2.4 Diversion of services

Where a service would be affected by the reservoir construction or subsequent use, it should be diverted outside the reservoir area. The route and method of diversion or the details of the reconstruction should be agreed prior to work commencing on site and this work is often best carried out prior to the main construction. In many instances, the diversion works will be carried out by the owners and a charge made accordingly.

10.3 EARTHWORKS

10.3.1 General

Earthworks, in terms of reservoir construction, will include the following:

- control of surface and groundwater by temporary works during construction

- site preparation
  - removal of vegetation, structures and topsoil from the borrow pit and foundation area of the embankment
  - removal of vegetation and structures below top water level in the reservoir area
  - excavation of any unsuitable foundation material and overburden in the borrow pit
- foundation preparation in the embankment area
- excavation for and construction of seepage control measures
- selection and excavation of fill in the borrow pit
- transportation of fill materials
- fill placing and compacting
- management of stockpiled materials, especially topsoil
- construction of permanent drainage works
- protection and care of the works
- reinstatement and rectification of damage to soils and vegetation
- possible reshaping works within and adjacent to the reservoir level.

It must be stressed that the earthworks are a controlled excavation and fill placing operation and not simply a muckshifting exercise to be completed at the earliest opportunity. A careful steady approach is required, with placing methods adjusted, as required, to the changing fill or weather conditions or other site specific features.

BS 6031 'Code of Practice for Earthworks'[65] gives useful information relating to the construction aspects of small embankments, whilst the ICE works guide on earthworks[95] contains useful practical advice.

10.3.2 Control of surface water and groundwater

Before earthworks are started on site, all necessary measures should be taken to avoid surface water flowing into the embankment and borrow areas from streams, ditches, etc. Where a watercourse is dry, flow should be expected in wet weather and measures should be provided to divert any possible flow in the watercourse. Groundwater should be expected in any excavation, particularly in the valley bottom, where the groundwater level is likely to be at or above the ground level. The presence of such water pressures will lead to a softening of the ground under plant, as well as to movement and instability and/or water inflow into excavations. A conservative approach is essential in dealing with water flows and the highest realistic water levels should be considered when assessing excavations and foundation preparation works. Sufficient and reliable pumping plant should be provided to deal with water in excavations, and where this is critical, a standby pump may be needed. This will also allow unexpected inflows to be dealt with and offers some security against vandalism or mechanical problems. CIRIA Report 113 'Control of Groundwater for Temporary Works'[96] and the ICE Works Guide 'Control of Water'[97] give useful straightforward advice on dealing with groundwater.

Where an impounding structure is to be constructed, whether a weir or a dam, the flow must be controlled during construction and this may be achieved in various ways:

- diversion of flows through a suitably sized pipe placed into original ground. A cofferdam is then formed between the site and the inlet to the pipe and used to divert the flows in the watercourse
- construction of a cofferdam of sufficient height to impound all the flow in the stream. This is only practical for a short restriction in flow on a small watercourse and must not be used for the construction of an impounding dam which may be subject to large flows
- passing the flow along the existing channel or an excavated diversion channel and forming the embankment on either side, followed by diversion of the flow by a cofferdam whilst the short section is infilled. This method is not recommended for other than the smallest dams as no facilities are available for dealing with flood flows
- construction of a bywash channel to pass construction flows occurring in addition to flood flow during subsequent operation. The design of the diversion/overflow arrangements for this option needs careful consideration and specialist advice is often required.

In practically all cases, the first option is the only satisfactory and safe approach and is often the most economic as the diversion pipe can generally be utilised as the low level drawoff pipe.

A small cofferdam can be formed by sandbags or similar, but most cofferdams will need to be constructed of earth fill, often by dozing the unsuitable superficial materials. The height of the cofferdam must be sufficient to allow flow in the watercourse to develop an adequate depth to pass the construction flood safely through the diversion pipe. Comments on the assessment of the construction flood are given in Section 7.5 and recommendations made for suitable diversion pipe sizes and heights of cofferdam. These recommendations should provide a reasonable minimum standard of protection, but where greater flows are expected, a larger pipe and cofferdam should be used. Alternatively two pipes may be utilised to reduce the height of the cofferdam. All pipes must be installed correctly to minimise potential seepage flows around them.

The diversion works should discharge into the watercourse downstream of the proposed dam site and should be orientated so that minimal erosion of the existing banks is caused by the discharge of flow. Some localised protection of tipped stone or hardcore may be necessary. Anti-seepage collars should be provided on the diversion pipe as discussed in Section 10.4.3. If the pipe discharges into a length of excavated channel and then into the watercourse, localised protection measures are likely to be required in the excavated channel, particularly at changes of direction or gradient.

10.3.3 Construction plant and equipment

Material has to be transported from the place of excavation and then compacted in the embankment. The most economic means possible should be adopted for excavating, transporting and compacting the fill considering:

- the period of time available for the works
- the transport (haul) distance
- the state of the ground
- the fill type and quantity
- anticipated weather conditions
- the expected response of the fill to the likely weather conditions.

The type of construction plant to be used for a small embankment type reservoir depends on a number of factors, including:

- type of embankment to be constructed, i.e. homogeneous, zoned or diaphragm design, see Section 9.2
- size of construction
- type of material to be used in construction
- distance of borrow pit/pits from the embankment
- possible available equipment on site and availability of this equipment over the construction period
- time scale for construction
- complexity of construction; especially structures, pipelines, etc.
- anticipated weather conditions.

Details of typical earthmoving plant are given in Tables 10.2 and 10.3. Earthworks do not necessarily require the use of specialist plant, and farm equipment may be suitable for the more straightforward or small-scale works. Certain specialist items of plant may be hired for short periods. Figure 10.2 shows a typical hydraulic excavator as frequently found on a farm, and Figure 10.3 shows a typical tractor and roller. Much of the construction work can be carried out with similar items of plant, but specialist plant will be required to spread and compact the fill materials. These are shown in Figures 10.4 to 10.7 which show a range of dozers and rollers used frequently for small reservoir construction. The safe use of construction plant is of paramount importance and particular attention should be paid to the safety aspects detailed in Section 10.12.3.

Table 10.2 Excavators and transportation

Purpose	Typical plant	Comments
Excavation plant	Hydraulic excavators	Normally tracked, but may be wheeled. Highly flexible plant. Readily available for hire
	Hydraulic face shovel	Less flexible in use. Normally tracked
	Tracked loaders (dozer shovels)	Limited use in excavation, but suitable for loading. Wheeled loaders often available on farms, but may be required elsewhere. Can also be used to spread fill
	Wheeled loaders (dozer shovels)	
Haulage plant	Tractor and trailer haulage vehicles	Available on most farms, it must be remembered that this equipment may be required elsewhere on the farm
	Small dumpers	Manoeuvrable and can often tip in several directions
	Dump trucks	May be articulated and rear-dumping. Normally used on larger projects
Excavation and haulage plant	Standard wheeled tractor scrapers	Scraper can be towed behind a bulldozer. Used only if time is important or if haulage distance is short for large amounts of material
	Motorised scrapers	Scraper has integral drive system. Restricted to larger projects
Dozing and spreading plant	Tracked dozers wheeled dozers	Used to spread and place fill. Blade only, no front bucket. Normally used to pull compaction plant
	Tracked loaders (drots)	Often used to spread and break up fill and pull compaction plant.

NOTE: Tracked plant is normally used for earthworks, although wheeled plant may be suitable in dry conditions. Tracked plant is normally slower than wheeled plant, but may be more manoeuvrable.

Figure 10.2 Hydraulic excavator

Figure 10.3 Tractor and rib roller

10.3.4 Site preparation

All structures including fencing, buildings, drainage features, roads, kerbing etc., should be cleared over the entire reservoir area. Where archaeological remains are encountered, the work must cease by law in accordance with the Ancient Monuments and Archaeological Areas Act[98], and the County Archaeological Department be informed. Depending on the remains, work may be required to cease for a period to allow investigatory or recovery work to be carried out.

Trees, scrub, roots, as well as all vegetation and organic matter should be removed below the proposed top water level of the reservoir, from the foundation of the dam and from the borrow pit area. Worthwhile timber should be recovered and the remainder burnt to ash in an area clear of the proposed dam. Topsoil should also be stripped and should be stored in separate temporary stockpiles as described in Section 10.3.13.

Works in the reservoir (Section 10.3.9) are often best carried out at an early stage and as part of the preparation work in the reservoir and borrow area.

10.3.5 Foundation preparation

The interface between the foundation and the embankment is a critical area in terms of stability and seepage control. The foundation preparation must include removal of all soft, loose or otherwise unsuitable material (Box 8.8). The exposed foundation should be scarified by ploughing or disc harrowing along the line of the dam to provide a 'key' between the foundation and the embankment material. This enhances the embankment stability and lessens the risk of seepage. A cutoff trench, as discussed in Section 10.4.1, is required to intercept any land drains, other shallow features or permeable layers and this should be excavated into sound material once the general foundation preparation has been carried out.

Where a watercourse crosses the foundation, the unsuitable materials in the channel should be dug out to a sufficient depth to expose suitable foundation material and any slopes into the excavation battered back to slopes of 1V:6H or flatter. The cutoff trench should be excavated below the foundation level which is exposed at the bottom of the excavation.

10.3.6 Unsuitable material

The use of unsuitable material in embankment construction or its presence in the foundation below the embankment can lead to stability or settlement problems and seepage, erosion or other difficulties with the completed structure. Unsuitable materials are discussed in Section 8.5.3 and Boxes 8.7 and 8.8 and also include:

- topsoil, including subsoil
- peat-rich or highly organic materials, including logs and tree stumps
- very wet materials
- material from borrow areas which is very soft or soft

- material with a liquid limit in excess of 90% or a plasticity index greater than 65%
- frozen soil
- colliery spoil or combustible material
- material having hazardous chemical or physical properties
- hardcore, concrete or other building materials.

Unsuitable material is normally utilised as low-grade fill for landscaping purposes and used for creating shallows, variations in the shoreline, islands and other features around the reservoir (Section 10.3.9) or disposed of in the borrow area. In certain instances it may be possible to incorporate it within the embankment as fill to areas of landscaping or infilling following specialist advice.

Polluted or hazardous materials should be excluded from any aspect of the reservoir creation and should be removed to a recognised licensed tip. Normally, however, their presence on site may preclude the satisfactory creation of a reservoir and may require specialist advice. Other unsuitable materials should be disposed of in the borrow pit, used to form islands and shallows or used as fill for landscaping purposes elsewhere on site. The material should be placed in layers not more than 300 mm thick and compacted by earthmoving plant.

10.3.7 Trenches and excavations

Excavations should be carried out in such a manner as to avoid damage or deterioration to the formation of the excavation or trench and to minimise disturbance to the adjacent ground. Slopes of shallow excavations should normally be no steeper than 1:1, but flatter slopes will be required in poor ground, when water is present or as required for access on other constraints. Excavation in excess of 2 m to 3 m requires special consideration and flatter slopes may be necessary. Guidance on good trenching practice is given in Reference 59. Where the ground cannot be battered back, some means of temporary support such as trench sheets must be employed. It should also be remembered that where access into the excavation is required and the depth is in excess of 1.2 m, the Health and Safety Executive requires protective measures to be taken and such excavation should not be entered unless these are complied with on site. Soft material should be removed from the bottom of excavations and replaced with concrete beneath structures or bedding material beneath drains or pipes.

The minimum clearance around structures should be sufficient to give adequate working space and excavations should normally be battered back to safe slopes or supported as discussed.

The minimum clearance on either side of any pipes or joints should be not less than the values given in Table 9.1. A larger clearance is often necessary for construction purposes.

Backfilling of trenches should be undertaken at the earliest opportunity to minimise the risk of collapse. Backfilling should not take place until structures have achieved a sufficient strength to withstand loading imposed by the backfill. Cohesive materials should be compacted in layers not exceeding 200 mm using tamping compactors or by controlled use of a hydraulic excavator bucket. Granular backfill should be compacted by plate compactor or other suitable vibratory means. In general, backfilling should be subject to the same controls and constraints as described later in this section for fill placing.

10.3.8 Excavation of fill materials from the borrow pit

The information gained from a site investigation should be assessed to determine the likely distribution of the potential fill material. Unsuitable fill material, as defined in Section 10.3.6, should not be used for the embankment construction. Selection of the borrow pit material should be made during excavation so that the appropriate material is placed and compacted in the most suitable zone. If a homogeneous construction is adopted, the material should be selected and placed so that the less cohesive material is in the shoulders with the more clayey cohesive material in the centre of the embankment.

Topsoil and unsuitable material should be removed as outlined in Sections 10.3.4 to 10.3.6. Inadequate stripping of unsuitable material can result in the inclusion of poor quality fill and the presence of a zone of weakness detrimental to embankment stability and watertightness.

Where the borrow pit is to be excavated below the water table, the groundwater level must be lowered by drainage. Perched water tables may have sufficient capacity to require drainage, even when the borrow pit is above the main water table. Surface drainage in a borrow area should also be carried out to reduce the likelihood of near surface material becoming saturated. Surface drains should normally be shallow trenches excavated to a level below the borrow pit floor, and extended from the point of entry of any water to a sump for removal by pumping or gravity drainage where possible.

Material in the borrow pit is likely to be prone to softening by rainfall, standing or running water and frost, or to excessive drying in dry or windy weather. These effects will be exacerbated by the passage of plant, excessive handling of the fill or poor stockpiling. Drainage measures, limiting the amount of borrow area opened up at any time and systematic working should help to minimise problems to prevent deterioration of the material. Work may have to cease when adverse weather conditions persist. Any material which becomes unsuitable should be disposed of as detailed in Section 10.3.6.

If soil requires reworking either to increase or decrease its moisture content this is best dealt with within the borrow pit. The moisture content of the material can be reduced by ploughing the material and allowing it to dry. Additional water can be introduced by hoses or bowsers if the material is too dry. Normally, however, fill should be used as-dug.

Following the winning of fill for construction, the surplus and unsuitable material should be dozed to the required final slopes and any necessary regrading works carried out around the borrow pit. Slopes above the top water level should be seeded where appropriate.

10.3.9 Reservoir works

Works may be required in or adjacent to the reservoir for a number of reasons including:

- increase or reshaping of the water surface of the reservoir
- formation of spits, bays, etc.
- formation of islands
- creation of deeper areas or shallows
- flattening of ground slopes within or outside the reservoir area
- disposal of unsuitable or surplus materials
- protective works to the shoreline
- specific features for the reservoir usage (e.g. excavated fishing stands).

The earthwork elements of these works are normally carried out as part of the main earthworks and can utilise the unsuitable and surplus materials arising from the embankment construction as a low-grade landscaping fill. In some instances, where more substantial excavation or regrading works are required, the excavated materials may be suitable as fill for the embankment.

10.3.10 Embankment construction

Embankment construction should start at the earliest opportunity to avoid degradation of the prepared foundation. The embankment should be raised at a uniform rate and in general kept longitudinally level-so far as is practical under the prevailing weather conditions-and consistent with the progress of other work on the embankment. The rate of construction should be limited to not more than 1 m in height per week. The fill should be spread in layers and fully compacted as discussed in Section 10.3.11. Each layer of soil should be placed and compacted along the entire length of the embankment in a continuous process and any openings for access should be kept to a minimum. This is to avoid creating discontinuities which could lead to differential settlement and areas of weakness and potential leakage. Where openings are left, these should be infilled at the earliest opportunity by removing any dried material on the fill surface, cutting back any longitudinal slopes to not steeper than 1V:6H and ensuring the fill is adequately keyed into the previously placed material.

An embankment should be constructed with a surface crossfall of 1V:20H to shed surface water. The surface should be left sufficiently even to prevent the ponding of rainwater in ruts and holes and should be rolled smooth to encourage drainage at times of inclement weather and at the end of each day. Prior to further fill placing, the surface layer should be opened up by harrowing or ripping to key the new fill into the previously placed material.

Embankment fill must never be placed into standing or running water and fill placing should cease when the material is likely to become softened during and after inclement weather. Any fill which has softened should normally be removed, although it may be possible to dry the fill by surface harrowing or ripping with a dozer. No fill should be placed when either the fill or the placing surface is frozen.

The embankment should be overfilled beyond the required profile and then trimmed to the required slopes and levels. If a section of embankment has been underfilled and additional fill is required, a section of embankment should be cut back and additional fill built up as a series of layers. Under no circumstances should fill be placed as a sloping layer on the side of the embankment; this will subsequently tend to soften and may lead to shallow slope instability.

Materials should not be stored on the embankment and care should be taken to avoid contamination of previously placed fill materials. Diesel spillages can damage the fill, whilst granular materials spilt on the core and not removed can lead to seepage through the dam when in operation.

10.3.11 Fill placing and compaction

Soil compaction is the process whereby soil particles are packed more closely together through a reduction in air content. The objective of compaction is to modify the behaviour of tipped soil to produce a fill which has the desired properties for the required application; these generally are to decrease the permeability to the required values detailed in Section 9.2.1 and increase the strength of the fill. Some compaction arises from the plant transporting, placing and spreading the fill, but specialist compaction plant is normally used.

The mechanical means by which fill is compacted may be either of the following, dependent primarily on soil type:

- rolling for cohesive materials
- rolling with vibration for granular materials.

Within the scope of this guide, the compactive plant is likely to be a smooth drum roller towed by a dozer (Figures 10.4 and 10.5). For granular fill, the roller is normally used with the vibration facility, built into the roller, to vibrate the individual fragments into a dense state. On larger jobs, a sheepsfoot roller (Figure 10.7) may be used for the cohesive material; this has a series of protrusions or feet over the surface of the drum which knead the material and produce a dense fill.

Where structures are included in the fill, care should be taken to ensure that the standards of fill placing and compaction around structures are not significantly different from elsewhere. This will minimise future settlements. The use of heavy plant may need to be limited around structures and smaller hand operated plant employed as shown in Table 10.3. Filling around and above pipework also needs particular care and normally a pipe trench should not be excavated nor the pipe laid until the fill level is not less than 500 mm above the intended level of the pipe crown.

Moisture in the fill will influence the effect of compaction. The moisture content can be measured by taking a fill sample and having a moisture content test carried out as discussed in Section 8.4.6. There is a particular moisture content, dependent on the nature of the fill material and the compaction plant, at which the material can be compacted to a maximum density. This is termed the optimum moisture content and typical values for various soil types can be estimated from Table 10.4 for the compactive plant and methods likely to be employed. Figure 10.8 shows the typical relationship between the dry density, moisture content, permeability and compactive effort. The permeability can be seen to increase dramatically with decreased moisture content below the optimum value. Although this may appear complex, the general implication of Figure 10.8 is that the material should be close to its optimum value to produce the most satisfactory fill and this is especially necessary for the clay core or the impervious element of the dam.

In practice, clay soil to be used at maximum density in the core or impermeable zone of a dam should contain sufficient water to remain intact when rolled in the hand to form a thread of 3 mm diameter. If it does not, then the soil is classified as dry and the addition of water is required. Also, if the measured moisture content is lower than that shown in Table 10.4 by more than 10% of the particular value, the fill should be watered slightly to increase the moisture content for placing. This must be added sparingly and excess moisture must be avoided as compaction plant will not be able to operate on the material. The material should be broken up as shown in Figure 10.9 and water added in a controlled manner, as shown in Figure 10.10. The material should be well mixed during and after watering. This is normally best carried out on the dam although, if water is available, this can be undertaken at the borrow area before the fill is transported to the embankment. If the measured moisture content is higher than that shown in Table 10.4 by more than 20% of the particular value, the fill should be left or assisted to dry out before subsequent placing and compaction.

Ideally, the moisture content of fill to be used in the shoulders should be no lower than the relevant values in Table 10.4 but, in the case of cohesive fill, should not be so dry as to form hard friable fragments. If the potential shoulder fill is very dry, some limited watering will be required to bring the moisture content closer to the values in Table 10.4 and allow the fill to be compacted adequately.

Figure 10.4 Large dozer with towed vibrating roller

Figure 10.5 Crawler loader with towed vibrating roller

Figure 10.6 Self-propelled roller

Figure 10.7 Sheepsfoot roller

Table 10.3 Recommended compaction plant for construction of small embankment-type reservoirs

Compaction plant	Comments	Typical use
Smooth-wheeled roller	Most versatile and normally used for small dam construction. Where vibration facility available, normally used on granular soils only	All zones and materials, except very soft cohesive fill
Grid roller	Most efficient on dry, stiffer cohesive soils and well graded granular materials. Acts in similar way to the sheepsfoot roller using a steel grid instead of projections	Cohesive shoulder fill
Sheepsfoot (tamping) roller	Has a regular array of projecting 'feet' on the roller to knead the soil together. Most suitable on soft cohesive soils when used in conjunction with dozer blade to mix and blend soil, especially if water added. Bonds compacted layers	Core and cutoff trench. Homogeneous fill dams. Cohesive shoulder fill unless very stiff and dry
Pneumatic tyred roller	Suitable for soft cohesive soils and well graded granular materials. Less suitable to assist in mixing and blending soils	Core and cutoff trench Homogeneous fill dams Shoulder fill unless stiff cohesive material or firm granular material
Vibratory smooth wheeled roller	Used for granular soils, both well and uniformly graded. Efficient in reducing air voids and compaction at depth in previously placed fill. Roller should be used initially without vibration to achieve compaction and avoid roller sinking into loose fill	All granular materials, little benefit from vibration in cohesive materials unless very stiff and dry
Hand-guided and self-propelled vibrating rollers	Smaller version of vibrating roller. Can have two rollers in tandem. Roller should be used initially without vibration to achieve compaction and avoid roller sinking into loose fill	Fill in restricted areas and backfilling adjacent to structures, pipework, etc.
Vibrating plate tampers	Manually guided plant used for compaction of small areas of granular fill, especially in trenches	Compacting fill in localised areas immediately adjacent to structures, pipework, etc. Not suitable for cohesive soils
Power rammers	Manually guided plant used for compaction of small areas of cohesive fill, especially in trenches.	Compacting fill in localised areas immediately adjacent to structures, pipework, etc. Not suitable for granular soils.

NOTE: The machines for bulk earthworks may be self-propelled or towed by dozers or loaders. Self-propelled machines are normally limited to the larger projects beyond the scope of this guide and most small embankments are generally constructed using a smooth roller towed by a suitable dozer.

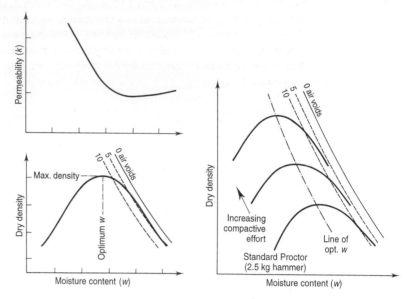

NOTE: All units in percent, bar dry density (kg/m³) and permeability (m/s). Air voids lines represent theoretical amount of air within the fill and compaction should seek to reduce these to below 10 percent.

Figure 10.8 Relationship of moisture content, dry density, compaction and permeability (Ref 37)

Table 10.4 Optimum moisture content: typical values for soil types

Soil	Optimum moisture content
Clay	15 – 20%
Silty clay	12 – 15%
Sandy clay	10 – 12%
Sand	7 – 10%
Gravelly sand	5 – 8%

The density to which a given soil type is compacted will depend primarily on the following factors:

- moisture content
- compaction equipment type, weight and contact pressures
- thickness of compacted layers – normally limited to 200 mm

- number of roller passes – normally not less than four passes
- roller speed.

The compaction requirements which should be observed for the different soil types to achieve optimum compaction at the specified moisture contents are given in Table 10.5 based on the standards required by the Department of Transport. This gives guidance on the maximum thickness of the layers and the minimum number of passes; if these guidelines are adopted, a relatively well compacted fill material should result. If there are any doubts regarding fill material or compaction requirements, specialist advice should be sought.

Table 10.5 Compaction requirements for typical soil types

Type of compaction plant	Category	Max depth of compacted layer (mm)	Minimum number of passes	Soil types
Smooth wheeled roller (or vibratory roller operating without vibration) Mass per metre width of roll:	over 2100 kg up to 2700 kg	125	8	cohesive
	over 2700 kg up to 5400 kg	125	6	or
	over 5400 kg	150	4	granular
Grid roller Mass per metre width of roll:	over 2700 kg up to 5400 kg	150	10	stiffer
	over 5400 kg up to 8000 kg	150	8	cohesive or
	over 8000 kg	150	4	granular
Sheepsfoot (tamping) roller Mass per metre width of roll:	over 4000 kg	150	4	see note
Pneumatic-tyred roller Mass per wheel:	over 1000 kg up to 1500 kg	125	6	softer
	over 1500 kg up to 2000 kg	150	5	cohesive
	over 2000 kg up to 2500 kg	175	4	or
	over 2500 kg	200	4	granular
Vibratory smooth-wheeled roller Mass per metre width of a vibratory roll:	less than 700 kg	100	unsuitable	granular
	over 700 kg up to 1300 kg	125	12	
	over 1300 kg up to 1800 kg	150	8	
	over 1800 kg up to 2300 kg	175	4	
	over 2300 kg	200	4	

Based on Table 6/4 of Department of Transport Specification for Highway Works (Ref 99)

NOTE: Sheepsfoot roller normally used on relatively soft cohesive soils. Generally used in conjunction with dozer blade to spread, mix and compact the soil, possibly with the addition of water. Subsequent passes of the roller, in conjunction with the blade, rework and compact the soil. Granular soils normally require vibratory compaction, especially if they are not well graded.

Figure 10.9 Use of plant to break up dry fill

Figure 10.10 Addition of water to dry fill

10.3.12 Damage to soil and vegetation by construction plant and equipment

The trafficability of any soil material is determined by its bearing capacity, which is related to the soil strength, moisture content, organic content and density. Wet and loose soils and soils with a high organic content all have low bearing capacities. Deformation of the soil surface then results in rutting, compaction and erosion of the surface as shown in Figure 10.1.

A single traverse by construction plant tends to impose short duration intensive forces on the soil, allowing recovery with little or no long-term effects. Where plant frequently traffics an area, compaction and smearing of soil will result, leading to loss of organic topsoil and compaction of the ground. This will cause increased runoff from the soil, with the attendant risk of drainage problems and erosion, and will lead to a reduced pore size in the ground between the soil fragments and decreased infiltration. Existing roots will be less able to develop and to extract water and nutrients whilst subsequent planting will find it difficult to become established and to flourish. Bare ground will tend not to recover and may only support isolated vegetation. Soil compacted during the construction stage should be harrowed or ripped to as great a depth as possible to break up the compacted upper layer and to provide improved conditions for plant growth.

Above ground and ground level vegetation can be badly damaged or destroyed by thoughtless or uncontrolled plant and equipment use and adequate precautions must be taken to minimise this. These should include:

- routing access away from vulnerable vegetation
- fencing off vulnerable areas
- avoiding maintenance work or fuelling to vehicles adjacent to vegetation
- avoiding excavation or stockpiling close to existing trees and shrubs.

Damaged vegetation should be cut back to sound undamaged growth. Where branches of areas of other vegetation are damaged, these should be cut back to sound wood, together with any other necessary pruning to produce a balanced tree or shrub.

10.3.13 Stockpiling materials

Temporary stockpiling of materials may be necessary as a result of the fill distribution in the borrow pit, material arising from foundation preparation or elsewhere which cannot be used directly or of imported materials which need to be stored before use. The materials will be liable to degradation if not handled and stored properly and they should be placed in formed stockpiles for subsequent excavation. The material should be spread in layers not exceeding 300 mm in thickness and compacted by earthmoving plant. The stockpiles should be shaped to shed rainwater at all times and should be located so as to avoid ponding or trapping surface runoff. The slopes should be stable and material should not be stockpiled to an excessive height.

Safe and convenient stockpile locations should be chosen carefully which comply with the following:

- stockpiles should be placed on a sound foundation and areas of weak, loose or otherwise unsuitable ground should be avoided
- sloping ground should normally be avoided
- stockpiles should not encroach within 10 m of excavations
- a sufficient margin should be left from adjacent structures, roads, fences etc.
- areas of trees and shrubs should be avoided and the stockpiles should not encroach within the spread of the branch canopy
- areas of future working or access routes should be avoided unless it is known that the material will be used before access to the area is required.

The stockpiles should be kept free of weeds, separate from other materials and prevented from becoming:

- compacted
- mixed with soil, rubbish, stone or hardcore
- contaminated with fuel, lubricants, lime or other chemicals
- buried.

Topsoil should not be stockpiled more than 2 m high to minimise the damage to and loss of the necessary soil organisms and structure. It should be handled to the minimum extent and be recovered from stockpiles without damage to the fabric of the soil. If any topsoil becomes unsuitable for re-use for any of the above reasons it should not be used in the embankment construction for topsoiling the slopes. If the stockpiles are kept for any length of time, e.g. over a winter, they should be seeded to help guard against deterioration and erosion.

10.4 SEEPAGE CONTROL MEASURES

10.4.1 Cutoff trench

The final depth of the cutoff trench should be determined on-site during construction and will depend upon the site conditions. The cutoff excavation must be finished within good quality clay or sound rock and it is essential that this condition is adequately met at the construction stage to prevent leakage through the foundation. Any land drains, which are intercepted by the excavation, should be cleaned out and the initial metre outside the trench filled with concrete. If there is evidence of loss of material into the drains outside the trench, greater lengths should be removed, the ends sealed with concrete and the removed sections replaced by good quality clay. Where land drains are encountered, they will normally need to be diverted to discharge clear of the embankment and should not be allowed to discharge into the embankment drainage system.

Dewatering of the cutoff trench may be required. Advice on methods for dewatering are given in References 96 and 97.

Filling and compaction of the cutoff trench should be carried out using the construction plant outlined in Tables 10.2 and 10.3. The work should be treated as part of the embankment construction and the fill placed in the cutoff trench should be continuous with the fill placed in the embankment. Work in any one area should be carried out without undue delay to avoid drying out or degradation of the sides of the trench.

10.4.2 Upstream clay blanket

The clay used to construct an upstream clay blanket should be similar to the material used within a central core design. Placement and compaction of the blanket material should be carried out using the construction plant outlined in Tables 10.2 and 10.3 and treated as part of the embankment construction work. Clay liners should be placed in the same manner, with particular care taken on slopes and adjacent to structures or pipework. Clay blankets and liners should normally be covered by granular material unless the clay will remain below the water level at all times.

During excavation of the clay material, whether from a borrow area within the reservoir area or from an external site, care must be taken to ensure suitable material is selected as detailed in Section 9.3.3. The contamination of clay material with silt or granular materials may have severe adverse effects on the integrity of the intended lining. In addition, double handling should be avoided as much as possible, and clay material should not be allowed to dry out prior to placing and compaction.

10.4.3 Anti-seepage collars

Anti-seepage collars should be provided around any pipe or culvert running beneath the dam as described in Section 9.3.3. Excavation for the anti-seepage collars should be made after backfilling and compaction of the backfill material in the pipe trench to existing ground or fill level. The collar should extend into the undisturbed ground on both sides and into the base of the trench by not less than 500 mm to reduce leakage. The pipe or culvert should pass centrally through the collar which must extend around the full perimeter of the pipe for a minimum distance of 500 mm . Thus the trench must be backfilled for a sufficient height above the top of the pipe or culvert to allow this to be achieved.

Anti-seepage collars should be constructed using concrete which can withstand any possible aggressive groundwater conditions. The collar must be formed in a single pour and be well compacted. The use of a vibrating poker is necessary to reduce the air voids and form a dense concrete around the full perimeter of the pipe or culvert.

10.5 DRAINAGE

10.5.1 Toe and other trench drains

Toe drains should be constructed in straight lines or regular curves, with careful excavation to minimise the disturbance to the adjacent ground. Where a pipe is included in the drain, the pipe and bedding should be carefully placed as detailed in Section 10.9.1. The trench fill material should be placed carefully into the drain from a height of not more than 1.5 m to avoid segregation. Figure 10.11 shows the construction of a curved toe drain containing a geotextile, permeable fill and a flexible pipe. The use of transverse horizontal sight rails to form an uniform gradient along the trench is also shown in this figure. This is achieved by sighting a traveller (a T-shaped timber guide) between the sight rails.

Adequate precautions should be taken to prevent damage to or contamination of the drain by the movement of plant or from material adjacent to the drain. Any section of drain or constituent material which is damaged during construction should be replaced.

Figure 10.11 Toe drain construction using sightline

10.5.2 Manholes

Where piped drains change direction or intersect, it is good practice to provide a manhole to allow the pipes to be rodded if they become blocked. A simple arrangement is shown in Figure 10.12.

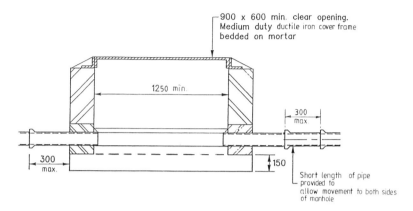

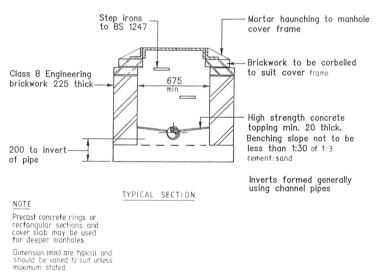

Figure 10.12 Simple manhole arrangement

The structure may settle slightly after construction. It is good practice to ensure some flexibility in the pipe by suitable arrangement of joints as shown. In the case of sealed drainage, as opposed to that discussed in the preceding section, a short length of soft, well compacted clay fill around the incoming pipe, immediately upstream of the manhole, will help to prevent seepage around the manhole, possibly leading to softening of the ground or flow outside the pipe.

10.5.3 Drainage blankets

Where a blanket of granular drainage material is to be placed on the foundation prior to the first layer of fill, or within the fill, this must be placed carefully to avoid contamination and segregation. The material should be end tipped and pushed forward by light plant to form a layer of required thickness. No compaction should be given to the layer. The succeeding fill layer should also be pushed forward over the blanket to minimise the passage of plant on the blanket and this work should be carried out at the earliest opportunity.

Care must be taken to ensure that the end of the blanket remains clear at all times and is able to discharge freely. Grading works and slope protection should be carried out carefully, to avoid damage to the blanket and to avoid sealing the surface outlet.

10.5.4 Drainage associated with structures

Drainage may be required adjacent to structures to prevent the development of water pressure and requires a sufficient thickness of material to be placed as backfill to such structures with a suitable outlet from the drainage backfill. Measures may also be required to deal with surface runoff collecting against the structures or splash from overflow channels and this drainage provision would be similar to the trench drains discussed in Section 10.5.1.

The drainage measures must be constructed such that no permeable passages are created through the embankment, resulting in loss of water and, more importantly, deterioration or erosion of the embankment itself. This requirement can be readily met by ensuring that the central portion of any structure passing through the embankment has well compacted clay backfill and that drainage is provided only over the downstream or outer length.

10.5.5 Headwalls

All drains discharging into a watercourse should have a headwall constructed at the outlet for ease of maintenance and to allow monitoring of the flows from the drain. Precast headwalls may be used, although a simple brick or reinforced concrete structure can be constructed. Flexibility of the pipework and a suitable localised clay backfill to the trench should be provided as detailed for manholes in Section 10.5.2.

The discharge level at the headwall should be located at the lowest possible level consistent with normal water levels in the watercourse and any monitoring arrangements. This normally requires the outlet pipe to be positioned not less than 150 mm above the normal water level.

10.5.6 Geotextiles

Geotextiles should be stored properly on site and protected adequately against damage from:

- plant
- sunlight
- mechanical damage
- vandalism.

Geotextiles should be placed, lapped and jointed in accordance with the manufacturer's instructions; a minimum lap length of 300 mm should be used. Where appropriate, the geotextile must be anchored adequately in accordance with the manufacturer's instructions.

Geotextiles should be installed in a clean and undamaged condition. In particular, separation fabric must be kept clean and must not be traversed by plant or labour during installation. Any geotextile damaged or soiled in any way during installation should be removed and replaced.

10.6 SLOPE AND CREST PROTECTION

10.6.1 Upstream slope

Broken stone or hardcore can be placed and spread on the upstream slope by simple farm or earthmoving plant. This should be tamped into place by the bucket of a hydraulic excavator or by hand to produce a dense stable mass. Pitching of regular concrete or stone blocks or other blockwork protection systems tend to be both expensive and labour intensive. Proprietary geotextile or composite geotextile/blockwork systems should be laid in accordance with the manufacturer's instructions. A grassed surface is normally required above any hard protection; this should extend up the slope to the crest.

Where vegetation is the primary protection against erosion, possibly in conjunction with other soft protection methods, it must be planted at the earliest opportunity to be afforded the best chance of establishment on the embankment. Comments on planting are included in Section 10.11.

10.6.2 Crest

The required crest protection should be established as soon as possible after completion of the embankment to minimise erosion by vehicles or foot traffic. Where a granular track is to be provided, the foundation should be adequately prepared and the material spread and fully compacted. A crossfall should be provided to shed rainwater, preferably towards the reservoir. Where any proprietary geotextile is used to strengthen the crest or to separate granular material from the fill below, it should be of the type specified for this purpose and should be adequately anchored in accordance with the manufacturer's instructions.

10.6.3 Downstream slope

The downstream slope will normally be topsoiled and grassed. A sufficient thickness of topsoil should be placed on the slope to allow a good covering of grass and vegetation to develop. Normally, this should not be less than 150 mm, but may be locally thickened where more substantial planting of trees and shrubs is required. Planting requirements and constraints are discussed in Section 10.11.3.

Where flow is to be passed over the downstream slope, a uniform sloping surface without undulations is required. Any reinforcing geotextile must be installed correctly, anchored adequately and covered by topsoil as detailed in the manufacturer's instructions or described in CIRIA Report 116[70].

10.7 RESERVOIR LINERS

Adequate site preparation and protection should be provided for all reservoir lining methods. Good general working practice must be adopted to ensure an acceptable lined reservoir, as described in Appendix L, and should be considered in conjunction with the manufacturer's instructions. The installation guidelines are for geomembranes and clay or bentonite linings (including bentonite mats). Guidance that applies to a particular lining method is given separately.

10.8 CONCRETE AND MASONRY

10.8.1 General

Concrete and masonry works are normally only a minor part of small reservoir construction falling within the scope of this guide, but a high standard of workmanship is necessary to ensure that the structure's function as envisaged. The service life will be many decades and thus the long-term durability of the concrete and masonry work is important. General advice on concrete construction is contained in References 100 to 102 and in References 86 and 103 for masonry.

Ready mixed concrete will be used on most small reservoir works. It should ideally be obtained only from suppliers who are accredited under the Quality Scheme for Ready Mixed Concrete (QSRMC) in accordance with Reference 104. Most suppliers and their depots are now accredited under this scheme.

On-site mixing should be considered only for very small structures containing up to 5 m^3 of concrete. BS 8000 Part 2 (94) gives guidance on mixing and transporting concrete, specifically for standard and prescribed mixes produced on site. It contains useful guidance on materials and their handling, and concreting in adverse weather (both hot and cold conditions).

A layer of blinding concrete of not less than 75 mm thickness should be placed below the base of any structure. This will allow the base reinforcement, formwork and any other features to be installed on a clean surface at the correct level and will facilitate the removal of mud and debris from within the formwork prior to concreting. In isolated structures, where seepage beneath a structure would not be problem, it may be possible to replace the blinding concrete by a layer of hardcore or stone of not less than 150 mm thickness to form a working surface. Where localised areas of soft or loose materials are present, these should be dug out and infilled as part of the blinding layer. The exposed foundation should be trimmed and prepared and the blinding layer placed at the earliest opportunity to minimise the risk of degradation of the surface.

10.8.2 Reinforcement

Reinforcement in concrete must be fixed accurately and securely and provided with sufficient spacers to ensure that the required cover is maintained. These may be plastic spacers or precast blocks; on no account should stones or brick fragments be used to support the reinforcement.

Reinforcement in masonry or composite masonry – concrete structures must also be adequately fixed and restrained to ensure that the required cover is provided.

The reinforcement must be free of any loose rust, scale or other substances before fixing, and further cleaned after installation to remove any mud or debris which may have collected. Fixing wires must be bent back and not allowed to extend into the depth of cover. Failure to ensure this or to use insufficient or unsuitable spacer blocks to hold the reinforcement securely is liable to lead to rapid corrosion. This will lead to surface deterioration, iron staining and possible spalling of the concrete or mortar cover.

The reinforcing bars or meshes must be adequately lapped to develop the required performance. This should be specified by the designer but must not be less than 300 mm .

10.8.3 Formwork

Formwork, or shuttering, must be suitably robust and secured adequately by means of ties, braces and props. The design and construction of formwork is detailed in References 105 to 107.

The appearance of the finished concrete will depend on the quality of the formwork facing and the concrete placing workmanship. Fair finish formwork is normally used where the concrete will be visible and should be plywood or other sheeting which is free from blemishes or other defects. Rough finish formwork is used for concrete which will be hidden by fill or by a surface covering where appearance is unimportant. The quality and robustness of the formwork facing and framework, however, must be sufficient to avoid deformation during concrete placing and hardening.

Sharp top corners to the concrete should be avoided by the inclusion of a timber triangular fillet, normally 25 mm × 25 mm, which also acts as a reference level for the concrete placing. Depending on the size, shape and location of the concrete element, timber fillets are often used to avoid sharp corners elsewhere on structures.

Concrete bases for smaller structures are normally formed in one concrete pour requiring formwork around the perimeter only. Larger bases may require the concrete to be cast in alternate strips or bays requiring intermediate formwork. The lowermost portions of concrete walls are normally cast as an integral part of the base slab to form a 'kicker'. This allows the wall formwork to be held securely in place against the previously cast concrete with a construction joint running along the length of the wall just above the base slab. Other horizontal joints in concrete walls should be avoided, if possible, by casting to full height in one pour.

10.8.4 Waterstops

Waterstops must be properly specified correctly located and jointed using methods approved by the manufacturer. The concrete must be thoroughly compacted around the waterstop to ensure that an effective barrier to flow is produced. A high degree of skill and care is necessary for building-in waterstops which will function correctly throughout the life of the structure. The use of waterstops is mandatory in accordance with the water-retaining code BS 8007[75] and is recommended in small reservoir construction where there is a possibility of water movement through the joints of a concrete structure.

10.8.5 Concrete placing and compaction

Concrete should not be placed when the shade temperature exceeds 30°C or is less than 5°C unless special precautions are taken. High temperatures are rarely a problem in the UK whilst low temperature concreting, if it cannot be avoided, may require accelerators or rapid-hardening cements. Insulation of newly placed concrete will be necessary to avoid frost damage, as will protection against strong drying winds or precipitation damage.

The concrete should be placed as quickly as possible and be consistent with good workmanship. Double handling should be avoided, not only to reduce time and effort, but to minimise segregation of the mix. With this in mind, the concrete should not be allowed to fall through more than 1.5 m. Chutes etc. should be used whenever possible to allow the concrete to flow into position.

On no account should extra water be added beyond that specified, as this will adversely affect the permeability and durability of the hardened concrete (Section 9.12.5). Thorough compaction of the concrete, using poker vibrators, is essential to ensure that there are no air pockets, to remove small air bubbles in the concrete and to work the concrete fully around the reinforcement and into the corners of the formwork. Failure to achieve thorough compaction will result in a lower strength, increased permeability and decreased durability. Care must be taken to avoid damage or movement of the reinforcement, spacer blocks, or waterstops during compaction. Over-compaction must also be avoided, when bleeding of fines and water to the surface of the concrete occurs leading to an unsatisfactory quality of finish, particularly in slabs. Where required, the concrete should be tested following the recommendations in BS 1881[108].

Immediately after compaction, the concrete should be protected by damp hessian, plastic sheeting or a spray-applied membrane. Over-rapid drying creates a poor-quality surface with reduced strength and durability and results in shrinkage cracks. If polythene sheeting is used, this must be securely anchored to avoid a wind-tunnel effect which can enhance the drying, whilst hessian should be kept damp. A spray-applied membrane is unlikely to be used, but its use must be checked for compatibility with any surface treatment or admixture used in the concrete. Further information on curing concrete is given in Reference 109.

10.8.6 Joint formation and concrete finishes

Sufficient and appropriate joints should be provided, as discussed in Section 9.12.7. Contraction joints can be formed by pressing a crack inducer into the wet concrete or cutting a groove in sheltered locations, when the concrete is a few days old or, in more exposed locations, after 24 to 48 hours. Information on joint construction is included in References 83 to 85.

Unformed top concrete surfaces should be finished by using a steel trowel to achieve a uniform, dense and durable surface. The trowelling has the effect of closing surface pores and cracks, making it more difficult for water to penetrate into the concrete. A smooth trowelled surface is likely to be slippery to walk on and thus where a concrete surface is to form a walkway, alternative treatments or finishes may need to be considered to produce a satisfactory surface. These may include a brushed or tampered surface without trowelling.

10.8.7 Masonry construction

The standard of construction will depend on the skill of the builder and the adequacy of supervision. Care must be taken to lay the bricks or blocks accurately and to maintain a uniform joint thickness and consistent alignment of the perpends or vertical joints. Verticality should be checked regularly, together with the alignment and level of courses. Once the first course has been laid, the corners should be built to about five courses and the lifts between corners placed subsequently. Masonry should not normally be raised by more than 1.5 m height in any one day and lifts between corners should be complete at the end of the day.

Mortar should be allowed to harden slightly and then finished to a specified profile by tooling, with all joints tooled where possible. Pointing of blockwork should be avoided, where possible, and smeared mortar on exposed faces should be minimised.

Masonry should be covered at the end of the day, particularly if hollow blocks are used, and should be protected from rain. Work should cease when the temperature drops below 3°C unless special precautions are taken and all work must be protected against frost.

Where walls are not properly restrained until completion, temporary propping may be required. This may also be necessary for high or long walls and especially during strong winds.

Movement joints should be incorporated at a maximum of 6m-spacing if not specified at the time of the design. It should be noted that panels of masonry are more prone to cracking if the length of panel exceeds twice the height and a closer spacing may be necessary. The position of joints should be considered in relation to other structural details and the progress of the work. Where openings are present in the masonry, consideration should be given to the provision of masonry reinforcement above and below the openings to prevent deformation.

10.9   DRAWOFF AND OTHER PIPEWORK

10.9.1   General

Pipes should be laid and jointed in accordance with the manufacturer's instructions. Particular care must be taken to ensure that the trench dimensions are appropriate and that sufficient space is available to form the joints. The bedding should be placed carefully and compacted prior to pipe laying. Additional pipe-laying bedding material should then be placed evenly on both sides of the pipe to avoid undue loading or lateral movement of the pipe. The material must be placed carefully to avoid damage to the pipe or the joints and must be compacted by suitable hand operated plant. Other fill materials should follow compacted by similar means, until the trench has been backfilled. Trench excavation for pipelaying in fill should not be commenced until the fill is sufficiently above the pipe crown to permit subsequent fill placing to continue, without detriment to the pipes.

Pipes beneath a dam must be installed within a clay backfilled trench with anti-seepage collars and with a high degree of workmanship. Construction of the pipework and anti-seepage collars should be completed before placement of the embankment fill commences. Pipe joints should be as flexible as possible to minimise the effects of subsequent movement.

Pipes elsewhere and clear of the dam need not normally be installed with anti-seepage measures, and a granular bedding is normally satisfactory. Where a pipeline is laid in sloping ground the backfilled trench may act as a channel for groundwater flow and erosion may result. This may also occur in ground with a high water table, where flow is able to move towards a lower level elsewhere along the pipeline. This can be prevented by filling intermittent sections with well compacted clay and, in the worse instances where substantial flows are encountered in the trench, by installing land drainage. Pipework adjacent to inlet structures, manholes or headwalls should be constructed so as to provide some flexibility if settlement of the structure occurs. This is normally achieved by the provision of short lengths of pipe with not less than two flexible joints, as shown in Figure 10.12.

10.9.2 Drawoff pipework

The low-level outlet pipe will be the lowest, and often the only, drawoff pipe. Its probable position in the foundation below the maximum height of dam will result in the greatest reservoir pressure and embankment loading on this pipe. The standard of installation of the pipework and seepage control measures must, by necessity, be high. The low-level outlet pipe often functions as the diversion pipe to pass flows in the watercourse during construction, and subsequently conveys compensation flows during reservoir operation. Apart from the downstream end, access for maintenance or other work will be essentially impossible and thus it is vital that the pipe is correctly and adequately installed.

Other drawoff pipework, if present, must also be installed to the same high standard although the reservoir pressures and loadings on the pipework will be less. Any pipes below the maximum reservoir level (including any flood rise) will be susceptible to water movement along the backfilled trench and appropriate anti-seepage measures must be provided.

10.9.3 Thrust blocks

Pipelines having unanchored flexible joints require restraint at changes of direction and at blank ends to resist the thrusts developed by internal pressure. Thrust acts radially outwards at a change of direction, on a line which bisects the angle of direction change i.e. the deviation angle. Thrust magnitude depends upon pipe diameter, internal pressure, type of flexible joint and the size of the deviation angle. Table 10.6 gives the thrust for various standard bends and fittings for pipe sizes up to 900 mm diameter.

The thrust must be resisted by the undisturbed ground in the sides of the trench and this is normally achieved by the provision of a block of concrete to cast insitu between the outside of the bend and the undisturbed trench side. The required area of contact against the trench side will depend on the thrust and the strength of the ground. A conservative value of one square metre of contact area per 50 kN of thrust may be used, with a minimum one metre length of concrete along the trench extending up to the pipe crown level. In many instances, the size of pipes and internal pressures appropriate for reservoirs within the scope of this guide, a reduced size may be possible, or the thrust block omitted. Specialist advice should be sought, however, before reducing or omitting any thrust block.

Guidance on the design of thrust blocks can be obtained from CIRIA Report 128 'Guide to the design of thrust blocks for buried pipelines'[131]

Table 10.6 Assessment of pipe thrusts

Nominal size DN (mm)	Thrust (kN) per 100 kN/m² internal pressure				
	Blank ends and junctions	90° bends	45° bends	22½° bends	11¼° bends
80	0.755	1.070	0.575	0.295	0.150
100	1.090	1.550	0.835	0.430	0.215
150	2.270	3.210	1.740	0.885	0.445
200	3.870	5.470	2.960	1.510	0.760
250	5.900	8.340	4.510	2.300	1.160
300	8.350	11.800	6.390	3.260	1.640
350	11.300	15.950	8.640	4.400	2.210
400	14.450	20.450	11.070	5.640	2.830
450	18.100	25.600	13.850	7.060	3.550
500	22.250	31.450	17.000	8.670	4.360
600	31.650	44.800	24.250	12.360	6.210
700	42.800	60.500	32.750	16.690	8.390
800	55.700	78.750	42.600	21.730	10.920
900	70.150	99.200	53.700	27.370	13.750

NOTE: Thrusts for other internal pressures may be calculated pro-rata.

10.9.4 Outlets

The downstream end of a pipe carrying compensation flow should discharge through a headwall, as detailed in Section 10.5.5, back into the watercourse downstream of the dam. The pipe should be orientated to discharge the flow in a downstream direction and protective measures may be required to prevent erosion of the watercourse. These could comprise hardcore or broken stone, but a small concrete apron may be preferable. This should extend for not less than 16 pipe diameters along the channel and should continue to a sufficient height up the sides of the watercourse.

10.10 OVERFLOW WORKS

10.10.1 Construction of drop inlet overflow weirs

A drop inlet overflow weir designed as a masonry or concrete weir chamber should be constructed in undisturbed ground at the edge of the reservoir. If the weir chamber is to be constructed of masonry, a competent bricklayer should be employed to build the structure with the appropriate bond to provide the strongest construction.

Hardcore or crushed stone should be placed as a protective layer around the structure to minimise erosion, particularly upslope of the chamber where surface runoff and locally increased wave action may occur.

10.10.2 Construction of crest weir overflow works

A crest weir structure should be sited on original ground, normally at, or just beyond, one end of the dam. In some instances, it may be sited further beyond the end of the dam if the topography, such as a very gently sloping abutment or side valley, allows this.

A high standard of concrete work is essential and good practice in design and construction is imperative. A cutoff, which should be formed by excavating through the blinding layer, should be designed and constructed as an integral part of the base slab of the structure. Reinforcement must be provided in the cutoff to ensure that cracking, and consequently seepage, cannot occur along the underside of the slab. Any areas of poor concrete should be made good or cut out and replaced. Additional joints should be kept to a minimum to avoid possible seepage paths, but it is good practice to provide an expansion joint between the downstream end of the structure and the remainder of the overflow channel or culvert. Further joints should be provided along the channel in accordance with good practice (Section 9.12.7).

The sill should be constructed to a constant level across the full width of the structure. It may be formed in reinforced concrete linked by reinforcement into the base slab or by setting precast kerbs or bricks to the required level. If this latter option is used, some form of linking element, e.g. short vertical bars cast into the slab, should be provided to tie the kerbs or bricks to the slab. These should be encased fully in mortar for corrosion protection to ensure that no unsightly discolouration occurs.

Where a precast or cast insitu culvert is used to pass the flow beneath the dam crest, particular care must be taken with the sealing and provision of the joints in the structure to avoid leakage into the fill.

Backfilling of the structure through the crest of the embankment is critical to the safety and satisfactory functioning of the dam. Selected clay should be used and be placed by hand or mechanical means and compacted fully into place to a standard not less than that for the remainder of the embankment.

10.10.3 Construction of auxiliary overflow works

The auxiliary overflow may be sited on undisturbed ground at or beyond one end of the dam but is often sited on the dam itself in view of the length of overflow normally required. The extent of the overflow should be clearly defined with its length graded to the required constant elevation and a uniform crossfall across the lowered crest. A line of kerbs set to the required level will provide a more definite controlling level.

The establishment of good low maintenance grass cover is an essential requirement for this form of overflow. Where vehicle access is required along the crest, other than for very occasional maintenance or access purposes, a surfacing of crushed stone or similar material is required. These two requirements are in opposition, but flow velocities over much of an overflow crest are normally low and the surfaced crest can adequately withstand these. A geotextile, if present, must be adequately anchored at its upstream edge to withstand the force of the flowing water. Normally this is achieved by extending the geotextile onto the upstream slope and terminating it in a trench as shown in Figure 10.13.

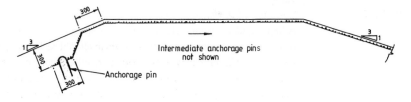

DETAIL AT CREST

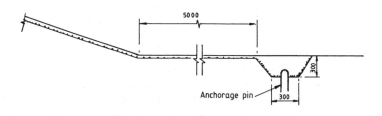

DETAIL AT TOE

Figure 10.13 Anchorage of geotextiles for erosion protection

Sufficient anchorages should be provided over the full extent of any geotextile; 500 mm lengths of small diameter reinforcement steel bar or similar bent to form U-shaped pins can be used as anchorages. Normally, an appropriate 2 m spacing is adequate to restrain the geotextile until the grass roots are fully established. The lower section of the downstream slope will experience the maximum flow velocity with the greatest erosion potential being at the junction with the original ground beyond the dam. It is important, therefore, that the geotextile is taken sufficiently far beyond the toe and adequately anchored, as also shown in Figure 10.13.

10.11 ENVIRONMENTAL/CONSERVATION WORK

10.11.1 Shallows and islands

As a general rule shallow margins for wildlife should be graded to at least 1V:10H down to a maximum depth of 1.5 m. These can be created by grading back the margins, or by depositing overburden. If the area available for shallows is limited they should be located where they will gain the most shelter from wind without being shaded.

10.11.2 Aquatic planting

Aquatic plants can be obtained from commercial suppliers although the cost can be prohibitive and care is needed to avoid exotic species. An alternative source could be from the spoil of recent dredgings. The RLA, County or District Council may be able to provide information on this. While transplants can be obtained from a nearby water body it is important to remember that wild plants are protected under the Wildlife and Countryside Act 1981[9] and it is necessary to obtain the permission of the landowner before material is collected. Some plants are specifically protected and advice should be obtained from English Nature if plants are to be collected from the wild.

Planting is best carried out in the spring to allow the plants a full growing season to become established, although rhizomatous species should be transplanted in February or March when the rhizomes are just at the end of their dormant period and about to shoot. Once the plants have been collected, the roots must not be allowed to dry out, and must be kept moist until they are re-planted.

Marginal plants should be placed just above the water line with the roots or rhizomes pushed into the soil. Emergents are planted in much the same way but in shallow water, or saturated ground, and it may be necessary to weigh them down. An additional layer of material may also have to be placed over a flexible liner, for some plants such as sedges have sharp growing tips to their roots that could pierce the membrane. Submerged and floating leaved plants need to be weighted and dropped into the water at the appropriate locations, whilst free-floating plants can simply be placed into the water for dispersal.

If different species are planted in groups, competition leads to the most vigorous becoming dominant. It is advisable to plant the reservoir margins with separate blocks of species which allows each group to become established and results in a varied waterside flora.

10.11.3 Terrestrial planting

Following construction, disturbed ground should be deeply ripped or harrowed. Topsoil should be replaced where necessary. Surfaces that are to be sown with grass seed should be reduced to a fine tilth and stones greater than 50 mm removed. A variety of grass seed mixtures are commercially available and can be selected to suit site conditions and specific requirements such as low maintenance or agricultural production. Specialist advice on appropriate seed mixes can be obtained from seed suppliers.

If it is desired to create a wildflower sward adjacent to the reservoir, the surrounding topsoil should be stripped. This is necessary because the higher fertility of topsoil and presence of a seedbank leads to the suppression of the desired plants by more vigorous species. Freshly exposed subsoil should be cultivated, instead, to create a fine, firm seedbed, with the selected grass and wildflower seed mix being sown by hand in late spring or early autumn. Conversely, the tall tussocky vegetation preferred by nesting wildfowl can be adequately created by mounding uncompacted subsoil and overburden and allowing it to revegetate naturally with grasses, teasels and thistles, etc.

Broadleaved trees and shrubs are best planted in the period November to March, although no planting should be done when the ground is frozen, waterlogged or very dry. Evergreens should be planted in October or in April when the soil is relatively warm. In nature conservation terms it is preferable if the trees and shrubs which are planted are native species and if the seeds and cuttings from which they are grown are of British, and preferably local, origin.

The success of any tree or shrub planting should be improved by ensuring that the roots are covered at all times and prevented from drying out, the soil is not compacted, the appropriate species are selected for site conditions, and the newly planted trees are thoroughly weeded and protected from grazers and rabbits. The most economical plantings are with bare-rooted transplants (200 to 400 mm) where a notch is made in the soil with a spade and then firmed back around the plant with the foot. For larger trees a pit must be dug, the hole being large enough to accommodate all the roots without constriction. Trees and shrubs must be adequately staked and protected as required. Planting mats should ideally be used to discourage weed growth around the new tree or shrub and regular weed control must be carried out to avoid competition.

10.12 CONSTRUCTION SAFETY

10.12.1 General

The construction of the reservoir must be in accordance with the Health and Safety at Work Act etc.[40] and the Construction Regulation[110, 111]. Prior to reservoir construction, the undertaker should take steps to become aware of the hazards which may exist, so that the operations can be planned in such a way as to eliminate risks by design, if possible, and to minimise and control any residual risks. The following paragraphs details some of the more common hazards that are encountered on sites involving earthworks. Various publications relating to safe working can be obtained from the HSE at the address given in Appendix D.

10.12.2 Safety related to the construction of earthworks

- The failure of temporary slopes is a common cause of accidents and this includes failure of slopes in cuttings, trenches and embankments. An appraisal of the stability of such slopes should be made if failure could cause a potential hazard and, if necessary, the slope cut to a safe angle or shoring installed. Special care should be taken where groundwater and/or soft or variable soils are encountered as this will substantially reduce the stability.
- Unless specifically allowed for in the design of an excavation, heavy plant should not be allowed to approach, nor excavated material placed near the edge of the slopes, whether cut or filled or near trenches or other excavations.
- Care should be taken when excavating from the base of a working face to ensure that the face is not overhanging or excessively high. If a localised failure occurs, there should be no risk to the driver and excavator or any other operatives or plant.
- Where bench working is being carried out, care should be taken to ensure that the higher benches are not later undermined or that there is not a risk of material falling or rolling from the upper levels.

10.12.3 Safety related to construction plant

- Heavy plant should be routed along distinct haul roads, preferably separate from pedestrians and light traffic. Wherever possible, the haul road should be formed so as to maximise sight lines by avoiding or smoothing bends and any high sections or humps along the route.
- Haul roads should be well maintained and kept in good condition as this will maximise vehicle control and minimise the braking distance. Gradients should be kept to a minimum and in dry weather the surface of the haul road should be watered to minimise dust clouds.
- Visibility from many large machines is poor, particularly when reversing, and it is advisable to keep well away from them. If this is difficult, such as when setting out or checking the fill, the drivers should be warned verbally or by roadside signs. Sitting in a parked vehicle may not be a safe position, as certain large construction plant can easily crush a light vehicle.
- Overhead obstructions such as cables, bridge soffits, or temporary works may cause clearance problems. These should be indicated clearly by signs or tapes.
- Drivers and operators of plant must be trained and fully competent in the use of the plant.
- Plant should be regularly maintained and not operated beyond its capability or capacity or as described by the manufacturer. This information should be obtained and understood prior to use of the plant.
- Plant should be operated with any appropriate guards in position and any necessary protective clothing or equipment should be worn when using the plant.

- Tipping of fill on the embankment should be carried out short of the edge of the advancing layer of fill and then the additional fill dozed forward.
- Vehicles that travel on public roads must, by law, be in a roadworthy condition. Where mud is spread on public roads by the wheels of site plant, road cleaning wheel washing equipment should be used to minimise the possibility of accidents involving the public. This is particularly important in the winter months.
- Wheeled vehicles working on slopes can slide out of control. Tracked vehicles are safer on steeper slopes but damage to the ground surface is more severe.
- Grass covered slopes are intrinsically hazardous and more likely to lead to sliding; the danger increasing with steeper slopes. Safe working methods depend on:
  - ground condition
  - weather
  - direction of travel
  - type and condition of equipment
  - weight being carried
  - care and experience of operator.

11 Contractual arrangements

11.1 CONTRACT REQUIREMENTS

Contracts for the construction of a small reservoir are normally made between two parties. The party who commissions the work is normally the landowner and the future undertaker (often referred to as the 'Client' or 'Promoter', or the 'Employer' in contractual terminology). The party who constructs the work or provides other construction services is usually referred to as the 'Contractor' and may be a civil engineering contracting firm or a local plant/earthmoving firm. A contract is also required between the parties when various aspects of work forming the site investigation are carried out or any specialist construction or other service is required during the construction from a further party.

The contractor undertakes to carry out the specified works for a sum of money in accordance with the employer's instructions, usually within a stated period of time.

The employer has an obligation to pay the contractor for the work he carries out and to give possession of the site to the contractor for the duration of the contract. The success of the contract is dependent, to a large degree, on the quality of the employer's instructions to the contractor, normally contained within the contract documentation. This is of legal significance and should include the following distinct elements:

- form of contract (or agreement) — forms the contract by reference to the required works and the other documentation
- conditions of contract — are the conditions and terms under which the contract is formed
- specification (including the drawings) — sets out how the work is to be executed
- method of measurement — sets out the financial arrangements

It is important that careful consideration is given to selecting the most appropriate form of contract documentation. Amendments to standard documentation should be included where necessary, but such changes should be kept to a minimum to reflect the particular requirements and minimise the potential consequences of inappropriate, incorrect or incompatible amendments and additions.

The place where a contract is made is the place where the contract is made binding by the communication of an acceptance. This means that a contract made and accepted 'in Scotland' is subject to Scottish Law.

Civil engineering contracts are usually made in writing in one of the forms discussed in Section 11.2 These have many advantages and their use is recommended. It is important to be aware that the making of a contract requires no formality. A binding contract may be made by the following methods:

- exchange of letters
- signing of contract documents
- verbal agreement.

A verbal agreement is not recommended as there is no record of the terms of the contract.

A number of Acts of Parliament affect civil engineering contracts. In general, however, civil engineering contracts are governed by the ordinary rules of the law of contract.

The basis of the contract is agreement. Agreement comprises an offer and acceptance. An offer must be distinguished from an attempt to negotiate. If it is accepted, an offer becomes a binding contract.

The acceptance of an offer must be unconditional and it must be communicated to the person who makes the offer. The terms of acceptance must correspond precisely with the terms of the offer to ensure that acceptance is unconditional. The words 'I accept your offer' constitute an unconditional acceptance and brings a binding contract into existence.

The terms of the contract do not have to be set out in full in the documents, letters or conversations which constitute the offer and acceptance. Terms contained in another document, a set of standard conditions for example, may be incorporated within the contract by reference.

An essential prerequisite of binding contract is that the Agreement must be supported by 'consideration'. In civil engineering contracts the consideration for the promise made by the contractor to carry out the works will usually be the promise by the employer to pay the price for the works. Reference to these financial arrangements should be clearly defined in the terms of the contract and the details specified or reference made to a standard method of measurement and payment.

Clarity and precision are essential requirements when drafting a contract. Standard forms of civil engineering contract clearly define the rights and duties of the contracting parties and their use is strongly recommended.

If the parties enter into a contract, but there is an error due to omission or misstatement in the contract documentation, the contract may be rectified by the parties given certain provisions. These are that both parties agree the contract is inaccurate and agreement on the form of correction to the contract is reached.

The contract should also allow for changes necessitated by conditions revealed during construction or changes in the undertaker's requirements. An adequate site investigation should minimise the former whilst the undertaker should clarify his requirements prior to the completion of the design insofar as is possible. Changes to the works agreed with the contractor should be conveyed clearly by means of a written instruction and amended drawings. The economic aspects of changes during construction are discussed further in Section 5.

The insurance requirements should be covered fully in the conditions of contract, together with the contractor's obligations to rectify faulty workmanship or outstanding works, both during the construction and for a specified time thereafter. Provisions for the settlement of disputes arising from the contract should also be included in the conditions of contract.

A contract is also recommended with any specialist advisers who may be required to provide advice on any aspect. This should refer to a brief or list of the employer's requirements in sufficient detail to enable the required services to be performed effectively.

11.2 CONTRACT OPTIONS

11.2.1 General

Two main options can be adopted by the employer to detail the conditions under which a contract is formed:

- The employer may negotiate with one or more contractors to carry out the required works shown on the drawings and described and defined in the specification. In some cases the contractor(s) may be involved in discussions at the design stage or carry out this work for the employer. Normally, the contractor will offer to carry out the required works to his standard terms and conditions, although the employer may amend these by negotiation.

- The employer may prepare contract documentation, normally specifying standard conditions of contract, and including the drawings and specification. Details of the method of measurement and payment will also be included. One usually or, more contractors are then requested to price the works.

The procedures are detailed more fully in Figure 11.1. The use of a non-standard or a standard form of contract is feasible for small reservoir construction; the choice should be the employer's, based on the size, scope and complexity of the proposed works, the degree of control that he is able or willing to exert and the financial arrangements for paying for the works. The contractor(s) should not be able to exert any pressure in the choice of contract and, ideally, the employer should clarify his preferred approach prior to any liaison with the contractor(s). Specialist advice should be taken where necessary.

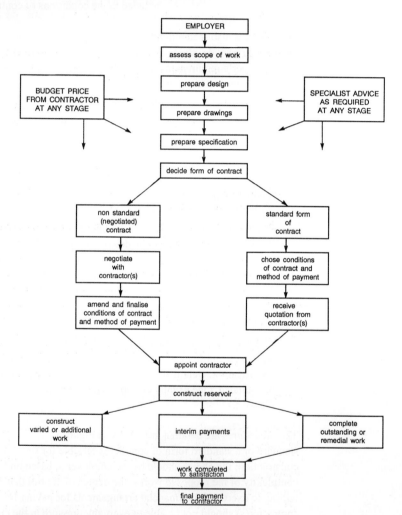

Figure 11.1 Contract requirements

11.2.2 Non-standard form of contract

A non-standard form of contract is normally based on a written quotation from a contractor and the contractor's standard terms and conditions. While the conditions included are designed to protect the contractor's interests rather more than the employer's, they will constitute rudimentary conditions of contract for the works. The employer can then negotiate with the contractor on the conditions included and clauses may be omitted or appended as agreed.

The advantages of entering into a contract on this basis are:
- provision of a ready-made template for the contract conditions
- speed of setting up in place
- reasonable statutory protections for the employer (see Section 11.6)
- ease of administration (lump sum payments).

It should be noted that amending the contractor's terms and conditions might reduce the significance of some of the statutory protections.

The disadvantages are:
- the conditions as laid out invariably favour the contractor
- the scope and quantity of the work required may not be well defined thereby giving scope for future disagreements
- the resolution of disputes may be disproportionately troublesome to the original amount of work required.

This approach is not recommended unless the works required are anything other than completely straightforward. A recognised standard form of contract normally offers a greater protection to the employer and lays out the rights and duties of all the parties. It is advisable to have the contractor's proposals checked by a specialist adviser to ensure they are acceptable. Any necessary additions or amendments should then be formulated and agreement sought with the contractor.

11.2.3 Standard form of contract

The standard form of contract recommended is the 'Conditions of Contract, Agreement and Contract Schedule' for use in connection with 'Minor Works of Civil Engineering Construction', First Edition[112]. Alternatively, the more comprehensive 'Conditions of Contract and Form of Tender, Agreement and Bond' for use in connection with 'Works of Civil Engineering Construction', Sixth Edition[113], may be used. Copies of these documents may be obtained from The Institution of Civil Engineers at the address given in Appendix D.

Where a ground investigation is to be carried out using a specialist contractor, the standard form of contract recommended is the 'Conditions of Contract for Ground Investigation', First Edition[114].

Should the undertaker wish to amend a standard form of contract, he should seek specialist advice when preparing the contract documentation and before requesting contractors to price the works.

11.3 SPECIFICATION

The specification forms part of the contract documents under the ICE standard forms of contract. It should be prepared in conjunction with drawings of the proposed work and should be concise and complete. The specification should describe the workmanship and materials required as well as indicating the position in the works of the various items if not indicated on the drawings.

The specification should also set out clearly any constraints on the contractor's freedom to do the work as and in the order he thinks fit.

11.4 DRAWINGS

The contract drawings should ideally detail all the contract work. They must give sufficient information to enable all parties to understand the requirements of the project. They provide the most accessible method to convey information. The drawing should be referenced with the title of the project, preferably the name of the employer and a specific title. Each drawing should be separately numbered and, if revised, should be given a letter suffix to delineate the latest version. Whenever information is updated the drawings should be amended and copies should be circulated to all parties to the contract.

The drawings should include:
- a large-scale plan of the works at a scale of 1:200 (or larger). This should have the main construction works marked on, together with immediately local access routes and any constraints on the works
- cross-sections across the proposed embankments including details such as slopes, internal zoning, excavation depths, etc.
- larger scale drawings with details of specific features of the embankment, including pipework, structures, slope protection, drainage, crest works, etc.

Additional drawings may be required to show works in the borrow pits or constraints on their use and extent, access routes where the works are distant from public roads or other specific requirements of the project.

Drawings are also required for approval of projects by the RLA and LPA. These authorities may require drawings at minimum scales. Where required, advice on the preparation and requirements of contract drawings should be sought from the specialist adviser.

11.5 FINANCIAL ARRANGEMENTS

11.5.1 General

It is essential to have the contractor's financial proposals for carrying out the work in a suitable form prior to commencement of the work. This will ensure that the cost implications of varying the quantity of work can be assessed adequately before the work is commenced and the effects of subsequent changes in the construction period.

11.5.2 Clarification of terms

It is important to clarify the financial terms of the contractor's proposal to carry out the work. It will usually take the form of one of the following:

- estimate
- budget estimate
- cost estimate
- budget price
- budget quotation
- quotation.

These may be presented as a single sum of money or broken down into a number of individual costs and/or items for specific items of work. Use of the following two terms only is recommended to avoid confusion:

BUDGET PRICE – An approximate best estimate of the final cost that has no contractual standing and may vary. A budget price may be sought at any time during the feasibility and design stages to give an indication of all or part of the cost of the proposed works.

QUOTATION – The actual cost of the work to be carried out that would be paid to the contractor and which is based on the requirements shown in the contract documents.

The contractor should be asked to submit his proposal as a 'Budget Price' or a 'Quotation' and the definitions of these two terms should be clearly explained to him before his proposal is submitted. A quotation is recommended with individual work items clearly defined with rates given. It should include provision for time-related charges where applicable.

The form of payment should be considered by the employer when seeking a quotation for the work prior to discussions with a contractor and should be specified if a standard Conditions of Contract is to be used. Payment normally takes the following forms:

- lump sum payment, with no itemised breakdown, for the works as shown in the contract documentation
- payment for clearly defined individual work items
- payment for work carried out on a time basis.

If a lump sum is proposed for the required works, with no itemised breakdown, problems may occur if unforeseen ground conditions are encountered or additional, alternative or remedial work is required. There is then no agreed basis on which to assess the cost of the additional work and this approach is not recommended other than for very small simple schemes.

Payment for clearly defined work items offers the fairest means of valuing the works carried out for both parties. The employer would normally prepare a list of work items required, together with a best estimate of the quantities based on the contract documentation (normally referred to as a Bill of Quantities). Appendix O shows a simple example of part of a Bill of Quantities for a small reservoir. The actual quantities are then remeasured at the end of the works and the contractor paid accordingly. If the quantities change, this is readily dealt with by the remeasurement, whilst additional or amended works can normally be priced by comparison with similar rates.

Payment for works carried out on a time basis can either be against previously agreed rates for labour and plant or against the contractor's actual costs plus an agreed increase for overheads, profit, etc. Materials are normally paid at cost plus an agreed handling charge. This method may appear attractive, but the employer is dependent on the contractor's method and speed of working and organisation to complete the works in the most efficient manner. Dates may be agreed to complete all or part of the works, but the employer is generally unable to have much control of the expenditure and this method is not recommended other than for very small simple schemes.

Payment for specialist advisery services would normally be on a time and expenses basis in view of the imprecise extent and nature of the work. A budget price, however, may be requested to give an idea of the likely expenditure.

11.6 STATUTORY PROVISIONS

The main Acts that provide statutory protection to an employer engaged in a simple negotiated contract are as follows:

- Sale or Supply of Goods and Services Act[115]
- Unfair Contract Terms Act[116]
- Misrepresentation Act[117]
- Latent Damages Act[118]
- Occupiers Liability Act[39].

11.7 INSURANCE

11.7.1 General

Provided the Conditions of Contract recommended in Section 11.2.3 are adopted, insurance requirements for the period of the contract are clearly defined. If a negotiated contract is adopted, the following definitions of liability apply for the period of the contract. Proof of the contractor's insurance and their current validity should be seen prior to the commencement of work.

11.7.2 Public liability

During the contract period, the liability of the contractor for causing injury or damage to third party persons or their property is covered by the contractor's public liability insurance (third party insurance).

11.7.3 Employees of contractor

The liability of the contractor for injury or accident to the contractor's employees is covered by the contractors employers' liability insurance. This insurance is compulsory for all contracts under the Employers' Liability Act[119].

11.7.4 Insurance of the works

The permanent works and any temporary works as well as any constructional plant should be insured by the contractor in the joint names of the contractor and employer for the duration of the contract. 'Joint names' means that both the employer and the contractor are covered by the insurance.

After the contract period, the employer should seek advice from his insurance company on the insurance requirements for the reservoir. The liability during the handover period, from the end of construction to the full taking over by the employer, requires special consideration.

11.8 SUPERVISION

Where a contractor is employed, the responsibility for organising the work is his, subject to any constraints of the contract contained in the specification or drawings, or statutory controls. The undertaker should keep a discreet 'eye' on the work without hindrance to the contractor. Where the undertaker is aware that the work is not being constructed correctly or that statutory or other requirements are being broken, a carefully phrased word with the contractor's senior member of staff should help to minimise unnecessary expenditure or difficulties for the contractor.

The general aspects to be monitored during the construction are outlined in Section 10.1.5 and these should be checked as necessary to ensure compliance.

Where other organisations are employed during the construction work, these should be similarly supervised and appropriate action taken. Where such organisations are employed directly as a sub-contractor by the contractor, any problems should be discussed directly with the contractor as there is no formal contract with a sub-contractor.

Part 4 Post-construction

12 Observations and monitoring

12.1 INTRODUCTION

Visual observations and monitoring (surveillance) should be carried out at regular intervals after completion of the dam to ensure problems associated with the dam, foundation and adjacent areas are identified at an early stage in their development. These problems are generally associated with the following:

- instability
- seepage
- erosion
- blockage (of channels and flow structures)
- theft, vandalism and damage
- deterioration.

Regular monitoring during the first filling is particularly important as many problems will only become apparent at this stage. The dam should be checked visually, at least daily, during this time and action taken if any seepages or other matters of possible concern become apparent.

Monitoring during construction is not included in this section but is discussed in Section 10.

Visual observations and monitoring should be considered under the following categories:

- reservoir level and flow measurements
- identification of defects
- problem monitoring
- reservoir margin areas
- embankment areas
- inlet, overflow, drawoff and ancillary structures
- downstream areas
- other aspects.

A plan of the reservoir and adjacent area at 1:200 scale or larger should be marked up with notes and observations on each occasion and kept as a record.

Photographs may be used to supplement visual observations; they should be processed promptly after being taken and clearly labelled with the date, brief notes about the contents of the photograph and other relevant matters. Ideally photographs should be taken from the same position to facilitate comparison.

It is good practice for the undertaker to retain a specialist adviser to carry out a periodic visit and review of the dam and reservoir. Monitoring observations and maintenance problems can be reviewed and a technical check carried out to assess the safety of the reservoir. The necessity for remedial measures can be discussed. An annual visit is normally recommended.

12.2 RESERVOIR LEVEL AND FLOW MEASUREMENT

Reservoir water level should be recorded on a weekly basis. It is good practice to install a reference gauge-post, accurately levelled with levels marked on it to facilitate rapid assessment of reservoir level. The requirements for reference posts are given in Appendix P and a typical example in use is shown in Figure 7.15.

Flows from drains and springs should be monitored regularly and ideally at weekly intervals. Details of vee-notch measuring weirs and other flow measuring devices are given in Appendix F. Use of a standard form for recording observations will help to improve the quality and consistency of the records.

12.3 IDENTIFICATION OF DEFECTS

The main surveillance indicators of possible defects are summarised in Appendix Q. 'An Engineering Guide to the Safety of Embankment Dams in the United Kingdom'[120], published by the Building Research Establishment also offers a useful summary of relevant matters.

Boxes 12.1 and 12.2 details the defects that are likely to be encountered and attempts to categorise these in terms of maintenance or remedial works. In general terms, maintenance work should be considered as work needed on a reasonably regular and routine basis to keep the reservoir functioning in a safe and satisfactory manner. Remedial works, on the other hand, should be considered as non-routine work necessitated by a specific event or failure of one or more elements of the embankment or reservoir. The distinction is clearly not straightforward and defects needing maintenance work which if deferred or dealt with in an inappropriate manner may make conditions become worse or more extensive. These may then necessitate more comprehensive remedial work. Most defects can be readily assessed into one or other category and the appropriate course of action selected.

Box 12.1 Problem identification – maintenance works

Embankment and reservoir perimeter
- visual inspection of reservoir, embankment and adjacent areas
- minor works to
    - reseeded bare patches on grass slopes
    - reinstate surface erosion
    - reinstate soft or hard slope protection
    - control minor surface deformation, desiccation or cracking
    - control/infill localised animal burrows
    - repair of localised damage to geotextiles and geomembranes
    - maintain constant control level of auxiliary overflow.

Structures and pipework
- operation of all moving parts, at least quarterly
- visual inspection of all pipework and structural elements
- removal of silt or debris adjacent to overflow and other structures
- lubrication of penstocks and valves and other moving parts (including pump maintenance)
- painting/surface treatment of exposed metal and timber elements
- minor works to repair superficial damage to concrete and other structures
- replacement of minor structural elements
- emptying of silt traps
- emptying of manholes and drain inlet/outlets
- check drains for free flow and rod as required.

Vegetation
- grass cutting and weed control
- removal of intrusive, damaged or inappropriate vegetation
- check stakes, tier and guards to vegetation
- aquatic vegetation control
- additional/replacement planting.

Miscellaneous works
- check access both on and off site
- check and lubricate all moving parts (e.g. hinges, gates, locks etc.)
- check availability/location of keys
- check fences and signs
- check condition of footpaths, access tracks
- minor works necessary as a result of vandalism.

12.4 PROBLEM MONITORING

The onset of a problem may develop slowly or may be rapid. The problem and the reason for its occurrence should be ascertained at the earliest opportunity and the cause modified, alleviated or removed as appropriate. The need for remedial work must be identified and the appropriate action and speed of response ascertained.

> **Box 12.2** Problem identification – remedial works
>
> **Embankment and reservoir perimeter**
> - works required to
>   - control erosion or improve existing protection measures
>   - control seepage or leakage by the installation of drainage or other measures to control or limit flow
>   - control developing spring flow or high groundwater levels by the installation of drainage
>   - control actual or potential instability by drainage, slope adjustment, addition of toe weight or other works
> - restore or increase freeboard
> - replacement or installation of geomembranes and geotextiles.
>
> **Structures and pipework**
> - repair of impact, spalling or severe corrosive damage to structures
> - reconstruction of damaged or distressed structures or elements
> - replacement of damaged pipework or other significant elements.
>
> **Vegetation**
> - tree and stump removal
> - work following toppling of trees.
>
> **Miscellaneous works**
> - more major works necessary as a result of vandalism
> - works resulting from changes upstream or downstream of the dam (e.g. increased flood risk from changes in the catchment).

Seepage may increase rapidly with time in particularly erodible materials or may be insidious and slowly worsen. Figure 12.1 shows a typical area of relatively long-term seepage with water-loving plants at the toe of a dam. An increase in seepage may be associated with a worsening situation, but may also reflect other influences, both internal and external to the dam. Table 12.1 summarises the actions that should be taken if seepage through or adjacent to the dam is detected or becomes worse or more extensive with time. Seepage flows should be monitored as detailed in Section 12.2. Seepage water should be examined for suspended solids and samples of seepage, and reservoir water should be taken for chemical analysis where a problem is suspected. Sample bottles can usually be obtained from the local water company and chemical analyses usually carried out there. Advice on chemical testing could be sought from the water company, although the RLA or ADAS may also be able to advise on the testing. Cloudy water may indicate erosion within the embankment and immediate advice should be sought. Changes in water colour and odour may be significant and should be monitored.

Sudden drops in reservoir level may be indicative of leakage and should be investigated immediately. If leakage is sufficient to cause a noticeable drop in the reservoir level, it will be major and so the identification and location should be straightforward.

Figure 12.1 Typical long-term seepage area at toe of a dam

Figure 12.2 Embankment instability

Figure 12.3 Surface movements at toe of dam

Figure 12.4 Instability of upstream slope following reservoir lowering

Table 12.1 Action on discovering instability or seepage

Nature of instability	Nature of seepage	Action needed	
		Non-impounding reservoir	Impounding reservoir
(a) Localised superficial movement only	(a) Localised seepage sufficient to dampen and soften soil often oil surfaces, but no visible flow	When reservoir is next emptied, investigate the cause and carry out remedial works. Meanwhile, inspect regularly for signs of further deterioration	Inspect regularly for signs of further deterioration
(b) As (a), but more extensive	(b) As (a) but more extensive	Empty reservoir at next available opportunity, but monitor the situation closely. Cease filling and consult specialist adviser	Monitor the situation weekly and consult specialist adviser. Consider opening drawoff pipe, where present, to lower reservoir level slowly and consult specialist adviser
(c) Localised deeper seated instability	(c) Localised seepage causing small amount of visible flow	Empty reservoir as soon as possible. Monitor every day to ensure leakage is not increasing. Ensure leakage is not masking or affecting any other monitoring. Cease filling and consult specialist adviser	Monitor the situation daily to ensure leakage is not increasing. Ensure leakage is not making or affecting any other monitoring. Open drawoff pipe, where present, to lower reservoir level slowly and consult specialist adviser
(d) —	(d) As (c), but more extensive		
(e) As (c), but more extensive	(e) Leakage, either local or more extensive, creating a significant flow	Empty reservoir immediately, taking advice from the RLA on disposal of water. Cease filling and consult specialist adviser	Monitor hourly. Consult specialist adviser immediately. Empty reservoir at earliest opportunity taking advice from Fire Brigade

Note: Superficial instability refers to movement of topsoil or surface layer of slope protection only.
Rate of emptying in cases (c) to (e) should be at a sufficiently slow rate to avoid creating instability of the upstream slope (Section 12.4).

Instability may be apparent from the deformation and cracking of an embankment or sloping ground. In general, cracks due to instability will tend to run along the slope, often with a change in level, and possibly gradient, across them.
Figure 12.2 shows cracking at the upper levels of a small dam, whilst Figure 12.3 indicates more substantial movement at the toe. Instability of the upstream slope can be initiated by a too rapid lowering of the reservoir and slipping of saturated fill which has been unable to drain sufficiently rapidly (Figure 12.4). Advice on the rate of lowering should be obtained from a specialist adviser. This should be defined both for normal operation, where relevant, and in the event of the discovery of a defect or problem which necessitates a lowering of the reservoir level. Regular monitoring of unstable or suspect areas may readily be achieved by the accurate installation of a straight line of pegs across the area and the use of a string line between the end pegs; these being located in stable ground well outside the problem area. Sighting along the line allows a rapid assessment of any ground movement to be made. Tilting of the pegs, provided they have not been knocked, also allows the depth of movement to be considered. Deep seated movement will tend to leave the pegs near vertical, unless substantial movement has occurred; superficial movement will tend to tilt the pegs. Measurements between nails set in pegs on either side of a crack allow the subsequent movement to be measured. If both nails are set at the same level, subsequent deformation can be measured with

a spirit level and tape. These methods can be used to monitor the embankment or potentially unstable ground downstream or around the reservoir. Table 12.1 summarises the actions that should be taken if instability is detected or becomes worse or more extensive with time.

Photographic records should be used to supplement visual observations and provide a means of monitoring changes with time.

The development of any changes in seepage or instability involving the embankment itself or areas immediately downstream should be referred to a specialist adviser, as detailed in Table 12.1. Instability upstream of the dam affecting or threatening dam structures or roads, buildings or other features around the reservoir perimeter should also be treated with caution. Elsewhere, these problems may be less critical, but any seepage or instability should be investigated carefully and remedial works put in hand.

12.5 RESERVOIR MARGIN AREAS

Poorly drained areas may indicate seepage or a raising of the groundwater level. These may be identified by any of the following:

- surface water ponding
- water emerging at the surface
- areas of excessive damage due to livestock, vehicular or human access
- ochreous (red-brown) deposits from iron staining
- vegetation indicative of poor drainage or dying off of previously established vegetation
- areas of abnormal vegetation growth
- areas where rapid snow melt occurs
- areas of localised ice accumulation
- changes in usage, e.g. cessation of burrowing by animals.

Vegetation indicative of poor drainage is summarised in Table 12.2, whilst Figure 12.5 shows an area of seepage on a reservoir perimeter and the typical growth of the associated vegetation.

The perimeter and protection measures around the reservoir should be examined for damage and signs of instability (Figure 12.6) or erosion (Figure 12.7). Necessary repairs should be included in the maintenance programme. If the damage is extensive, remedial measures involving an alternative form of bank protection may be required. Gulley erosion may occur around the reservoir perimeter where periodic inflow has not been controlled. These areas should be located and remedial measures adopted to intercept and control the water source before it reaches the reservoir perimeter.

Instability should be monitored using sight lines, as described in Section 12.4.

Table 12.2 Vegetation indicating poor drainage – (Ref 123)

Common Name	Scientific Name	Description
Common Reed	*(Phragmites australis)*	Occurs in dense stands on areas of base rich groundwater
Willow	*(Salix)*	Prefers wet land
Alder	*(Alnus)*	
Flying Bent Grass	*(Molinia caerulea)*	Indicates wet ground usually with a mat of decomposed humic material overlying the mineral soil
Meadow Sweet	*(Filipendula ulmaria)*	Indicates wet ground in hollows
Glaucous Sedge	*(Carex flacca)*	Indicates wet calcareous soil
Panicled Sedge	*(Carex paniculata)*	Often marks an issue of water from a perennial spring or seepage or perhaps a blocked drain
Common Sedge	*(Carex nigra)*	Indicates wet acid soil
Carnation Sedges	*(Carex panicea* and *Carex flacca)*	Indicates wet soil conditions. Carex panicea flourishes in acid conditions, the latter prefers base rich conditions
Oval Sedge	*(Carex ovalis)*	Likes damp meadows
Great Willow Herb	*(Epilobium hirsutum)*	Marsh, fen, swamp edges – very wet conditions
Hairy Willow Herb	*(Epilobium parviflorum)*	Often found on damp ground and at ditch edges
Bog or Marsh Willow Herb	*(Epilobium palustre)*	Wide range of wet sites
Silver Weed	*(Potentilla anserina)*	Often occurs at intermittent seepage areas
Horse Tails	*(Equisetum species)*	
Persicarias	*(Polygonum species)*	These fleshy-stemmed annuals, often with a dark blotch on the lanceolate leaves, are commonly called redshank. They are indicative of wet conditions
Water Grass	*(Agrostis stolonifera var.polustis)*	Wet land species which occur in less tractable areas of otherwise sound arable fields
Marsh Foxtail	*(Alopercurus geniculatus)*	
Tussock Grass	*(Deschampsia caespitosa)*	Likes wet conditions, when found in a sward the moisture regime probably needs improvement. Either drainage or sward management needs to be considered
Rushes	*(Juncus species)*	Often mark an issue of water from a perennial spring or seepage or perhaps a blocked drain
Soft Rush	*(Juncus effusus)*	Grows in really wet neutral to acid conditions
Common Rush	*(Juncus conglomeratus)*	Grows in somewhat drier soils
Hard Rush	*(Juncus conglomeratus)*	Likes base rich wet conditions often in calcareous areas
Sharp-Flowered Rush	*(Juncus acutiflorus)*	Occurs mostly on wet land. They spread by rhizomes and form exclusive patches in wet and acid soils
Jointed Rush	*(Juncus articulatus)*	
Creeping Buttercup	*(Ranunculus repens)*	Common, but less obvious, broad-leaved indicators of wet land
Lesser Spearwort	*(Ranunculus flammula)*	
Marsh Thistle	*(Cirsium palustre)*	
Marsh Ragwort	*(Senecio aquaticus)*	

Figure 12.5 Seepage around reservoir perimeter

12.6 EMBANKMENT AREAS

12.6.1 General

Regular visual examination of an embankment may provide an indication of the internal condition and behaviour. Evidence of seepage may be provided by damp or softened areas, springs, change in vegetation or sound (Appendix Q). This may be relatively minor, but will become potentially more serious with time, and if left to develop, may result ultimately in major leakage and/or failure of the embankment. Deformation may be indicated by cracking, slumped material, loose soil or disturbed or tilted structures (e.g. fences, signs etc.) and trees. Deformation may arise from many causes, but may be indicative of dam instability, particularly if this is associated with the onset or increase in seepage. The development of seepage and potential or actual instability must be considered as potentially serious. The comments and recommendations of Section 12.4 on problem monitoring should be followed. Further comments on deformation are included in the following sections concerning specific areas or features of an embankment.

Figure 12.6 Instability around reservoir perimeter

Animal burrows may develop in any area on or adjacent to an embankment and their positions and development should be noted. They will tend to become more extensive with time as the animals colonise a larger area, and may lead to sudden changes in the embankment behaviour and appearance as burrows collapse or interlink. In extreme cases, the burrows may allow a seepage pattern through the embankment to become established. Appropriate control measures and remedial works should be put in hand at the earliest opportunity and the burrows monitored regularly until this work has been completed. The possible problems with crayfish activity are described in Section 9.14.6.

Figure 12.7 Erosion around reservoir perimeter

Vegetation on and adjacent to the embankment should be examined regularly. The Forestry Commission has published a useful guide to the recognition of hazardous trees and the action to be taken[121] whilst Hoskins[122] gives guidance on the particular hazards to dams. Where a tree is damaged or in poor health, it should be pruned, lopped, trimmed or removed as part of the regular maintenance programme. Figure 12.8 shows a large willow on an old dam which has suffered the loss of the upper part of the trunk.

The tree appears to be regenerating, although in need of surgery, but a close inspection from the far side shows the true rotten state. Any inappropriate vegetation, particularly if woody, must be removed from areas where it could become detrimental to the satisfactory operation of the dam. These include:

- on the upstream slope where it could affect lining or slope protection works
- on the crest affecting similar works or access
- on the downstream slope at the position of the auxiliary overflow
- other positions which may be unsuitable as a result of the reservoir usage
- adjacent to drains and ditches
- adjacent to structures
- locations where it precludes access.

Figure 12.8 Unhealthy vegetation

Figure 12.9 shows a young sycamore sapling becoming established immediately adjacent to a main overflow inlet. If left in place, such trees will continue to grow and ultimately cause damage to the structure as shown further down the same channel in Figure 12.10.

Bare patches arising from grass cover loss, which could be subsequently eroded must be identified (Figure 12.11). An unsurfaced crest is particularly vulnerable to erosion and Figure 12.12 shows the crest of an embankment in a public park where severe erosion has occurred, resulting in the loss of freeboard of the fill and in seepage through the upstream brick wall.

12.6.2 Embankment crest and upstream shoulder

The upstream face and crest should be examined for cracks and deformation. These may develop following a drawdown of the reservoir level and will tend to result in longitudinal cracking of the fill. In more extreme cases, rapid large scale movements of the upstream slope may occur. Cracks may also develop in unsurfaced crests or crest tracks. Any cracking which extends continuous or intermittently for more than a few hundred millimetres should be investigated carefully and an explanation sought. Cracking may also occur as a result of settlement of the fill or sub-base under surfaced crest tracks, but such settlements should not occur after the reservoir has been completed and operational for some years. Cracking of cohesive fill may also occur as a result of desiccation and shrinkage of the surface layers of fill. The cracks can allow water or rainfall to enter rapidly into the surface layers which may promote surface deformation.

Variation in the water line along the upstream face may be indicative of deformation and possible instability, but could be the consequence of poor control during construction which resulted in the creation of an undulating profile.

The slope protection should be inspected and any damage or deterioration noted. Figure 12.13 shows an area of poorly engineered and constructed slope protection at the waterline immediately adjacent to a dam where the lack of vegetation and water flow over a poorly prepared surface has led to erosion and loss of support for the geotextile. The area where an embankment runs into natural ground is particularly prone to erosion damage and should be monitored regularly. The loss or poor health of vegetation which functions as slope protection should also be noted. Well established and healthy vegetation, as shown in Figures 9.4, 9.35 and 9.36, will control erosion where wave action is not significant. More substantial wave attack may be controlled by the roots of larger trees and shrubs, but these may not be sufficient to control erosion adequately, as shown in Figures 12.14 and 12.15. Where exposure conditions are severe and protection other than grass is absent, rapid erosion can result. Figure 12.16 shows the effect of unchecked erosion, where much of the crest width has been lost and only a minimal width of embankment remains. The few reeds which have become established upstream, in the foreground of the figure, provide no resistance to wave action.

Figure 12.9 Tree becoming established adjacent to overflow

The condition and adequacy of the slope protection lining and the protection works to the liner should be examined regularly. The deterioration of these features may lead to severe erosion problems within the dam, and ultimately failure, or loss of water from the reservoir. Liners are prone to damage from a number of causes, as detailed in Section 13.4 and prompt action is required. A damaged unprotected liner is shown in Figure 12.17.

12.6.3 Downstream shoulder and embankment toe

These areas should be examined for indications of seepage, and their positions marked on the site plan used to record the visual observations. Measures are likely to be required to control seepage or surface water flows and to direct these to a suitable watercourse.

Figure 12.10 Damage to channel by trees

Erosion or deterioration of the downstream face may arise from overtopping, weather effects, livestock, burrowing animals, pedestrians, vehicles or vandalism. Features on the face may tend to concentrate surface water flow and lead to localised problems. Geotextiles beneath the topsoil should be checked to ensure they have not become exposed or damaged.

Instability and deformation may be evident which may be shallow or more deep-seated. Specialist advice should be sought unless the cause can be ascertained without doubt. Instability may be initiated by or be associated with the development of seepage.

Drains should be checked to ensure that they function properly and flows should be estimated and noted. Drain flows may be measured using a simple measuring structure, such as a vee notch or rectangular weir. Details for the installation and use of measuring weirs are given in Appendix F. Increased flows may be a result of leakage from the reservoir, local rainfall, flow from an external source (for example a burst water main) or a change in the embankment behaviour. Changes in flow or cloudiness may be indicative of hidden damage as detailed in Section 12.3. A poorly engineered and constructed drain is shown in Figure 12.18; the lack of a headwall and uncontrolled surface flow have resulted in substantial erosion at the drain outlet. The steep-sided ditch has also suffered from instability and erosive flow.

Figure 12.11 Slope erosion

12.7 INLET, OVERFLOW, DRAWOFF AND ANCILLARY STRUCTURES

All reservoir structures that affect the safe operation of a reservoir should be examined. Any signs of deterioration of materials, complete or partial blockage by debris and loss or damage should be noted. Obstruction in channels will result in increased water levels and in a reduction of carrying capacity, as shown in Figure 12.19. Debris at any overflow, especially if trapped against a screen, can cause a substantial increase in water level, as shown in Figure 7.17. Immediate clearing, repair and replacement should be carried out as part of the maintenance programme. Major clearing, repair and replacement may be implemented as remedial measures. All elements of the structure, including concrete and masonry, as well as metalwork and timberwork, should be examined for damage or deterioration. Any materials that have deteriorated should be replaced or repaired.

Damage to or distress of structures may result in deformation, tilting or cracking. Figure 12.20 shows the settlement and break-up of an overflow slab as a result of erosion of the underlying ground. Such events should be examined carefully and photographs taken periodically of cracks or any other signs of movement (with a scale for relative comparison). Where major deformation is observed, specialist advice should be sought to assess the need for suitable remedial measures. All formed slopes adjacent to structures should be examined for instability and erosion. If these aspects are a problem, remedial measures should be implemented to increase stability and provide protection against erosion.

The sill or weir level of overflow structures should be monitored to see that no undue settlement or tilting has occurred; variation in the depth of outflow will indicate whether this is happening. If this becomes excessive or the sill construction itself has deteriorated or is damaged, the required level should be reinstated. This is particularly necessary for auxiliary overflows because their extended length and very infrequent use can allow settlement or deformation to pass undetected. This can affect their safe functioning at a future date.

The areas adjacent to overflow structures should be monitored for erosion of the channel or embankment. Areas susceptible to debris accumulation should be examined and the positions marked on the site plan. These may be indicative of settlement of the structure or shortcomings in the design of the works.

Pipework may suffer damage or fracture due to deterioration, vandalism or accident fracture. A check on pipework should include any pipework in chambers, which may not be readily visible, and the condition of any fittings (i.e. valves etc.), manholes, headwalls or other structures on the pipeline. Care should be taken where entry into confined spaces is required and appropriate precautions should be taken.

Where the pipework meets, passes through or is on the embankment, the local area should be monitored for possible evidence of leakage if differential movement or settlement is suspected.

Figure 12.12 Severe crest erosion

Figure 12.13 Damage to geotextile slope protection

Figure 12.14 Erosion to upstream slope between roots

Figure 12.15 Erosion to upstream slope undercutting vegetation

Figure 12.16 Severe upstream erosion

Figure 12.17 Damage to exposed liner

Figure 12.18 Erosion damage to drain outlet

Figure 12.19 Blockage of channel restricting flow

Figure 12.20 Settlement and distress of overflow structure resulting from erosion beneath

12.8 DOWNSTREAM AREAS

12.8.1 General

A reservoir may influence downstream areas in a number of ways. Leakage or raised groundwater levels may result in flowing or standing water in poorly drained areas. Alternatively, the construction of the dam may lead to a reduction in subsurface flow and reduced groundwater levels. This will cause a decrease of water-loving vegetation species, localised deposition of material and/or the deterioration of the watercourse.

Discharge from the reservoir due to large flows or poor design may cause scour of the downstream channel. Realignment of watercourses may lead to localised erosion until steady conditions become established. The release of silt from low-level drawoff works may deposit material in the downstream channel and restrict flow in the watercourse.

Erosion may also occur in areas downstream of the auxiliary overflow. These areas may be associated with localised instability or deformation. It is essential that these are identified and reinstated.

12.8.2 Seepage

Surface features indicating seepage have been described in Section 12.4. The positions of these features should be marked on the site plan, together with descriptive notes and photographs. The appearance of springs adjacent to the dam during or after filling the reservoir may signify a serious risk of imminent failure of the embankment, particularly if the flow is cloudy; a specialist adviser should be contacted immediately. Springs may merely be indicative, however, of groundwater changes. Flows may be monitored by installing a simple measuring device as described in Appendix F. Any seepage that is present should be controlled and drainage measures should be installed in the downstream area following the guidance provided in Sections 9 and 10.

12.8.3 Instability

Surface features indicating instability have been described in Section 8. The position of these features should be marked on the site plan, together with descriptive notes and photographs. The appearance of a slip adjacent to the dam either during or after filling the reservoir may signify a serious risk to the embankment; a specialist adviser should be contacted immediately. Instability should be monitored using sight lines, as described in Section 12.4.

12.9 OTHER ASPECTS

12.9.1 Public safety

It is necessary to consider the implications of the surveillance monitoring on safety. Specific requirements to be checked include ensuring that all safety signs and safety equipment (for example life preservers, life lines) are provided and are in a satisfactory working condition. Specialist advice should be taken to clarify those requirements which depend on the reservoir use. Spare signs and equipment must always be readily available as replacements. All areas of restricted access, such as control structures or confined, should be checked to see that they remain secure. Protective fencing, gates and locking systems should be checked and damage or deterioration noted and repaired.

12.9.2 New development

When new development is planned or carried out adjacent to or in close proximity to a reservoir, particularly downstream, an assessment of the potential implications on the reservoir should be made as discussed in Section 7. If there is any doubt about the implications, particularly where changes to the catchment may affect the flood situation, specialist advice should be sought.

12.9.3 Appraisal of visual observations and monitoring records

The visual observations and monitoring should be assessed immediately after the information is obtained. Where appropriate, flows, deformations, etc., should be plotted on graph paper and trends identified. Reservoir level, rainfall and any other relevant data should also be added to the plot. A considerable saving can be made if observations and records are clear, concise and accurate. The information should be made available for the annual review by the specialist adviser as discussed in Section 12.1.

Where seepage or instability has been identified, the problem monitoring procedure detailed in Section 12.4 should be followed and specialist advice sought as detailed in Table 12.1

13 Maintenance works

13.1 INTRODUCTION

Maintenance is an essential requirement for every reservoir scheme, regardless of size. Maintenance may require regular clearing, cleaning or operation of structures and replacement of deteriorated or defective materials and equipment as required. More substantial deterioration or damage, or correction of a more fundamental problem will necessitate remedial works, as detailed in Section 14.

The main causes of deterioration are:

- use
- ageing
- long-term changes within the embankment
- access and use by humans and livestock
- plant and vehicle access and use
- weather
- uncontrolled vegetation
- burrowing animals and pests
- changes in external factors.

The maintenance requirements should be considered at an early stage so that the cost implications in terms of plant, labour and materials can be assessed adequately. Pollution of the reservoir has not been included within the scope of maintenance works, but is discussed in Section 14.

13.2 EMBANKMENT AREAS

13.2.1 General

Earth embankments and their foundations are subject to settlement and deformation which may continue for an extended period, possibly measured in years, after their completion. They are also susceptible to the influences and effects of the reservoir and the groundwater and will be liable to deterioration from internal and surface flow. These may cause erosion of the fill or material within and beneath the dam or long-term chemical changes from the internal water movements. Instability may result from the development of high water pressures within the dam as a result of the construction methods, settlement and deformation, inappropriate usage or subsequent works (e.g. excavation at or into the toe of the downstream shoulder). Uncontrolled growth of vegetation or the development of unsuitable species on or near to the embankment may also cause problems.

13.2.2 Internal erosion and deterioration

Maintenance options arising from internal erosion and deterioration are limited; normally more major remedial works are required.

13.2.3 Surface erosion

Bare areas on the embankment should be reseeded using a grass seed mix compatible with that used for the original seeding. Where regular access across the embankment leads to worn areas, the provision of a footpath or an access track should be considered. Vehicle access should be minimised or prohibited unless a suitable surface has been provided.

Erosion on the upstream slope may require an enhanced or alternative form of protection if the existing protection has proved to be inadequate or the soft protection has not developed as expected. In certain minor cases, however, simple localised protective works using sandbags filled with 1:8 cement:sand may be utilised. Figure 13.1 shows this form of protection provided around a large tree on an old dam which had formerly been subjected to increased wave attack and erosion. Pockets can be left in such works to allow suitable vegetation to establish itself to improve the appearance. Unless the extent of the works is limited, however, such work should normally be considered as remedial work within the scope of Section 14.

Figure 13.1 Sandbag protection to upstream slope

Where vegetation on the upstream slope provides a protective function, this must be maintained in a healthy and stable condition. Work may be needed to prune or lop the vegetation or remove unsuitable species or growth in undesirable locations as described in Section 13.10. In many instances, it is desirable to limit growth to less than 2 m, although isolated larger shrubs or trees in appropriate locations and of a suitable type may be allowed to develop subject to the comments in Section 9.2.8.

13.2.4 Surface deformations and shallow instability

Shallow instability may develop from a number of causes. The reasons for the onset of movement should be ascertained prior to carrying out any work. In many instances, remedial measures will be needed, but where the deformation can be traced to a minor problem or change in conditions, limited maintenance works may be sufficient to deal with the situation. These may include areas where minor cracks develop, animal burrows are formed or toppled vegetation creates a hole. Such areas should be filled immediately with suitable fill and/or topsoil, depending on their depth and extent. Any settlement at the crest of the embankment which exceeds that allowed for at the design stage should be reinstated immediately with similar material. The embankment should be examined carefully after a severe storm and any eroded or settled areas should be reinstated. The slopes and crest of a clay embankment may be subject to excessive cracking during prolonged drought conditions, necessitating maintenance works to deal with these effects.

13.3 RESERVOIR PERIMETER

Any protection measures that have been constructed must be maintained and materials replaced as required. Where the existing protection has proved to be inadequate, more extensive works will be required and these are considered further in Section 14.6. Vegetation adopted as a bank protection measure should be managed, as outlined in Section 13.10. Footpaths, and their immediate surroundings, should be adequately maintained and granular material should be available for reinstating damaged areas as required.

13.4 GEOMEMBRANES AND GEOTEXTILES

13.4.1 Geomembranes

Exposed geomembranes provide an irresistible target for vandals, and may require regular maintenance unless public access is prohibited or controlled strictly. Maintenance may involve repairs to punctures or torn welds if these are of limited extent and any loss of the protective layer should be replaced. Puncturing of geomembranes by livestock or pests may result in excessive maintenance or remedial works. Accidental damage may also occur from careless use of plant, tools and equipment. These risks will be minimised if the guidance given in Appendix L is followed. Advice and recommendations for the repair of geomembranes should be obtained from the supplier at the installation or pre-contract stage.

13.4.2 Geotextiles

Exposed geotextiles have also been prone to vandalism and liable to damage by plant, vehicles, equipment, foot traffic, pests or livestock, unless the particular recommendations are adopted. For minor repairs to geotextiles, a minimum 300 mm overlap between adjacent sheets should be adopted. When repairs or replacement of geotextiles are carried out, it is important to ensure that the same, or similar, grade of geotextile is used as differing grades may behave differently in the ground. Advice and recommendations for the repair of geotextiles should be obtained from the supplier at the installation or pre-contract stage.

13.5 INLET STRUCTURES

Silt traps need to be cleaned out regularly to maintain their efficiency. The type and size of the silt trap will dictate the method of emptying. In general, backactors are used for the more accessible excavated silt traps and sludge gulpers for more restricted concrete 'box' type structures.

Where silt traps are not incorporated, problems may occur due to silt accumulation. Regular desilting may be required if the quantities of silt entering the reservoir are high.

13.6 DRAWOFF WORKS

All pipework should be checked regularly to ensure it is clear of debris. The regular operation and lubrication of all control elements, such as penstocks and valves, is essential to ensure that they will function as required when needed. Generally, all control elements should be operated at least six-monthly and lubricated annually.

13.7 OVERFLOW WORKS

The effectiveness of the main and auxiliary overflows depends upon proper, regular maintenance. Debris must not be allowed to accumulate. Grass auxiliary overflows warrant particular care and attention and should be kept clear of unwanted vegetation, other than grass, and any damage repaired without delay. Erosion can be minimised by maintaining a good grass cover. Grass should be cut at regular intervals on the downstream slope beyond the auxiliary overflow to maintain the grass at the required length. Care should be taken during cutting to ensure that the ground surface is not rutted nor the grass torn or damaged. Access by livestock and pedestrian traffic should be limited and the risk of vandalism or unauthorised traffic minimised on the embankment. Where a geotextile is present, particular care is needed to avoid damaging or exposing the material during maintenance works, including grass cutting.

Control weirs may contain timber, plastic, metal or concrete stoplogs to regulate the flow. These should be repaired regularly and replaced when necessary. The control level of an auxiliary overflow should be checked regularly to ensure that it remains level and no low areas develop.

Under no circumstances must the sill level of any overflow structure be raised, even temporarily, by introducing timbers or blocks to raise the overflow level. This practice reduces freeboard and will result in a much reduced overflow capacity. The possibility of overtopping is then enhanced substantially with the risk of embankment erosion and breaching.

13.8 ANCILLARY STRUCTURES AND DRAINAGE

Bywash channels should be regularly cleared and any unstable slopes battered back to a stable slope angle. Fish passes must be regularly cleared of debris, particularly at those times of year when fish will be passing through the reservoir.

The routes of toe and other drainage should be checked and intrusive vegetation removed. Any settlement or loss of granular material should be made good, but the reasons for the settlement or loss should be investigated and appropriate remedial action taken if necessary (i.e. settlement adjacent to the drain may indicate loss of ground necessitating local reconstruction of the affected length of drain). Where pipes are present in drains, these should be checked to ensure that they are free-flowing; this may require the drains to be rodded or a more substantial flow of water to be pumped through to clear debris. Manholes and headwalls should be cleaned out as necessary.

13.9 BURROWING ANIMALS AND PEST CONTROL

Earth embankments are susceptible to damage from burrowing animals, as detailed in Section 9.14.16. They may also affect vegetation on the embankment and be detrimental to the use of the reservoir. If a problem arises from excessive damage by burrowing animals or other pest, advice should be sought from the local ADAS adviser on pest control. He will be able to advise on the appropriate measures to control, remove or exterminate the problem.

It is important to be aware that certain animals are protected by law, (see Section 9.14.6), and this must be clarified with MAFF before any pest control measures are implemented.

After burrowing animals have been removed or exterminated, the burrows should be backfilled with similar material to the rest of the embankment. The burrows should be cleared out, where possible, and filled with thoroughly rammed and well compacted material. The possible design measures discussed in

Section 9.14.16 may need to be implemented to limit the burrowing and damage. Close monitoring after backfilling has taken place should be undertaken to ensure that the problem has been eradicated.

13.10 VEGETATION MANAGEMENT

13.10.1 General

The vegetation around a small reservoir must be managed to provide continuing value to wildlife and visual amenity. If it is not, natural succession occurs whereby the open water reverts to marshy and, ultimately, to dry ground. No reservoir habitat creation scheme should be designed without considering the long-term maintenance requirements and preparing an outline site management plan that identifies the major tasks and objectives of management. Information on the practical aspects are contained within the guides issued by the British Trust for Conservation Volunteers, namely:

- Waterways and Wetlands[124]
- Hedging[125]
- Woodlands[126].

Further advice is also available from ADAS and the Forestry Commission (References 127 and 128).

Vegetation management should also attempt to maximise the establishment and growth of the vegetation at and around the reservoir. The main reasons for vegetation, particularly trees and shrubs, failing to become established or thrive is often as a result of:

- compaction of the ground during construction and landscaping operations, leading to dense ground and associated waterlogging
- drying out of roots before planting
- trying to plant trees and shrubs of excessive size
- inadequate or inappropriate staking
- lack of protection against mechanical damage (e.g. mowers) or grazing wild or domestic animals
- lack of control of weeds around the tree or shrub.

Vegetation management on the embankment is discussed in Sections 13.2.3 and 13.10.4.

13.10.2 Aquatic vegetation

The aim of aquatic vegetation management should be to maintain the vegetation zonation described in Section 4.4 and prevent aggressive species becoming dominant with the loss of less competitive plants and species diversity. Emergent plants that are likely to require particular attention are reedmace and common reed although canadian pondweed and hornwort can cause large-scale problems over open water. reservoirs which are overgrown with vegetation provide poor conditions for fishing and wildfowl, and are also visually less attractive. Unwanted vegetation can be cut and removed, either manually or mechanically, although the spoil should be left to drain so that aquatic invertebrates can return to the water. Herbicides can be used to control aquatic vegetation; they should not prove too detrimental to nature conservation interests, provided they are species specific and used correctly. Only those chemicals approved for use in or near water should be used as detailed in the MAFF 'Guidelines for the use of herbicides on weeds in or near watercourses and lakes'[129] and, if there is a chance the chemical may enter a watercourse, consent must be obtained from the RLA.

13.10.3 Grassland

The ecological value of grassland is best maintained by grazing or cutting, although livestock should be controlled to avoid trampling and fouling of the water. Sheep are preferred to cattle because they are less harmful to vegetation and the often soft ground adjacent to the reservoir. Embankment slopes should never be grazed by animals larger than sheep as the relatively steep sloping ground is particularly prone to deterioration, surface abrasion and the formation of hoof marks and holes.

A wildflower sward should be cut at least three times in the year following establishment, with the cuttings removed to reduce nutrient levels. Subsequent management depends on site conditions and the effect required, one cut in late summer for example, would produce an open sward suitable for wildflowers, provided the cuttings were removed. Where it is desired to maintain tall tussocky vegetation for nesting wildfowl, management will only be necessary every two to four years to prevent scrub encroachment.

Where regular grass cutting is to be carried out on an embankment or reservoir perimeter, hand-held equipment should be used on slopes of up to 1V:2.5H. Flatter grassed slopes below approximately 1V:4H may be cut using tractor-towed equipment.

13.10.4 Scrub and trees

Newly planted shrubs and trees require careful maintenance to ensure successful establishment. Regular weed control must be carried out to avoid competition and they must be protected adequately from grazers and rabbits. Once fully established, scrub maintenance work should involve mainly cutting back to encourage different stages of growth and improve structural habitat diversity. A rotational system of management is best so that there is always some mature cover available. Trees can be lightly pruned four or five years after establishment to encourage a leading, growing shoot. Once they mature, management should focus on preventing their growth, shading the water surface too much and ensuring the continuing health of the tree. Trees which become too large, out of balance or unhealthy, will require selective lopping and pruning, and may ultimately require removal and replacement if they die or become too unhealthy or mis-shaped.

Trees on the reservoir perimeter, especially on the southern side, may need to be cut back to avoid shading or limit the amount of leaf debris entering the water. Saplings growing in unwanted locations on and near the embankment or near structures (Section 9.2.8), should be removed at an early stage before they have become fully established and a significant problem.

Maintenance work to scrub should involve mainly cutting back by pollarding, pruning or coppicing (cutting to ground level) the shrubs to encourage different stages of growth and improve structural habitat diversity. A rotational system of cutting is best so that there is always some taller mature cover available.

Vegetation and particularly trees, will not always continue to perform their necessary function and may develop various weaknesses. Dealing with hazardous trees is an important aspect of maintenance works. The Forestry Commission guide[119] describes the hazards and recommended courses of actions. In certain instances, mature trees may have reached the end of their useful life and will require felling (Section 14.5). It should be noted that trees damaged or weakened by natural or man-made agents do not die off immediately. They will tend to linger on, with little new growth and be prone to increased pest and disease attack. Thus they will gradually weaken and die, typically four of five years after the real cause of their decline.

Disease is often a secondary effect following physical damage, extreme climatic conditions, insect or animal attack, pollution or general poor health. Older trees and clumps of single species are more prone to attack, but most diseases and pests attack a specific species, e.g. Dutch Elm Disease. All trees have a natural span which may range from a few decades for birches to several centuries for the oaks; ultimately all trees must die. Toppling is associated with trees of all ages, but an older tree is more likely to be affected as are trees left adjacent to felled areas which may be subject to substantially changed wind loadings.

13.10.5 Timing of management work

Aquatic vegetation management also needs to be carefully planned. Winter management work will reduce the amount of food available to wildfowl at a time when food supplies are already depleted, and is likely to disturb dormant animals such as frogs. Similar work in spring or summer could disturb the breeding and development of amphibians or invertebrates.

The timing of vegetation management is a crucial factor to consider if a reservoir is being managed for wildlife benefit. Wildflower grassland for example should not be cut during the flowering season and any undue disturbance should be avoided during the bird nesting season.

Ill-timed or inappropriate trimming or pruning of hedges and trees may be detrimental to nesting birds and small mammals. Consequently, the timing of any vegetation management should be planned to avoid disturbance at important times of the year, an aspect which is also discussed in the practical guides (References 124 to 126).

Regular grass cutting on the embankment should be carried out and this should facilitate rapid inspection although the requirements during the colder months are drastically reduced as the growth rate slows down.

13.11 SAFETY

The regular maintenance of all safety signs and life saving equipment is essential. Spare signs and life saving equipment should be immediately available if damaged, lost or stolen. During desilting works the reservoir level may be required to be drawn down and any areas of silt must be fenced off from the public, with warning signs erected for the duration of the works.

Stockproof fencing and fencing to prevent public access should be checked regularly and repairs carried out immediately. Locks and gates should also be checked and lubricated and the availability of keys checked to ensure that access is readily available, if required, in an emergency.

Where a reservoir is used for fire fighting purposes regular maintenance must be carried out to ensure the required access to the reservoir is not restricted in any way. Liaison with the local fire prevention officer should ensure that any additional requirements can be incorporated as part of the regular maintenance programme.

14 Remedial works

14.1 INTRODUCTION

Regular visual observations and maintenance works are intended to ensure that the reservoir operates as required and to identify problems that may warrant remedial works. These works must have no adverse effects on the safety or stability of the reservoir or the surrounding areas. In many cases, other than very localised problems, specialist advice should be sought before work is carried out to minimise effects on adjacent areas or to avoid adopting an inappropriate approach.

The need for remedial works must be identified and the appropriate action and speed of response ascertained. Section 12 gives guidance on the appropriate response to problems which may develop and the necessary monitoring and action to be taken.

Specialist advice must be sought on the best approach to carrying out remedial works which involves the following:

- loss of freeboard, whether temporary or permanent
- full or partial breaching of the dam, including dismantling
- major works on the main overflow
- works necessitated by internal erosion or deterioration
- works necessitated by seepage or instability (Table 12.1)
- excavation or regrading works in or just beyond the downstream toe
- work on pipelines beneath the dam
- works around the reservoir which could affect adjacent roads, structures or slopes.

Where remedial works are carried out at a place of work, as discussed in Sections 3.1 and 3.6.2, there is likely to be a Health and Safety Executive involvement.

14.2 EMBANKMENT AREAS

14.2.1 General

Remedial works to embankments involving earthworks are usually carried out to reduce and control leakage, improve stability or provide additional protection to embankment slopes and the reservoir perimeter.

14.2.2 Leakage reduction and control

Measures may be necessary to reduce or control leakage beneath, around or through the embankment. Leakage may be reduced by locally increasing the impermeability of the upstream end of the seepage path by placing cohesive fill as a blanket or in a trench or by installation of a lining, either as a geomembrane, or more rarely by using bentonite. This approach is not straightforward unless the treatment extends over relatively large areas or a specific feature is located and dealt with (e.g. a pipe left in situ). Specialist advice is often needed to provide a safe economic solution. Under no circumstance should measures be attempted to stop leakage by placing cohesive fill at the point of issue. This will not stop the flow but merely cause the leakage to break through elsewhere, possibly resulting in enhanced difficulties. Alternatively, the flows may lift off and soften the placed material, resulting in localised instability and/or deteriorated fill.

In all instances when leakage is identified, the reservoir water level should be lowered to well below the level of the leakage and maintained at that level until the remedial works have been completed. Monitoring of leakages should be carried out to assess the effectiveness of the repairs as discussed in Section 12.2. It is important to monitor the reservoir level at the same time as leakage. If leakage continues, alternative remedial measures should be considered and specialist advice sought.

No dam is fully watertight, but the detrimental effects of the leakage may be reduced by adequate leakage control. This may be achieved by the installation of drainage, as described in Section 9.3.7. Flows from remedial drainage, together with reservoir level and rainfall, should be monitored to establish whether leakage is increasing or decreasing with time.

14.2.3 Internal erosion

The onset of leakage should be regarded as a possible indication of internal erosion, especially if this is accompanied by cloudy or discoloured flow. Leakage flow will tend to be constant at a given water level, whereas internal erosive flow will tend to increase with time at a given water level. If left unattended, the flow will continue to increase and ultimately lead to a washing out of the embankment. The process may be very slow, but in dams with fine sand in the fill or foundation, the process may be very rapid.

Any works to deal with internal erosion require specialist advice at the earliest opportunity; inexperienced attempts to control or stop the flow are likely to worsen the situation. Depending on the quantity and position of the flow, suitable drainage measures and minor works may be applicable, but in many instances, more extensive remedial works, possibly including partial breaching of the dam, may be required. Such work should only be carried out by an experienced contractor in conjunction with specialist advice.

14.2.4 Stability improvement

The slope stability of an embankment or a natural or excavated slope in close proximity to a reservoir can be improved by the following methods.

- Regrade the slope to a flatter gradient.

    Regrading a slope increases the resistance to sliding and may reduce the forces causing instability. This is particularly appropriate if earthmoving plant is readily available. Care must be taken to ensure that the regrading work itself does not affect the stability of a wider area. Where existing mature vegetation would be affected by the regrading work, the loss of the reinforcing effects of the roots should be considered before work is carried out.

- Provide weight to the toe of a slope using fill materials.

    Earthfill and rockfill is commonly used to provide a stabilising weight over the toe and lower portion of an unstable slope to increase the restraint. Similar comments as above apply regarding the stability of a wider area and the loss of vegetation.

- Reduce water pressures in the slope by installing drainage.

    The strength of the soil within an unstable slope can be increased by installing a drainage system to decrease the water pressures within the soil. Adequate discharge facilities into a watercourse must be provided to deal with the drainage flows.

Earthworks involving the removal of material or excavation in and adjacent to the downstream shoulder should only be carried out after specialist advice. Whatever remedy is adopted, care must be taken to preserve the safe functioning of other features of the embankment. In particular, drains or natural drainage channels must not be obstructed and safe loading on structures or pipes must not be exceeded.

14.2.5 Slope protection

Small scale repairs can be made to upstream slope protection as part of a regular maintenance programme, but more extensive damage or deterioration will require remedial action. Where inadequate protection of the slope is evident it may be necessary to adopt more suitable alternative protection measures. These are described in Sections 9 and 10. During the work on the slope protection, the reservoir level must be held down at a sufficiently low level by means of the low-level outlet or pumping to allow the works to be completed safely.

Where erosion of the crest or downstream face occurs due to overtopping, the cause of the overtopping must be considered immediately and specialist advice sought. This may require the reservoir level to be lowered to minimise the problem and reduce the likelihood of further overtopping. Raising the embankment to increase the freeboard is another option, although this must be considered fully as the crest width will be reduced and/or the slopes steepened. Rarely, further overtopping may be accommodated by the provision of suitable protection.

14.2.6 Surface erosion

Surface runoff from the embankment or the abutments immediately downstream of the dam may cause surface erosion and gulleys: the junction of the fill and natural ground is particularly vulnerable to these effects. A collecting drain should be installed to intercept this drainage and lead the flow into the watercourse. Figure 14.1 shows a typical geotextile wrapped drain installed to control seepage adjacent to one end of a small dam. The upper surface of the geotextile has been protected against damage by a covering of the permeable fill. Design and construction details of drainage measures are given in Sections 9 and 10.

Figure 14.1 Geotextile wrapped drain to control seepage flows

14.3 INLET, DRAWOFF, OVERFLOW AND ANCILLARY STRUCTURES

Structures may deteriorate or be subject to damage beyond the scope of routine maintenance. These may include:

- impact damage (floating debris, falling trees, vandalism, boats, etc.)
- concrete spalling (weathering of poor concrete, rusting of reinforcement)
- severe corrosion (pipes, valves, handrailing, etc.)
- cracking and deterioration of pipes (settlement, ground instability).

The reservoir must be drawn down to allow access to the structure. Measures will be needed to control inflow unless a low-level outlet is present. Work to the main overflow is particularly hazardous if this involves partial demolition and reconstruction; the timing and method of carrying out such works must be fully considered and specialist advice sought.

Major changes to the catchment or area downstream of the reservoir may affect the size of the design flood and necessitate an enlarged overflow. Experience in operating the reservoir may also indicate that the existing overflow capacity is inadequate and improved or additional overflow facilities are needed. This aspect is not straightforward and specialist advice must be sought.

14.4 GEOMEMBRANES AND GEOTEXTILES

14.4.1 Geomembranes

Areas of extensive damage to a geomembrane used either as an upstream liner on an embankment or as a reservoir liner should be repaired as soon as they are identified. The reservoir water level should be lowered below the level of the damaged area whilst repair work is undertaken. Care should be taken not to lower the reservoir too quickly to prevent bank instability as a result of rapid drawdown as described in Section 9.3.

Recommendations for the repair of geomembranes are described in Appendix L.

14.4.2 Geotextiles

Damage to geotextiles is not likely to be extensive if they are well installed and maintained, as described in Sections 10.5.6 and 13.4.2. Extensive remedial works are only likely to become necessary as a result of external damage or if an unsuitable geotextile has been used. Complete replacement is often required and the existing geotextile should be removed and a replacement installed, as described in Section 10.5.6.

Geotextiles in drains may suffer hidden damage or deterioration which may become evident from the behaviour of the flow. Changes in flow or colour may indicate damage or loss of adjacent ground or fill and should be investigated immediately. Alternatively, malfunctioning of a drain may be indicated by surface settlement of the adjacent ground. Excavation of the drain and replacement with more suitable materials should normally be carried out to avoid the risk of long-term problems.

14.5 VEGETATION

Where tree removal or works on a well wooded embankment dam are planned, it is essential that the effects of tree loss are fully assessed. The timing of the works must also be considered with respect to the time of year and the vegetation condition and the type and extent of vegetation to be removed must be known. The potential effects of removal can only then be assessed. This must consider any changes in stability and the effect on other vegetation and trees and the likely consequence of increased runoff. The felling of trees is best carried out by an experienced person who can control the operation and fell a tree into the intended location. Any other vegetation damaged during the felling should be cut back and all felled or cut vegetation removed. Where appropriate, replacement planting should be carried out.

Failure to fell mature trees when they cease to perform their intended function or become a hazard may be as serious as ill-considered removal. It should be noted that a felling licence from the Forestry Commission is required and certain trees may be protected by a tree preservation order or a blanket protected woodland order. Where weaknesses are noted and felling is required, it is essential that the major roots adjacent to the trunk are also removed and the hole backfilled with compacted material of a similar type to the rest of the embankment.

Any tree which has toppled, heaving up a large disc of earth and larger roots, should be removed and the hole inspected. Loosening is possible down to two or three metres, depending on the species, but the hole typically may be less than one metre deep. Remote from the crest and upper levels of the embankment, the hole can be refilled with compacted material. Adjacent to the crest, all disturbed material should be removed to sound fill and the hole refilled; this may require temporary protective measures against inflow from the reservoir.

Where excavation adjacent to trees is planned, care should be taken to minimise the effect on the tree itself both from excavation or construction plant or by ill-considered stockpiling of material.

14.6 RESERVOIR PERIMETER

14.6.1 General

Remedial works around the reservoir are normally limited to slope protection or the control of springs and surface water inflows. Problems of instability are less likely, but may occur if areas of naturally unstable ground are reactivated or construction activities have created slopes which are too steep or encroach towards existing features. Instability of carelessly placed fill around the reservoir perimeter may occur if the material has not been compacted adequately or has been placed with slopes which are too steep.

14.6.2 Wave erosion and slope protection

Measures are often required to control erosion in areas not protected at the time of construction. In many cases some erosion can be tolerated or can be controlled by suitable vegetation or soft landscaping (Box 9.1). In other cases, where erosion is severe, more extensive and durable measures will be required and a suitable form of hard protection could be installed. Gabions, as Figure 14.2, stone or coarse gravel, may provide an adequately enhanced protection which is environmentally acceptable.

Figure 14.2 Use of gabions to control wave erosion

14.6.3 Springs and surface water flow

Overland flow into the reservoir from springs or surface water around the perimeter may cause gulley erosion. Suitable drainage should be installed, discharging into the reservoir at a suitable headwall, as detailed in Sections 9 and 10.

14.6.4 Stability improvement

The addition of toe weight to buttress unstable areas is often not possible and slopes are normally either regraded or their drainage improved, as discussed in Section 14.2.4.

14.7 CHANGES IN STORAGE CAPACITY/WATER AREA

If the storage capacity is to be increased, it is essential that the revised storage of the enlarged reservoir is considered at an early stage. If this is less than 25 000 m^3 above the natural level of any part of the land adjoining the reservoir, the reservoir will remain outside the scope of the Reservoirs Act 1975. If the storage capacity is required to be greater than 25 000 m^3 then reference should be made to Appendix A to assess the implications of the Act concerning the undertaker. Specialist advice should be sought at an early stage as construction may only proceed under the control of a qualified engineer, as detailed in Appendix A. If the reservoir remains outside the Act, specialist advice will generally still be required as an inappropriate size or type of embankment raising may threaten the integrity of the entire dam.

Methods to increase the storage capacity include raising the level of the dam and the overflow structures, as well as desilting and excavating material from within the reservoir area. If a small increase in level is required, additional freeboard against wave action may be gained by incorporating a wave wall, normally of concrete, along the upstream side of the dam crest. The capacity of the overflow structures must be reassessed in view of the amended reservoir area and level.

Methods to decrease the storage capacity include lowering the level of the overflow structures, which usually increases the freeboard, or partially filling the reservoir with material. The capacity of overflow structures must be reassessed in view of the amended reservoir area and level.

Raising or lowering the top water level will change the position of the water line along the embankment and around the reservoir perimeter. This may severely affect established vegetation and existing slope and bank protection measures. It may also necessitate revised reservoir access requirements.

Changes to the reservoir volume, water area or top water level are likely to need planning permission from the LPA. The agreement of the RLA will also be necessary and changes will normally require revised consents and licences. This is likely to apply, whether the reservoir is enlarged or reduced and many of the comments in Section 3 are likely to be applicable. Dismantling (i.e. removal of the embankment and making provision for the safe passage of flood flows) may also necessitate planning permission.

14.8 DISMANTLING RESERVOIR STRUCTURES

When a reservoir becomes redundant for whatever reason, it is necessary to ensure that the embankment and associated structures are dismantled safely. Specialist advice should be sought to ensure this, and the Health and Safety Executive will also have an involvement. Provision must be made to control and pass inflows for impounding reservoirs. The reservoir level must be drawn down as low as the drawoff works will allow and a diversion channel or pipe should be installed. The stability of any excavations is particularly important and temporary excavations should not have slopes steeper than 1V:3H. Flatter slopes may be required if working within material deposited in the reservoir. Care should be taken if working with silt from within the reservoir area which will be very soft and liable to flow when excavated and may possibly present a gas hazard.

The work will need planning permission from the LPA and the agreement of the RLA in view of the changes to the flow conditions in the watercourse. The licences and consents from the latter authority may require measures to ensure that any material deposited within the reservoir area is left in a safe condition and not liable to be washed downstream at times of flood flows and that such flows can be passed safely by, or through, the dismantled structure.

Removal of the reservoir may invite action by downstream parties. This aspect should be considered and appropriate action taken at an early stage if removal is planned. Specialist advice is often required.

References

1. HMSO
   Reservoirs Act 1975

2. HMSO
   Water for Irrigation
   Bulletin 202, 1977

3. HMSO
   The Agriculture Act 1986

4. HMSO
   The Town and Country Planning Act 1990

5. HMSO
   The Town and Country Planning General Development Order 1988,
   SI No 1813, 1988

6. HMSO
   The Town and Country Planning General Development Order 1991,
   Amendment No 2, SI No 2268

7. HMSO
   The Town and Country Planning General Development Order 1992,
   Amendment No 3, SI No 2805

8. HMSO
   National Parks and Access to the Countryside Act 1949

9. HMSO
   Wildlife and Countryside Act 1981

10. HMSO
    Countryside Act 1988

11. EEC
    The Assessment of the Effects of Certain Public and Private Projects on
    the Environment 1985
    EEC Directive 85/337/EEC

12. HMSO
    The Town and Country Planning (Assessment of Environmental Effects)
    Regulations 1988, SI No 1199

13. HMSO
    Environmental Assessment
    DoE Circular 15/88, Welsh Office Circular 23/88

14. DEPARTMENT OF THE ENVIRONMENT WELSH OFFICE
    Environmental Assessment: A Guide to the Procedures 1989

15. DEPARTMENT OF THE ENVIRONMENT MAFF
    Planning Permission and the Farmer
    HMSO, 1989

16. FRANEY & CO LTD
    The Planning Fact Book
    Edited by Thomas, P

17. HMSO
    Water Resources Act 1991

18. HMSO
    Water Resources (Licences) Regulations 1965

19. HMSO
    Land Drainage Act 1991

20. HMSO
    Salmon and Freshwater Fisheries Act 1975

21. WATER RESEARCH CENTRE
    Water Supply Byelaw Guide
    Ellis Horwood Ltd, 1st Edition, Second Edition 1989

22. HMSO
    Diseases of Fish Act 1983

23. HMSO
    Registration of Fish Farming and Shellfish Farming Business Order 1985

24. HMSO
    Salmon Act 1986

25. HMSO
    Fish Health Regulations 1992

26. HMSO
    Theft Act 1968

27. HMSO
    Forestry Act 1967

28. HMSO
    The Water Act 1989

29. HMSO
    The Control of Pollution Act, Part 1 1974

30. HMSO
    The Control of Pollution Act, Part 2 1984

31. DEPARTMENT OF THE ENVIRONMENT MAFF
    Code of Good Agricultural Practice for the Protection of Water
    PB 0587, 1991

32. HMSO
    The Control of Pollution (Silage, Slurry and Agricultural Fuel Oil) Regulations 1991

33. HMSO
   The Environmental Protection Act 1990

34. HMSO
   The Public Health Act 1936

35. HMSO
   The Public Health (Recurring Nuisance) Act 1969

36. DEPARTMENT OF THE ENVIRONMENT MAFF
   Code of Good Agricultural Practice for the Protection of Soil
   PB 0617, 1993

37. MASON, P A
   Farm Waste Storage – guidelines for construction
   CIRIA Report 126, 1992

38. NATIONAL RIVERS AUTHORITY
   Policy and Practice for the Protection of Groundwater
   NRA, 1992

39. HMSO
   The Occupier's Liability Act 1957

40. HMSO
   The Health and Safety at Work Act etc. 1974

41. HMSO
   The Control of Substances Hazardous to Health Regulations 1988

42. HMSO
   Management of Health and Safety at Work Regulations 1992

43. HMSO
   Local Government Act 1972

44. ANDREWS, J and KINSMAN, D
   Gravel Pit Restoration for Wildlife
   RSPB/Tarmac Quarry Products, 1990

45. NORFOLK FWAG/THE ENVIRONMENTAL RESEARCH FUND
   Farm Pond Management – a practical guide for farmers and advisers

46. PROBERT, C
   Pearls in the landscape
   Farming Press, Ipswich, 1989

47. BRITISH ASSOCIATION FOR SHOOTING AND CONSERVATION
   Ponds and Lakes for Shooting and Conservation
   BASC, 1986

48. GAME CONSERVANCY
   Ponds and Lakes for Wildfowl
   Game Conservancy Booklet No 3, 1993

49. ANGLING FOUNDATION
The Creation of Low-Cost Fisheries: A guide for angling clubs, landowners and local authorities, 1981

50. INSTITUTE OF HYDROLOGY, BRITISH GEOLOGICAL SOCIETY
Hydrological Summary for Great Britain (published monthly)
Institute of Hydrology

51. INSTITUTE OF HYDROLOGY, BRITISH GEOLOGICAL SOCIETY
Hydrological Data UK (published annually)
Institute of Hydrology

52. NATIONAL WATER COUNCIL
Design and Analysis of Urban Storm Drainage from the Wallingford Procedure
DoE/National Water Council Standing Technical Committee Report No 28, 1981

53. INSTITUTE OF HYDROLOGY, NATIONAL ENVIRONMENTAL RESEARCH COUNCIL
Flood Studies Report
NERC, 1975

54. THE INSTITUTION OF CIVIL ENGINEERS
Floods and Reservoir Safety: an engineering guide,
Thomas Telford, 2nd Edition, London, 1989

55. FARQUHARSON F A K, LOWERING M J AND SUTCLIFFE J V
Some Aspects of Design Flood Estimation
BNCOLD/University of Newcastle, Symposium on inspection, operation and improvement of existing dams, September 1975

56. FARQUHARSON, F A K, MACKNEY, D, NEWSON, M D and THOMASSON, A J
Estimation of Runoff Potential of River Catchments from Soil Surveys, (including map)
Soil Survey of England and Wales, Harpenden 1978

57. BRITISH STANDARDS INSTITUTION
Code of Practice for Site Investigations
BS 5930: 1981

58. WELTMAN, A J and HEAD, J M
Site Investigation Manual
CIRIA Special Publication 25 1983

59. IRVINE, D J and SMITH, R J H
Trenching Practice
CIRIA Report 97, 1983

60. MAFF/SOIL SURVEY OF ENGLAND AND WALES
Soil Texture: assessment of soil texture and of soil organic matter
Leaflet 895, 1984

61. BRITISH STANDARDS INSTITUTION
    Methods of Test for Soils for Civil Engineering Purposes
    BS 1377: 1990

62. HEAD, K H
    Manual of Soil Laboratory Testing
    Vol 1: Soil classification and compaction tests
    Vol 2: Permeability, shear strength and compressibility tests
    Vol 3: Effective stress tests
    Pentec Press, 1992

63. TAVENAS, F, LEBLOND, P, HEAN, P and LEROVEILS, S
    The Permeability of Natural Soft Clays, Part 1: Methods of laboratory measurement
    Canadian Geotechnical Journal 20, No 4, 1983

64. COPPIN, N J and RICHARDS, I G
    Use of Vegetation in Civil Engineering
    CIRIA and Butterworths, London, 1990

65. BRITISH STANDARDS INSTITUTION
    Code of Practice for Earthworks
    BS 6031: 1981

66. BRITISH STANDARDS INSTITUTION
    Code of Practice for Foundations
    BS 8004: 1986

67. TOMLINSON, M J
    Foundation Design and Construction,
    Longman Scientific and Technical, 5th Edition Harlow, 1986

68. INSTITUTION OF STRUCTURAL ENGINEERS
    Earth Retaining Structures
    In: *Civil Engineering Code of Practice* No 2, reprinted 1987

69. PADFIELD, C J and MAIR, R J
    Design of Retaining Walls Embedded in Stiff Clays
    CIRIA Report 104, 1984

70. HEWLETT, H W M, BOORMAN, L A and BRAMLEY, M E
    Design of Reinforced Grass Waterways
    CIRIA Report 116, 1987

71. YOUNG, O C, BRENNAN, G and O'REILLY, M P
    Simplified Tables of External Loads on Buried Pipes
    TRRL, 1986

72. COMPSTON, D G, CRAY, P, SCHOFIELD, A N and SHANN, C D
    Design and Construction of Buried Thin-Wall Pipes
    CIRIA Report 78, 1978

73. BRITISH STANDARDS INSTITUTION
    Precast Concrete Pipes, Fittings and Ancillary Products
    BS 5911, Part 114: 1992 Specification for porous pipes

74. BRITISH STANDARDS INSTITUTION
Structural Use of Concrete
BS 8110, Part 1: 1985 Code of practice for design and construction

75. BRITISH STANDARDS INSTITUTION
Code of Practice for Design of Concrete Structures for Retaining Aqueous Liquids
BS 8007: 1987

76. BRITISH STANDARDS INSTITUTION
Concrete
BS 5328
Part 1: 1991 Guide to specifying concrete
Part 2: 1991 Methods for specifying concrete mixes
Part 3: 1990 Specification for the procedures to be used in producing and transporting concrete
Part 4: 1990 Specification for the procedures to be used in sampling, testing and assessing compliance of concrete

77. BUILDING RESEARCH ESTABLISHMENT
Sulphate and Acid Resistance of Concrete in the Ground
BRE Digest 363, 1991

78. BUILDING RESEARCH ESTABLISHMENT
Code of Practice for Design of Concrete Structures
Concrete Part 1: Materials
Concrete Part 2: Specification, design and quality control
Digests 325 and 326, 1987

79. BRITISH STANDARDS INSTITUTION
Specification for Aggregates from Natural Sources for Concrete
(including granolithic)
BS 882: 1992

80. BRITISH STANDARDS INSTITUTION
Testing Aggregates
BS 812, Part 1: 1975 and Part 2: 1975

81. BRITISH STANDARDS INSTITUTION
Concrete Admixtures
BS 5075, Part 1: 1982 Specification for accelerating admixtures, retarding admixtures and water reducing admixtures

82. BRITISH STANDARDS INSTITUTION
Guide to Selection of Constructional Sealants
BS 6213: 1992

83. CIRIA
Manual of Good Practice in Sealant Application
CIRIA Special Publication 80, 1991

84. CIRIA/WRC
Civil Engineering Sealants in Wet Conditions
CIRIA Technical Note 128, 1987

85. BLUNDELL, R and O'LEARY, E
   Joints in In-situ Concrete
   Concrete Society Digest No 10.

86. MASON, A P *et al*
   Standard Method of Detailing Reinforced Concrete
   Concrete Society Technical Report No 2, 1986

87. BRITISH STANDARDS INSTITUTION
   Code of Practice for Use of Masonry
   BS 5628, Part 3: 1985 Materials and components, design and workmanship

88. BRITISH STANDARDS INSTITUTION
   Precast Concrete Masonry Units
   BS 6073, Part 1: 1981 Specification for precast concrete masonry units

89. BRITISH STANDARDS INSTITUTION
   Specification for Clay Bricks
   BS 3921: 1985

90. BRITISH ASSOCIATION FOR SHOOTING AND CONSERVATION
   Ponds and Lakes for Shooting and Conservation
   BASC, 1986

91. BARRINGTON, R
   Making and Managing a Trout Lake
   Fishing News Books, Blackwell Scientific Publishing Ltd, Oxford, 1983

92. HMSO
   The Safety Signs Regulations 1980

93. BRITISH STANDARDS INSTITUTION
   Specification for Indicator Plates for Fire Hydrants and Emergency Water Supplies
   BS 3251: 1976

94. BRITISH STANDARDS INSTITUTION
   Workmanship on Building Sites
   BS 8000
   Part 1: 1989 Code of practice for excavation and filling
   Part 2: Section 2.1, 1990 Code of practice for mixing and transporting concrete
   Part 2: Section 2.2, 1990 Code of practice for sitework with insitu and precast concrete

95. HORNER, P C
   Earthworks
   ICE Works Construction Guides, Thomas Telford, London, 1981

96. SOMERVILLE, S H
   Control of Groundwater for Temporary Works
   CIRIA Report 113, 1986

97. QUINION D W and QUINION G R
Control of Groundwater
ICE Works Construction Guides, Thomas Telford, London, 1987

98. HMSO
Ancient Monuments and Archaeological Areas Act 1979

99. DEPARTMENT OF TRANSPORT
Specification for Highway Works
HMSO, 1992

100. SHIRLEY, D E
Introduction to Concrete
BCA Publication 45.028, 1987

101. GRANT, M
Use of Ready-Mixed Concrete
Concrete Society Digest No 8, 1988

102. HEALTH AND SAFETY EXECUTIVE
Site Safety and Concrete construction – A guide for small contractors
Health and Safety Executive HS(G) Series 46: 1989

103. BRITISH CEMENT ASSOCIATION
Concrete Masonry for the Contractor

104. BRMCA
Quality Scheme for Ready-Mixed Concrete
BRMCA Advisory Sheet 502, 1986

105. BRITISH STANDARDS INSTITUTION
Code of Practice for Falsework
BS 5975: 1992

106. CIRIA
Formwork – A guide to good practice
Concrete Society: 1986

107. WILSHERE, C J
Falsework
ICE Works Construction Guide, Thomas Telford, London, 1987

108. BRITISH STANDARDS INSTITUTION
Testing Concrete
BS 1881: 1983
Part 101: Method of sampling fresh concrete on site
Part 102: Method for determination of slump
Part 107: Method for determination of density of compacted fresh concrete
Part 108: Method for making test cubes from fresh concrete
Part 111: Method of normal curing of test specimens

109. BIRT, J C
Curing Concrete
Concrete Society Digest No 3: 1984

110. HMSO
The Construction (General Provisions) Regulations 1961
No 1580

111. HMSO
The Construction (Working Places) Regulations 1966
No 94

112. INSTITUTION OF CIVIL ENGINEERS
Conditions of Contract, Agreement and Contract Schedule for Use in Connection with Minor Works of Civil Engineering Construction,
ICE, Thomas Telford, 1st Edition 1988

113. INSTITUTION OF CIVIL ENGINEERS
Conditions of Contract and Form of Tender, Agreement and Bond for Use in Connection with Works of Civil Engineering Construction,
ICE, Thomas Telford, 6th Edition 1991

114. INSTITUTION OF CIVIL ENGINEERS
Conditions of Contract for Ground Investigation,
ICE, Thomas Telford, 1st Edition 1983

115. HMSO
Sale or Supply of Goods and Services Act 1987

116. HMSO
Unfair Contract Terms Act 1962

117. HMSO
Misrepresentation Act 1962

118. HMSO
Latent Damages Act 1986

119. HMSO
Employers' Liability Act 1965

120. JOHNSON, T A, MILLMORE, J P, CHARLES, J A and TEDD, P
An Engineering Guide to the Safety of Embankment Dams in the United Kingdom
BRE Report 187, 1990

121. FORESTRY COMMISSION, DOE
The Recognition of Hazardous Trees
Forestry Commission Leaflet, 1992

122. HOSKINS, C G
Embankment Dams, Vegetation and Engineers
Dams and Reservoirs Vol 3, No 3, BDS, 1993

123. CASTLE, D A, McCUNNALL, J and TRING, I M
Field Drainage – Principles and practices
Batsford Academic and Educational, London, 1984

124. BRITISH TRUST FOR CONSERVATION VOLUNTEERS
Waterways and Wetlands
BTCV, 1975

125. BRITISH TRUST FOR CONSERVATION VOLUNTEERS
Hedging
BTCV, 1978

126. BRITISH TRUST FOR CONSERVATION VOLUNTEERS
Woodlands
BTCV, 1980

127. ADAS/FORESTRY COMMISSION
Farm Ponds: Management and Maintenance
Leaflet P3025, 1986

128. ADAS
Pond Management for Wildlife
Leaflet P3242, 1990

129. MAFF
Guidelines for the Use of Herbicides on Weeds in or near Watercourses and Lakes
MAFF, 1985

130 CIRIA
Early-Age Thermal Crack Control in Concrete
CIRIA Report 91, 2nd Edition, 1992

131 CIRIA
Guide to the design of Thrust Blocks for Buried Pipelines
CIRIA Report 128, 1994

Appendix A Reservoirs Act 1975

A1 INTRODUCTION

The 1975 Act came into force in 1986 to '...make further provision against escapes of water from large reservoirs or from lakes or locks artificially created or enlarged.' It follows on from an earlier Act, entitled the Reservoirs (Safety Provisions) Act 1930, and essentially extends and strengthens this earlier Act.

The Act, like the 1930 Act, provides a legal framework within which qualified engineers, who are members of various panels set up by the Secretary of State, inspect reservoirs at intervals not exceeding ten years and recommend measures to ensure safety. It is the responsibility of the undertaker to arrange that both inspections and the recommendations are carried out and of the enforcement authorities to ensure that they are implemented. The technical functions of the engineer are virtually unchanged from those defined by the 1930 Act but the duties and powers of the authorities to ensure that the Act is enforced have been considerably strengthened.

The Act improved on the 1930 Act with the following new features:
- the local authorities have duties and increased powers to enforce the Act
- creation of public, locally available registers of reservoirs
- non-compliance is a criminal offence
- panel appointments last five years instead of life
- reservoirs must be supervised, and a new panel of engineers has been created for this purpose. When under construction or alteration there is no requirement for a supervising engineer and the reservoir is under the direct control of a construction engineer
- the inspecting engineer must now be independent, both of the undertaker and the construction engineer for the reservoir
- provisions for enforcement authorities to take emergency action
- provisions for the abandonment and discontinuance of reservoirs.

The Act does not extend to Northern Ireland.

A2 DEFINITIONS

- Large Raised Reservoirs (LRR) – reservoirs designed to hold or capable of holding more than 25 000 m^3 (roughly 5.5 million gallons) of water as such above the natural level of any part of the land adjoining the reservoir (including the bed of any stream)
- a qualified civil engineer – a member of the appropriate panel appointed by the DoE (see Section A3.5 and 3.6).

A3 RESPONSIBILITIES, POWERS AND DUTIES OF PARTIES INVOLVED IN THE ACT

A3.1 General

Responsibilities, powers and duties under the Act are held by the following persons or groups of persons:

- the Secretaries of State for the Environment, for Scotland and for Wales
- enforcement authorities
- undertakers (generally owners and lessees of reservoirs)
- panel of engineers appointed by the DoE for design, construction and inspection
- panel of engineers appointed by the DoE for supervision.

Current lists of panel engineers can be obtained from the Water Directorate of the DoE as listed in Appendix D.

A3.2 Duties of Secretary of State

Within their respective territories the Secretaries of State for the Environment, for Scotland and for Wales each have a general oversight of local authorities' functions under the Act. They receive reports at two-yearly intervals from enforcement authorities and have powers to cause an inquiry to be held into any question of whether an enforcement authority has failed to perform any of its functions under the Act. The Secretaries of State, after consultation with the Institution of Civil Engineers, are required to set up panels of engineers to carry out technical functions under the Act. The Secretaries of State have determined that four panels should be formed. The duties which their members may undertake are set out in Sections A3.5 and A3.6.

A3.3 Enforcement authorities

Enforcement authorities under the Act are:

- in England and Wales (outside the Greater London and metropolitan county areas), the county councils
- within the Greater London and metropolitan county areas, the London boroughs and metropolitan district councils
- in Scotland, the regional and island councils.

They have a duty to ensure that undertakers comply with the Act in all respects, and must send reports at two-yearly intervals to their respective Secretaries of State on the steps they have taken in this regard. Under the Act, nearly all technical duties are the prerogative of the expert panel engineer and enforcement authority duties are mainly of an administrative nature with some legal content. They must maintain a register containing prescribed information on large raised reservoirs wholly or partly within their area.

This will include:

- name and location of reservoirs
- name and address of undertaker
- names and addresses of panel engineers concerned with the reservoirs
- dates of most recent reports and certificates (under both the present and previous Acts).

The register is open to the general public and, as is shown above, will contain sufficient information to enable any person to check that the provisions of the Act concerning inspections, appointment of qualified civil engineers, etc. are being observed. It does not contain information of a commercial nature or anything that might affect security.

A3.4 Undertakers

For the purposes of the Act, the word 'undertaker' in relation to a reservoir means the owner or lessee, and ultimate responsibility for the safety of a reservoir rests with the undertaker. Where a public highway crosses a dam, the highway authority is also deemed to be an undertaker under the Act. Other owners of part of the dam or reservoir may also be deemed to be undertakers.

A3.5 Panel engineers (excluding supervising engineers)

The professional aspects of reservoir safety are the responsibility of civil engineers, qualified for the task by their appointment to panels by the Secretary of State, and the terms qualified civil engineer and panel engineer are synonymous under the Act. All technical matters relating to safety rely on panel engineers' expertise and judgement. The first three panels are qualified to design, supervise the construction of and inspect the different types of reservoir and their powers are set out in outline below

- All Reservoirs Panel (AR) – Qualified to act in relation to all reservoirs and as referees under the Act
- Non-Impounding Reservoirs Panel (NIR) – Qualified to act in relation to all reservoirs that do not impound directly from a catchment (generally pumped storage)
- Service Reservoirs Panel (SR) – Qualified to act in relation to all non-impounding reservoirs constructed of concrete, brickwork or masonry (generally service reservoirs in a supply system).

Engineers in all the above panels can also act as supervising engineers and for emergencies under Section 16 of the Act on reservoirs within their own capacity.

A3.6 Supervising engineers

The fourth panel comprises supervising engineers and is required under Section 12 of the Act. Its purpose is to ensure that there is continuous professional supervision of a reservoir at all times, and not just at inspections, when material changes in the condition of the reservoir, or catchment that could affect the safety of the former, might become apparent. Supervising engineers are not empowered to design or to carry out supervisory works under the Act that pertain to safety.

A4 SUPPLY OF INFORMATION AND RECORDS

Information on all existing reservoirs must be supplied to the Enforcement Authorities. The information required includes the following:

- name and address of undertaker
- situation of reservoir
- final certificate for the reservoir (or preliminary certificate if the final one is not available)
- any certificate on the execution of the works for the construction or alteration of the reservoir, including the annex to the certificate
- report on the latest inspection
- name of the last Inspecting Engineer, Supervising Engineer and others in Schedule 1 of the prescribed form of record.

The undertaker must keep a record in the form prescribed by the regulations and which must include the following:

- the water level in the reservoir and the depth of water flowing over the overflow weir, at intervals to be prescribed by the inspecting engineer
- names and addresses of the last inspecting and the supervising engineers including those of any other panel engineers concerned with the reservoir and its construction
- details of direct and indirect catchments and standard average rainfall as provided by the meteorological office
- details of the dam
- information on overflows, etc.
- details of drawoffs and other methods of lowering the water in the reservoir
- means of access to the reservoir.

Details are set out in Statutory Instrument 1985 No. 177 The Reservoirs Act 1975 (Registers, Reports and Records) Regulations 1985 obtainable from HMSO. Thomas Telford Ltd publish a booklet (*Prescribed Form of Record for a Large Raised Reservoir*) in the prescribed layout which assists in setting out the information required.

This may seem a lot of information, but the technical information and most of the remainder are likely to be required by the inspecting engineer during his inspection and to help him make any recommendations that he considers necessary concerning the safety of the reservoir. Its availability may be of interest to him if he is asked to quote for an inspection. Information such as access to the site and the names of various engineers are of use to all parties in an emergency.

A5 APPOINTMENT OF SUPERVISING AND INSPECTING ENGINEERS

The undertaker is required to appoint a supervising engineer to each reservoir to ensure that it is under continual supervision (one engineer can cover more than one reservoir). He must also appoint qualified civil engineers from the appropriate panel to inspect the reservoir every ten years or such lesser period as the previous inspecting engineer may have recommended. The undertaker is responsible for:

- notifying the enforcement authority of such appointments
- affording those appointed all reasonable facilities, including making available the records of the reservoir, certificates, reports, drawings, etc.
- paying the fees of these engineers (he may wish to obtain a quotation or quotations before appointing them).

The duties of the Supervising Engineer include:

- to inform the undertaker of any matter that might affect safety or any failure to keep the records of the reservoir correctly and, in particular, to ensure that certain sections of the Act are complied with
- to report to the undertaker not less than once a year on any matter the inspecting or construction engineer has noted in his report or annex to the final certificate for the supervising engineer's attention
- if he considers an inspection is required, to notify the undertaker who must then arrange for it to be carried out.

The Supervising Engineer must also send to the enforcement authority a copy of any advice to the undertaker that the reservoir be inspected, any failure to keep the correct records or to take any other action. Once appointed, frequency of visits by the supervising engineer to the reservoir is a matter for that engineer to decide and he will wish to satisfy himself that it is adequate for safety purposes, bearing in mind the circumstances (which may alter over the years) and his statutory duties, without incurring unnecessary expense for the undertaker. Factors to be considered might include size and type of reservoir and dam, the age of the dam, previous history, proximity of dwellings, etc. The Act provides in Section 12 that the supervising engineer shall report to the undertaker once a year on any matter the inspecting engineer has noted for the supervising engineer's attention. While it is not mentioned in the Act, an inspecting engineer in his inspection report may wish to suggest the frequency and times of the year of the supervising engineer's visits to the reservoir.

The Inspecting Engineer must be independent of the undertaker (i.e. only employed in a consultant capacity) and must not be the construction engineer or his partner or fellow employee in the same business. The Supervising Engineer may be an employee of the undertaker. It should also be noted that if the inspecting engineer, subsequent to his inspection, acts as a construction engineer under the Act on an alteration to the same reservoir, he may not carry out the next statutory inspection of that reservoir.

An engineer carrying out an inspection under the Act is required to supply the undertaker with a report and certificate. If he recommends work in the interests of safety, this work must be supervised by a member of the appropriate panel (AR, NIR or SR) who certifies that the work has been undertaken. Copies of all certificates and reports (including those reports of the supervising engineer) are sent to the enforcement authority, to enable them to check that the requirements of the Act are being fulfilled. Details of reports and certificates are given in SI 1986 No. 467 and SI 1986 No. 468.

If an undertaker is aggrieved at a recommendation made by an engineer in the interest of safety (or as to the time of the next inspection) he may refer his complaint to a referee who must be a panel engineer appointed by mutual agreement between the undertaker and inspecting engineer or, in default of agreement, by the Secretary of State. The referee may then modify the report as he considers necessary. It should be noted that only recommendations in the interests of safety are enforceable under the Act.

A6 NEW, ENLARGED AND RESTORED RESERVOIRS

Where an undertaker proposes to construct a reservoir falling within the provisions of the Act he must employ a qualified civil engineer from the appropriate panel to supervise the design and construction of the works and inform the local enforcement authority (or authorities if it lies across a boundary) of his intention to do. The information prescribed under Regulation 5 on page 13 of Statutory Instrument 1986/468, The Reservoirs Act 1975 (Certificate, Reports and Prescribed Information) Regulations 1986 must be provided. The information can be divided into four groups:

- name and situation of reservoir
- nature of proposed works and timing
- name and address of panel engineer (construction engineer)
- type of reservoir (impounding, non-impounding), construction and height of dam, capacity and surface area of reservoir.

Where construction of a Large Raised Reservoir (LRR) within the meaning of the Act (see Section A1) involves the alteration of an existing reservoir, the same provisions apply to the alterations as to the construction of a new large raised reservoir.

These include:

- increase the capacity of a LRR
- alter a non-LRR so that it becomes a LRR
- bring back into service a LRR that has been abandoned or discontinued

Various certificates and drawings and descriptions are required at various stages of the construction, as follows:

- **Preliminary** Certificates – Issued at the stage of construction when the construction engineer considers that the reservoir can be properly filled, wholly or partially with water. The certificate will specify the level of filling and any conditions. Further preliminary certificates can be issued to supersede the previous certificate.

- **Interim Certificate** – Issued in connection with additional works where, at any time during work, the construction engineer considers that filling should not take place to a level previously given in a Preliminary Certificate. The subsequent certificate will specify a lower level of filling and any conditions prior to the issue of a further preliminary certificate. Subsequent interim certificates can be issued which supersede the previous certificate.

- **Final Certificate** – Issued when the construction engineer is satisfied that the reservoir is sound and satisfactory and may safely be used for water storage. The certificate will specify the level up to which water may be stored and the conditions of storage. This will also recommend certain monitoring and may contain matters to be watched specifically by the supervising engineer.

- **Certificate as to Efficient Execution of Works** – Issued after completion of the works when the construction engineer is satisfied that the reservoir is sound and satisfactory and may be safely used for water storage. Detailed drawings and descriptions giving full information of the works actually constructed must be annexed to the certificate.

A7 DISCONTINUANCE AND ABANDONMENT

Under the Act, an owner having no further use for his reservoir and wishing to reduce his costs must use either Section 13 'Discontinuance' or Section 14 'Abandonment'. The latter involves the use of the appropriate panel engineer to report on measures that must be taken to ensure that the reservoir is incapable of filling accidentally or naturally above the natural level of the adjoining ground or does so only to an extent that does not constitute a risk. The undertaker must then have these measures carried out and obtain a certificate to state that they have been done from the engineer concerned. If the measures involve alterations to the reservoir then the 'discontinuance' provisions of the Act apply. An abandoned reservoir remains on the register and the undertaker must employ a supervising engineer and also have the reservoir inspected under Section 10. Enforcement procedures are included in Section 14.

If an owner wishes his reservoir to be removed from the register so that the inspection and supervision provisions of the Act no longer apply to it, he must carry out the provisions of Section 13 of the Act. These involve employing the appropriate panel engineer to design and supervise the carrying out of works to ensure that the reservoir is incapable of holding more than 25 000 m^3 of water above the natural level of the adjoining ground and obtain a discontinuance certificate from the panel engineer concerned to this effect. This normally requires that the dam be breached or demolished.

A8 PENALTIES AND ENFORCEMENT POWERS

Penalties are provided for:

- failure to carry out the provisions of the Act regarding construction, supervision, periodical inspection, discontinuance and abandonment or, if the undertakers fail to comply with a notice from the enforcement authority
- failure to provide the enforcement authority in due time with any notice required under the Act
- failure to provide a person, generally a qualified civil engineer, with facilities required under Section 21(5) etc.
- using or providing false documents or information.

Proceedings for an offence in England and Wales may be instituted by the enforcement authority in whose area the reservoir is sited, by the Secretary of State or by others with the consent of the Director of Public Prosecutions. The enforcement authority also has considerable reserve powers under the Act relating to, inter alia, design and supervision of construction, alterations, supervision, inspection, discontinuance and abandonment.

When these are used by the Enforcement Authority they are entitled to recover reasonable expense from the undertaker. In Scotland the normal Procurator Fiscal system applies.

The enforcement authorities have considerable powers to assist them in their duties. If they become aware that a qualified civil engineer is not being employed, where required, under the Act, they may require the undertaker to appoint one within 28 days. If the undertaker fails to do so, the undertaker must reimburse the authority any reasonable expense incurred in the matter. If an enforcement authority considers a reservoir is unsafe and that immediate action is needed to protect persons or property, they are empowered to take steps at the reservoir to prevent or mitigate the risk using a qualified civil engineer. They must inform the undertaker as soon as possible of their proposals and are entitled to recover reasonable costs from him.

Appendix B Capacity and maximum depth of water

B1 **GENERAL**

The capacity and the maximum depth of water in a reservoir can only be assessed accurately once the levelling survey (Section 8.3.2) has been carried out and a contour plan of the proposed reservoir area produced. A preliminary appraisal can be made at the feasibility stage by approximate methods which should give an estimate to within 20% and probably closer. Clearly an individual valley or area of land may have specific features which are unusual and fall outside the scope of the suggested methods of assessment; these must be considered and the potential capacity amended accordingly. Such situations might include valleys which are curved in plan, or land which has variable ground surface gradients in direction, extent or amount. In many instances, however, these can be allowed for by reasonable averaging and a sufficiently accurate preliminary estimate of the likely capacity made for a given maximum depth of water at the proposed dam.

Not all the stored water will be available for abstraction, as described in Section 6.3.1, and the equivalent depths of water should be considered taking into account the following typical losses:

•	annual evaporation	600 mm
•	annual seepage	300 mm
•	dead storage	500 mm
	TOTAL	1400 mm

Non-impounding reservoirs are likely to be filled in the winter only therefore if flow into an impounding reservoir is likely to cease in the summer, allowance should be made although some reduction in the depths for evaporation and seepage. Where summer inflow into an impounding reservoir is available by means of a perennial flow in a watercourse, this will normally far exceed the actual evaporation and seepage and no allowance for these is usually provided.

Reservoirs for amenity, conservation, recreation and sporting uses often require a minimum area although the depth may be important. If summer inflow is unavailable, allowance for evaporation and seepage may be required and minimum areas and depths may need to be considered after assessment of summer losses. A reducing summer level, if too excessive, may also affect the proposed reservoir purpose.

Fill for the embankment will usually be obtained by excavation within the reservoir area, as described in Section 9.3.8. This results in excavations being entirely or mainly below top water level and increases the potential storage available. The increased volume can be maximised by careful assessment of the relative levels and the position of the borrow pit, as described in Section 9.3.8. The constraints on the position and extent must be considered at an early stage and any potential additional storage only assessed conservatively at the feasibility stage.

Assessment of reservoir shape, surface area and volume can best be carried out using a readily available Ordnance Survey plan of the area, enlarged to a reasonable scale, unless the information from the levelling survey is available. The largest available scales are 1/2500 in many areas and 1/1250 for urban areas and these should ideally be enlarged to 1/500 for a preliminary estimate purposes; this can usually be carried out at most commercial printing and copying organisations. Crown copyright exists on all Ordnance Survey material and it should be borne in mind that reproduction in whole or in part of such material by any means is prohibited without the prior permission of the Ordnance Survey. The address of the OS is included in Appendix D. These scale drawings are sufficient to show the major topographical and development constraints and have limited spot height information, but are not contoured. Consequently, it is necessary to obtain a limited number of ground levels at an early stage to give an indication of the ground profile. For an impounding reservoir, these should comprise a minimum of a line of levels along the proposed dam centreline (at least one in the valley bottom and one towards each abutment) and a similar number along the valley to beyond the proposed top water level. Sufficient levels must be taken for a non-impounding reservoir to indicate the general ground profile along the dam centreline and typical ground levels within the reservoir area.

B2   NON-IMPOUNDING RESERVOIRS

Non-impounding reservoirs have traditionally been constructed in a rather stark rectangular configuration as shown in Figure 2.1. Increasing environmental awareness now requires these to be created with a more variable embankment alignment and to be landscaped into their surroundings. This presents a more complex shape to be assessed, but an irregular reservoir can usually be represented by a series of mathematical shapes (i.e. rectangles, triangles, etc.) and a representative depth of water assigned to each. Greater accuracy can be obtained by superimposing a rectangular grid over the plan of the reservoir by use of squared graph paper and assigning a representative depth of water for each square. If the proposed reservoir is sketched on a 1/500 scale plan, a one centimetre grid will give a reasonable accuracy; each grid square then representing 25 m^2.

The excavation for the fill would theoretically give a considerable amount of storage below the original ground level compared to that retained by the dam. In the past, attempts have been made to balance the amount of excavation and filling (excluding topsoil and unsuitable material) to minimise costs and maximise the below-ground storage. This approach is generally not tenable due to the constraints on the excavation, the required fill volume and variable embankment alignment and additional fill may be required from outside the reservoir area. The likely increase in below-ground storage is clearly dependent on the shape of the reservoir and the excavation, the height of the embankment and the depth of the excavation but is unlikely to exceed 25% of above-ground storage and may possibly be somewhat less. This increase should be ignored for the preliminary estimate of volume and only considered at a later stage when the likely size of the excavation is known.

An approximate method for estimating the required volume if the reservoir can be represented by a square or rectangular shape is as follows:

Water volume (m^3) = $d(l_{e1} - 3d)(l_{e2} - 3d)$

where l_{e1}, l_{e2} = embankment lengths of two adjacent sides of the reservoir measured along centrelines (m)

d = maximum depth of water, measured at the dam (m)

This can be solved by trial and error to find a value for the water depth, d.

The approximate fill volume for a non-impounding reservoir constructed on a level, stripped ground surface can be estimated as follows:

Fill volume (m^3) = $l_p(Z^2 + 3Z)$

where l_p = reservoir perimeter (m) = $2(l_{e1} + l_{e2})$ for a rectangular reservoir)

Z = embankment height above stripped ground level (m)
  = $d + 1$ (m)

d = as defined in above

The one metre addition to the water depth is to give a preliminary estimate of the dam height to allow for wave and flood rise above the normal top water level.

This preliminary assessment of the fill volume does not include any allowance for a cut-off trench or any variations resulting from sloping ground along or across the dam and assumes side slopes of 1V:3H. If these are considered significant, further allowance should be included. An additional 30% of the calculated fill volume should be included for wastage and unsuitable materials to assess the amount required from the borrow pit within the reservoir or elsewhere.

B3 IMPOUNDING RESERVOIRS

A preliminary estimate of the volume and surface area of an impounding reservoir can readily be assessed as follows:

Water volume (m³) = $0.52\, snd^3$ or $0.26\, l_r\, d(l_e - 2n)$

Surface area (m²) = $1.4\, snd^2$ or $0.7\, l_r\, (l_e - 2n)$

Where l_e = embankment length, measured along centreline (m)

l_r = length of reservoir, measured as a straight line from the embankment to the upstream end of the reservoir (m)

d = maximum depth of water, measured at the embankment (m)

n = valley side slope (expressed as 1V:nH)

s = valley bed slope (expressed as 1V:sH)

If the valley side slopes differ substantially, the water volume can be calculated by replacing 'n' with $(n_1 + n_2)/2$ where n_1 and n_2 are the differing side slopes.

These formulae assume that the valley is evenly graded and parabolic in section and that the fill volume is excavated from within the reservoir area.

The approximate fill volume for an impounding reservoir constructed on a level stripped ground surface can be estimated as follows:
Fill volume (m³) = $1.3 l_e\, Z(1 + Z)$

Where parameters are as defined before.

This preliminary assessment of the fill volume does not include any allowance for a cut-off trench or any variations resulting from sloping ground across the embankment. If these are considered significant, as described in Section B2, further allowance should be included. An additional thirty percent of the calculated fill volume should be included for wastage and unsuitable materials to assess the amount required from the borrow pit within the reservoir area.

B4 EXAMPLES

B4.1 Non-impounding reservoir for abstraction purposes

The depth of water and capacity are required for an unlined reservoir having an available capacity of 16 000 m³. The site will allow a reservoir which can be approximated to a rectangle with sides of 100 m and 80 m length respectively. Water can only be abstracted during the winter and allowance must be made for evaporation and seepage losses from the reservoir.

As the reservoir can be approximated to a rectangular shape, the depth of water can be found by trial and error, as described in Section B2.

Thus $d(100 - 3d)(80 - 3d) = 16\,000$

whereby d is just in excess of 2.35 m, say 2.4 m. Allowing for losses and dead storage as Section B1.2, the actual depth required is 3.8 m and the volume is $3.8(100 - 3 \times 3.8)(80 - 3 \times 3.8) = 23\,096\text{ m}^3$, say $23\,100\text{ m}^3$.

The perimeter (lp) is $2(80 + 100)\text{ m} = 360\text{ m}$
The embankment height above stripped ground level (Z) is $d + 1 = 4.8\text{ m}$.
Thus the approximate fill volume is $360(4.8^2 + 3 \times 4.8)\text{ m}^3 = 13\,478\text{ m}^3$
$$\text{say} \quad 13\,500\text{ m}^3$$

It may be noted that the fill volume (before addition for wastage etc) is about 84% of the usable stored water volume or about 58% of the total stored volume. These represent a water-to-earth ratio of 1.18 and 1.71 respectively. This figure would be improved if detailed appraisal subsequently allows some below-ground storage to be considered.

B4.2 Impounding reservoir for abstraction purposes

The maximum depth of water and capacity after allowing for losses are required for a reservoir having an available capacity of $16\,000\text{ m}^3$. The valley has an average bed slope of 1V:50H and side slopes of 1V:20H. The stream flow is perennial and adequate to more than balance the evaporation and seepage losses and provide sufficient water to maintain the required compensation flow. Sufficient water must remain in the reservoir following the maximum extraction to provide an adequate area for wildlife.

The maximum depth of water at the embankment is given by:

$$d^3 = \frac{16\,000}{0.52 \times 20 \times 50}\text{ m}^3$$

$$d^3 = 30.77\text{ m}^3$$

hence $d = 3.13\text{ m}$

Only dead storage needs to be provided and sufficient area needs to be provided following the maximum drawdown. The water surface area with the recommended minimum dead storage depth of 0.5 m would extend for $2n \times 0.5$ across the valley and $s \times 0.5$ along the valley bottom. These would result in a minimum water area of about 20 m by 25 m. This should be sufficient in the short term prior to refilling with a water surface area of the order of 250 m^2 and probably slightly greater as a result of the flatter ground in the valley bottom. If a greater surface area was required, a one metre minimum depth would extend to approximately 40 m by 50 m.

Adopting the minimum dead storage, the maximum depth of water is 3.13 + 0.5 m = 3.63 m. Hence the actual volume is $0.52 \times 20 \times 50 \times 3.63^3$ m^3 = 24 872 m^3, say 24 900 m^3.

Maximum height of the embankment (Z) 3.63 + 1 m	=	4.63 m say 4.7 m
Length of the embankment (l_e) is 2 × 20 × 4.7 m	=	188 m
Length of the reservoir (l_r) is 50 × 3.63 m	=	182 m

Thus the approximate fill volume is $1.3 \times 188 \times 4.7 \,(1 + 4.7) = 6547$ m^3, say 6600 m^3. Prior to the addition for wastage, etc, this fill volume represents about 40% of the available water or 26% of the total, i.e. a water-to-earth ratio of 2.42 and 3.77 respectively.

These water-to-earth ratios compare with the values of 1.18 and 1.71 respectively for the comparable situations for a non-impounding reservoir for the same usable volume of 16 000 ^{3}. They clearly indicate that the earthworks quantities for a non-impounding reservoir are typically at least two, and often more, times greater for a given water storage in most situations (i.e. the water-to-earth ratio is general significantly greater for an impounding reservoir).

B4.3 Amenity/sporting reservoir with no abstraction

The depth of water and capacity are required for an impounding reservoir which will form a water area for wildfowling and fishing. A water surface area of not less than one hectare is required, exclusive of any variations in the reservoir perimeter which will be formed by localised earthworks. The ground slopes in the valley and flow in the watercourse are as Example B4.2.

The approximate depth of water for the required surface area is given by:

$$d^2 = \frac{1 \times 10\,000}{1.4 \times 20 \times 50} = 7.14 \text{ m}^2$$

hence $d = 2.67$, say 2.7 m

The stream flow will balance any losses and the reservoir will function at or above the top water level; consequently no additional storage is required. Hence the actual maximum depth of water is as calculated and the volume of stored water is $0.52 \times 20 \times 50 \times 2.7^3$ m^3 = 10 235 m^3.

The maximum height of the embankment (Z) is 2.7 + 1 m	=	3.7 m
The length of embankment (l_e) is 2 × 20 × 3.7 m	=	148 m
The length of the reservoir (l_r) is 50 × 2.7 m	=	135 m

Thus the approximate fill volume is $1.3 \times 148 \times 3.7 \times (1 + 3.7) = 3345$ m^3, say 3400 m^3. This represents about 33% of the total stored volume and a water to earth ratio of 3.06.

This last example shows how a smaller dam in the same valley has a greater volume of fill to the total stored water (i.e. a lesser water-to-earth ratio). In simple terms, this illustrates that the unit cost of the stored water can be minimised by maximising the storage subject to the influences of all the other constraints.

Appendix C General requirements for various recreational and sporting purposes

Requirements	Water	Water edge
Canoeing	Suitable for instruction, casual recreation and racing. Sprint racing canoeists require minimum 1km straight course, 95 m wide and 2 m deep with still water in sheltered conditions. The casual canoeist has few requirements apart from access to an area of fairly calm water. Sprint canoeing has similar requirements for water conditions as rowing, but both require very large areas and are likely to be outside the scope of this guide.	Access points for easy launching. Canoes can be launched easily from the bank, but a landing stage is useful – boat storage is often required.
Hydro-planing and power boating	Minimum area of water needed is 6.5 ha and preferably larger. Unobstructed area is essential with a minimum water depth of about 1 m. The water must be weed free. A large bay in the reservoir is suggested for pits for underwater access to the boats. Due to dangers to other users, motor boat racing requires exclusive use of water, and the only satisfactory way in which facilities can be shared is by the introduction of zoning (on very large areas) or timetabling. This use is not likely to be suitable for the size of reservoirs falling within the scope of this guide.	Good road access. Firm slipways, jetties and fuel points. Storage for engines and boats.
Rowing	Preferably still water in sheltered conditions. The minimum for competitive events is a straight stretch 800 m long and 30 m wide.	Special boathouse and landing platforms required for long boats. Launching is by ramp or steps. Landing stage: 18 m long for sideways launching of 'eights'.
Sailing	Minimum depth of 1.5 m, preferably 20 ha or more, though sailing does take place on much smaller areas of water. For recreational sailing, 7 to 15 boats/ha is acceptable, but not more than 7 boats/ha for racing. Dinghy sailing needs relatively sheltered water.	Firm slipways. Adequate access roads. Dinghy parks
Sub-aqua Swimming	Basic requirements are an area of still, clear and unpolluted water at least 20 m depth (9 – 12 m according to some authorities) with a clear sub-surface visibility. The depth required is not likely to be available at most reservoirs within the scope of this guide.	Easy entry into water

Requirements	Water	Water edge
Water skiing	Exclusive use of an area of relatively calm water, not close to vertical banks, as this activity tends to build up a wash. Some doubt as to minimum water area, but not less than 10 ha is needed. In practice, it appears that a number of clubs operate on smaller areas. The space requirement for recreational skiing is very much less, since a number of boats, each with a group of skiers, may operate together. The sport can generate strong local opposition. Apart from possible danger to other users, the wash created may damage banks and disturb anglers, while the high noise level associated with the use of motor boats, especially those powered by outboard as opposed to inboard engines, is often considered undesirable on environmental grounds. This use is not likely to be suitable for the size of reservoir falling within the scope of this guide.	Firm slipways and easy entry into water.
Model boat sailing	Small stretch of sheltered weed-free water, although reservoirs of 0.5 ha and upwards are to be preferred for large yachts and radio-controlled power boats. A water depth of 0.5 m would be sufficient for model use, but a greater depth is advisable to limit possible weed and rush growth.	Easy and safe access to water edge is necessary around full perimeter of the reservoir for launching and retrieving boats.
Fishing	The reservoir needs to be unpolluted and relatively undisturbed. The depth of water does not appear to be critical for coarse fishing although a variable depth is suggested as giving optimum conditions, with ideally one third of the water area less than 2 m depth, one third between 2 m and 3 m depth and one third in excess of 3 m. Some fish require colder water with localised areas of deep water. Value decreases with substantial areas of deep water and high acidity.	20 m of bank per fisherman or less during match fishing. Defined pathways, stands or 'benchings'. Strong or strengthened banks.
Shooting	A water body designed for wildfowling should be as large as possible, but not less than 0.2 ha. The shape is more important, and an irregular shoreline is essential with a gentle bank grading down to 2 m depth in the centre of the water area. The water body should ideally be located beneath an existing flightline.	Associated wetland areas are as important as the water area itself.

Boating and other interests will require car parking, a club house with changing rooms, toilets and first aid facilities; possibly with secure storage.

Appendix D List of useful addresses

D1 GENERAL ADDRESSES AND TELEPHONE NUMBERS

ADAS
Headquarters
Oxford Spires Business Park
The Boulevard 01865 842742
Langford Lane
Kidlington
Oxford
OX5 1NZ

ADAS
Soil and Water Research Centre
Anstey Hall
Maris Lane 01223 840011
Trumpington
Cambridge
CB2 2LF

The Association of Consulting Engineers
Alliance House
12 Caxton Street
Westminster 0171 222 6557
SW1H 0QL

The Association of Geotechnical Specialists
PO Box 250
Camberley
Surrey 01276 678949
GU15 1UD

British Geological Society
Keyworth
Nottingham
NG12 JGG 01602 363100

British Trust for Conservation Volunteers
36 St Mary's Street
Wallingford
Oxon 01491 839766
OX10 0EU

British Association for Shooting and Conservation
Marford Mill
Rossett
Wrexham 01244 570881
Clwyd
LL12 0HL

British Association for Shooting and Conservation (Scotland).
Croft Cottage
Trochy 013507 23226
Dunkeld
Tayside
PH8 0DY

British Association for Shooting and Conservation
(Northern Ireland)
The Courtyard Cottage
Galgorm Castle
Ballymena
County Antrim
BT42 1HL
01266 652349

British Association of Landscape Industries
Landscape House
9 Herry Street
Keighley
West Yorkshire
BD21 3DR
01535 606139

Building Research Estabishment (BRE)
Geotechnics Division
Garston
Watford
Herts
WD2 7JR
01923 894040

Council for National Parks
4 Hobart Place
London
SW1W 0HY
0171 924 4077

Council for the Protection of Rural England
Warwick House
25 Buckingham Palace Road
London
SW1W 0PP
0171 976 6433

Council for the Protection of Rural Wales
Ty Gwyn
31 High Street
Welshpool
Powys
SY21 7JP
01938 552525

Country Landowners Association
16 Belgrave Square
London
SW1X 8PQ
0171 235 0511

County Wildlife Trust
(See Royal Society for Nature Conservation
for local addresses)

Countryside Commission
John Dower House
Crescent Place
Cheltenham
Gloucestershire
GL50 3RA
01242 521381

Countryside Council for Wales
Plas Penrhos
Penrhos Road
Bangor
Gwynedd
LL57 2LQ
01248 370441

Department of the Environment (DoE)
Romney House
43 Marsham Street
London
SW1P 3PY
0171 276 3000

DoE for Northern Ireland
North Land House
3 Frederick Street
Belfast   01232 244711
BT1 2NS

Department of Agriculture Northern Division
Watercourse Management Division Headquarters
Hydebank
4 Hospital Road   01232 647161
Belfast
BT8 8JP

English Nature (Nature Conservancy Council for England)
Northminster House
Northminster Road
Peterborough   01733 340345
PE1 1UA

Farming and Wildlife Advisory Group (FWAG)
National Agricultural Centre
Stoneleigh
Kenilworth   01203 69699
Warwick
Warwickshire
CV8 2RX

FWAG (Scottish Branch)
Rural Centre
West Mains
Ingliston   0131 335 3982
Newbridge
Mid Lothian
EH28 8NZ

Federation of Civil Engineering Contractors
Cowdray House
6 Portugal Street
London   0171 404 4020
WC2A 2HH

Flexible Revetment Association
c/o Ardon International Ltd
PO Box 111
Tunbridge Wells   01892 355777
TN4 0PZ

Forestry Commission (HQ)
231 Corstorphine Road
Edinburgh
EH12 7AT   0131 334 0303

Forestry Authority for England
Great Eastern House
Tenison Road
Cambridge   01223 314566
CB1 2DU

Forestry Authority for Scotland
Portcullis House
21 India Street
Glasgow   0141 248 3931
G2 4PL

Forestry Authority for Wales
North Road
Aberystwyth
SY23 2EF   019070 625866

The Game Conservancy Trust
Burgate Manor
Fordingbridge
Hampshire   01425 52381
SP6 1EF

Health and Safety Executive
Information Centre
Broad Lane
Sheffield   01742 892345
S3 7HQ

Her Majesty's Inspectorate of Mines
St Anne's house
Stanley Precinct
Bootle   0151 951 4136
Merseyside
L20 3RA

Her Majesty's Land Registry
Lincoln's Inn Fields
London
WC2 A3PH   0171 405 3488

Her Majesty's Stationery Office
49 High Holborn
London
WC1V 6HB   0171 873 0011

Institute of Fisheries Management
Balmaha
Coldwells Road
Holmer   01432 276226
Hereford
HR1 1CH

Institute of Hydrology
Maclean Building
Crowmarsh Gifford
Wallingford   01491 838800
Oxon
OX10 8BB

Institution of Civil Engineers
1 – 7 Great George Street
London
SW1P 3AA   0171 222 7722

Institution of Water and Environmental Management
River Engineering Section
15 John Street
London   0171 831 3110
WC1N 2EB

International Geotextile Society, UK Section
c/o Netlon Ltd
Kelly Street
Blackburn   01254 262431
Lancashire
BB2 1PJ

London Gazette
Room 418
51, 9 Elms Lane
London   0171 873 8300
SW8 5DR

Meteorological Office
London Road
Bracknell
Berkshire   01344 420242
RG12 2SZ

Ministry of Agriculture, Fisheries and Food (MAFF)
Whitehall Place
London
SW1A 2HH   0171 238 5782

Ministry of Agriculture, Fisheries and Food (MAFF)
River and Coastal Engineering Group
Eastbury House
30 – 34 Albert Embankment   0171 238 0000
London
SE1 7TL

National Association of Agricultural Contractors
Huts Corner
Tilford Road
Hindhead   01428 605360
Surrey
GU26 6SF

The National Farmers Union
Agricultural House
25 – 31 Knightsbridge
London   0171 235 5077
SW1X 7NJ

National Farmers Union of Scotland
17 Grosvenor Crescent
Edinburgh
EH12 5EN   0131 337 4333

Ordnance Survey
Romsey Road
Maybush
Southampton   01703 792 000
SO9 4DH

River Purification Board
See Scottish Environment Protection Agency (SEPA)

Royal Agricultural Society for England
Stoneleigh Park
Kenilworth
Warwickshire   01203 696969
CV8 2LZ

Royal Institute of Chartered Surveyors
Great George Stree
Westminster
London   0171 222 2605

Royal Forestry Society for England, Wales and Northern
Ireland
102 High Street
Tring   01442 822028
Hertfordshire
HP23 4AF

Royal Society for Nature Conservation
The Green
Witham Park
Lincoln   01522 544400
LN5 7JR

Royal Society for the Protection of Birds
The Lodge
Sandy
Bedfordshire
SG19 2DL
01767 680551

Scottish Natural Heritage
(Nature Conservancy Council for Scotland)
12 Hope Terrace
Edinburgh
EH9 2ES
0131 447 4784

Scottish Office
27 Perth Street
Edinburgh
EH3 5RB
0131 556 8400

Soil Survey and Land Research Centre
Cranfield University
Silsoe
Bedfordshire
MK45 4DT
01525 860428

Thomas Telford Ltd
Thomas Telford House
1 Heron Quay
London
E14 4JD
0171 987 6999

Welsh Office
New Crown Building
Cathays Parks
Cardiff
CF1 3NQ
01222 825111

WRc Evaluation and Testing Centre
Water Byelaws Advisory Service
Fern Close
Pen-y-Fan Industrial Estate
Oakdale
Gwent
South Wales
NP1 4EH
01495 248454

D2 ENVIRONMENTAL AGENCY (ENGLAND & WALES)

Headquarters
Rivers House
Waterside Drive
Aztec West
Almondsbury
Bristol
BS12 4UD
01454 624400

Regional Addresses

ANGLIAN	Anglian Region Kingfisher House Goldhay Way Orton Goldhay Peterborough PE2 0ZR	01733 371811
NORTH WEST	North West Region Richard Fairclough House Knutsford Road Warrington WA14 1HG	01925 53999

SEVERN TRENT	Severn Trent Region Sapphire East 550 Streetsbrook Road Solihull Birmingham B91 1QT	0121 711 2324
SOUTHERN	Southern Region Guildbourne House Chatsworth Road Worthing West Sussex BN11 1LD	01903 820692
SOUTH WEST	South Western Region Manley House Kestrel Way Exeter EX2 7LQ	01392 444000
THAMES	Thames Region Kings Meadow House Kings Meadow Road Reading RG1 8DQ	01734 535000
WELSH	Welsh Region Rivers House/Plas yr Afon St Mellons Business Park St Mellons Cardiff CF3 0LT	01222 770088
NORTHUMBRIAN AND YORKSHIRE	Northumbrian and Yorkshire Region Rivers House 21 Park Square South Leeds LS1 2QG	01532 440191

ENVIRONMENT AND HERITAGE   01232 254754
SERVICE (N IRELAND)
Calvert House
Castle Place
Belfast
BT1 1FY

SCOTTISH ENVIRONMENT PROTECTION AGENCY   01786 457700
(SEPA)
Head Office
Erskine Court
The Castle Business Park
Stirling
FK9 4TR

West Region HQ	North Region HQ	East Region HQ
Rivers House	Graesser House	Cleawater House
Murray Road	Fodderty Way	Heriott Watt Research Park
EAST KILBRIDE	DINGWELL	Avenue North
G75 0LA	IV15 9XB	Riccarton
		EDINBURGH
		EH14 4AP

D3   SITE INVESTIGATION

Archive Sources of Information

(i) Aerial Photographic Maps

ADAS Aerial Photography Unit Brooklands Avenue CB2 2DR Cambridge	01223 455780
Aerofilms Limited Gate Studios Station Road Borehamwood Hertfordshire WD6 1EJ	01602 363100
British Geological Survey Photographic Department Kingsley Durham Centre Keyworth Nottingham NG12 5GG	0181 207 0666
Cartographical Services (Southampton) Limited Landford Manor Landford nr Salisbury Wiltshire SP5 2EW	01794 390321
Earth Images PO Box 43 Keynsham Bristol BS1 82TH	01275 839643
Engineering Surveys Ltd Clyde House Reform Road Maidenhead Berkshire SL6 8BU	01628 21371
Geonex UK Ltd 92 94 Church Road Mitcham Surrey CR4 3TD	0181 685 9393
Central Register for Air Photographs Department of Environment Lambeth Bridge House Albert Embankment London SE1	0171 276 0900
Central Register of Air Photography for Wales The Welsh Office New Crown Buildings Cathays Park Cardiff CF1 3NQ	01222 825111

Central Register of Air Photographs for Scotland
Scottish Development Department
St Andrews House
St James Air Photographic Centre   0131 244 4045
Edinburgh
EH1 3SZ

Public Record Office of Northern Ireland
66 Balmoral Avenue
Belfast
BT9 6NY   01232 661621

(ii) Soil Survey Information

Soil Survey and Land Research
Silsoe College
Silsoe
Bedfordshire   01525 860425
MK45 4DT

The Macaulay Institute for Soil Research
Craigiebuckler
Aberdeen
AB9 2QJ (for the Soil Survey of Scotland)   01224 318611

(iii) Government Publishers

Her Majesty's Stationery Office (HMSO)
Enquiries, Room 308
Publications Centre (PC51D)
51 Nine Elms Lane   0171 873 0011
London
SW8 5DR

(iv) Topographical Information

British Map Library
British Museum
Great Russell Street
London   0171 323 7700
WV13 3DG

Ordnance Survey
Romsey Road
Maybush
Southampton   01703 792000
SO9 4DH

(v) Geological Information

British Geological Survey
Keyworth
Nottingham
NG12 5GG   01602 363100

(vi) Archive Mining Records

British Coal Opencast: Headquarters
200 Lichfield, Berry Hill
Mansfield
Nottinghamshire   01623 22681
NG18 4RG

British Coal
Mining Records Centre
Bretby
Burton Upon Trent   01283 550606
Staffordshire
DE15 0QD

British Coal
Mining Records Centre
Newton Grange
Dalkeith   0131 654 2777
Midlothian
EH22 4NZ

Her Majesty's Inspectorate of Mines
St Anne's House
Stanley Precinct
Bootle   0151 951 4136
Merseyside
L20 3RA

Institution of Mining and Metallurgy
44 Portland Place
London   0171 580 3802
W1N 4BR

Appendix E Waterborne diseases

A number of diseases can be transmitted by water. Those of major importance include:

Causal Organisms	Disease
Bacteria	Dysentery
	Gastro enteritis
	Enteric fever
	Weils disease (Leptospirosis)
Viruses	Infectious hepatitis
	Poliomyelitis (infantile paralysis)
Intestinal parasites	Cryptosporidium

Disease causing organisms are discharged in human and animal excreta. Water supplies should be protected from contamination.

The risk of water being contaminated can be limited by controlling the discharge of possible contaminants into inland watercourses. Care should be taken to ensure the purity of the supply by all possible means, usually by protection of the source. Stream/river water used for impounding and non-impounding reservoir supply is likely to have been monitored by the RLA and information upon its purity should be sought if water is to be extracted from this source. Care should be taken with any runoff from agricultural land so that contamination from animal excreta is limited. Fences should be erected to control the movement of livestock near a reservoir and any reservoir feeder watercourses.

Water quality is of paramount importance for the protection of public health for sports, amenity and recreational purposes and any water abstraction usage for human or animal consumption. The development of certain forms of potentially dangerous algae associated with enhanced temperatures, amongst other factors, have also proved to be a problem to humans and animals in recent years.

Appendix F Flow gauging
(After Ref 2)

F1 **90 DEGREE VEE-NOTCH WEIR FLOW GAUGE**

Details of the vee-notch weir flow gauge are shown in Figure F1. It is important that the notch angle is 90° and that the dimensions are complied with, including the bevel in the notch. The flow over the weir can be estimated from the following table:

Head over notch (mm)	Flow (litre/sec)
10	0.02
20	0.08
30	0.22
40	0.46
50	0.79
60	1.25
70	1.83
80	2.54
90	3.41
100	4.42
110	5.60
120	6.95
130	8.48
140	10.2
150	12.1
160	14.2
170	16.5
180	19.0
190	21.7
200	24.7
210	27.9
220	31.3
230	34.9
240	38.8
250	42.9
300	67.5
350	98.9
400	137
450	184
500	239

NOTE: Flow = $1336h^{2.48}$ l/s
Where h = head over notch in metres

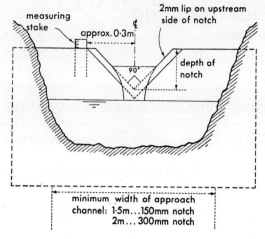

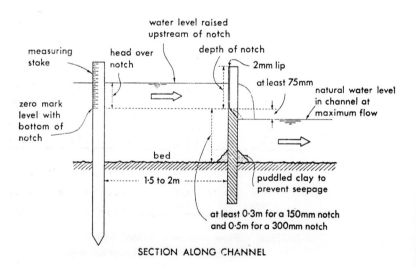

Figure F1 Installation of 90° vee-notch weir flow gauge

F2   RECTANGULAR WEIR FLOW GAUGE

Details of the rectangular flow gauge are shown in Figure F2. It is important that the dimensions shown are complied with, including the bevel to the weir. The flow over the weir can be estimated from the following table for varying weir length. Flows over intermediate lengths of weir can be interpolated or calculated directly from the formula.

Head over weir (mm)	Flow (l/s) for weir length			
	0.6m	1.0m	1.3m	1.6m
20	3.07	5.13	6.67	8.21
30	5.64	9.41	12.2	15.1
40	8.66	14.5	18.8	23.2
50	12.1	20.2	26.3	32.4
60	15.9	26.4	34.6	42.6
70	20.0	33.4	43.5	53.6
80	24.3	40.8	53.1	65.5
90	29.0	48.6	63.3	78.1
100	33.9	56.9	74.1	91.4
110	39.0	65.6	85.5	105
120	44.4	74.6	97.3	120
130	50.0	84.1	110	135
140	55.8	93.9	122	151
150	61.8	104	136	167
160	67.9	114	149	184
170	74.3	125	163	202
180	80.8	136	178	220
190	87.4	148	193	238
200	94.3	159	208	257
210	101	171	224	276
220	108	183	240	296
230	116	196	256	316
240	123	209	273	337
250	131	221	290	358
300	170	290	379	469
350	212	363	476	589
400	257	441	579	717
450	304	524	688	853
500	353	610	803	996

NOTE: Flow = $1817(L - 0.1h)h^{1.5}$ l/s

Where h = head over weir in metres

L = length of weir in metres

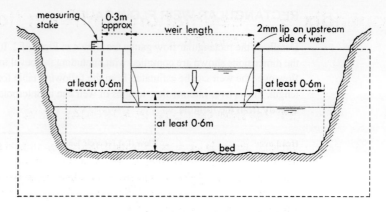

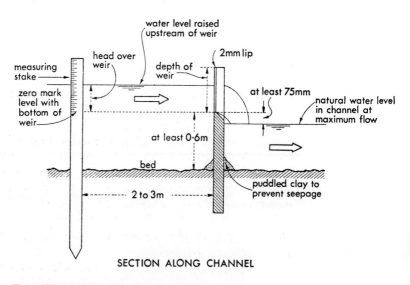

Figure F2 Installation of rectangular weir flow gauge

Appendix G Estimation of stream flow
(After Ref 2)

G1 ESTIMATION OF FLOW IN A TYPICAL YEAR

Table G1 shows, in tabulated form, the daily readings taken at a stream gauge every day during the winter months. The figures are equivalent to a site in eastern England with low rainfall where irrigation facilities might be considered. The readings are used as a basis for the estimation of the probable run-off from the stream catchment area in a dry winter.

For any stream gauging, the daily readings of head or depth of water over the flow gauge, as indicated on the measuring stake, should be tabulated day by day on a chart similar to Table G1 and the corresponding daily flows found from Appendix F according to which type of flow gauge is being used. The daily flows in each monthly column are totalled and multiplied by the factor indicated to give the total flow in each month in cubic metres per month and the monthly flows are totalled to give the winter flow. If, by chance, some daily readings are missed, an estimate of the flow on the omitted days can be made by comparing the flows on the day before and on the day after the omission; such estimated flows are shown in brackets in Table G1. Heads, as indicated on the measuring stake, should be read to the nearest 10 mm up or down, and any heads less than 10 mm shown as zero.

G2 ESTIMATION OF FLOW IN A DRY YEAR

If the flow in a stream in any winter is known, as above, then the flow in the same stream in the driest winter in 20 years (refer to MAFF Guide 'Water for Irrigation (Bulletin No. 102)) can be estimated in the following manner. The method described should only be used when more sophisticated hydrological methods are not available to the reader. As with all simple empirical methods, the results can in some cases differ from reality by an appreciable amount and if the calculated dry winter flow does not exceed the amount of water required by a sufficient margin, then alternative sources should be investigated.

The following information is required to calculate the dry winter flow:

- gauged stream flow in any winter
- rainfall in that winter
- average winter rainfall
- rainfall in previous summer
- average summer rainfall

Available from the Meteorological Office or ADAS for the area concerned

The method is based on the assumption that the winter rainfall and runoff have a simple relationship if the rainfall in the previous summer is equal to the summer average. This relationship, once established, can be used to estimate the dry winter runoff. Thus the calculations involve two adjustments to the actual gauged winter runoff. The first adjusts for the rainfall occurring in the previous summer, producing a corrected winter runoff based on the average summer rainfall. The second adjusts the corrected winter runoff to give an estimate of the runoff occurring in the driest winter in 20 years. An error, which tends to over-estimate the dry winter runoff, will be introduced if storm flows form a large proportion of the total winter runoff, but for streams which are gauged once a day, it is likely that most storm flows will be missed.

G3 EXAMPLE USING GAUGED RUNOFF

G3.1 Available information

Assume the following parameters for a catchment in eastern England within National Grid Reference Square TL:

Gauged runoff in a winter	=	118 mm (Table G1)
Rainfall in that winter	=	361 mm
Average winter rainfall	=	297 mm
Rainfall in the previous summer	=	343 mm
Average summer rainfall	=	290 mm
Average annual rainfall AAR	=	297 + 290 mm = 587 mm

G3.2 First adjustment for rainfall in previous summer

The 'surplus' or 'deficit' of the rainfall in the previous summer is 'rainfall in previous summer' minus 'average summer rainfall':

= 343 − 290 mm
= + 53 mm

Therefore, there is a 'surplus' of 53 mm of summer rainfall. If the summer rainfall had been below average the answer would have been a 'deficit' with a negative (−) sign.

The proportion of the summer rainfall 'surplus' or 'deficit' which influences the winter runoff is:

$$\text{surplus or deficit} \times \frac{\text{gauged winter runoff}}{\text{winter rainfall}} \times 2$$

$$= +53 \times \frac{118}{361} \times 2$$

$$= 35 \text{ mm}$$

Corrected winter runoff = gauged winter runoff − (+35) mm
= 118 − 35 mm
= 83 mm

If there is a summer rainfall 'deficit', the corrected winter runoff would be greater than the gauged winter runoff.

The corrected winter runoff of 83 mm is that produced by 361 mm of winter rainfall if 290 mm of rainfall had occurred in the previous summer. Assuming 80% of the flow occurs in the winter months, this suggests a total average annual runoff of 104 mm and losses (see Section 6) of 483 mm (587 − 104 mm) or approximately 82% of the average annual rainfall.

G3.3 Second adjustment for rainfall in dry winter

As the rainfall occurring in the driest winter in 20 years is about 70% of the average winter rainfall, dry winter rainfall = 70% of 297 mm
= 208 mm

$$\text{Estimated dry winter runoff} = \text{corrected winter runoff} \times \frac{\text{dry winter rainfall}}{\text{winter rainfall}}$$

$$= 83 \times \frac{208}{361} \text{ mm}$$

$$= 48 \text{ mm}$$

Thus the runoff occurring in the driest winter in 20 years is estimated at 48 mm and is only 8% of the average annual rainfall.

As the stream catchment area is 40 hectares, the estimated total dry winter stream flow is 19 200m^3, which is equivalent to an average flow throughout the six months of winter of 1.2 l/s.

It can be seen from the above calculations that, in this particular case, the estimated dry winter stream flow is less than half of the gauged winter stream flow. This shows that what may appear to be an abundant source of water in some winters may well become seriously deficient in a dry winter.

Table G1 Typical stream monitoring records and assessment of water flow

Day	October Head (mm)	October Flow (l/s)	November Head (mm)	November Flow (l/s)	December Head (mm)	December Flow (l/s)	January Head (mm)	January Flow (l/s)	February Head (mm)	February Flow (l/s)	March Head (mm)	March Flow (l/s)
1	60	1.25	60	1.25	90	3.41	80	2.54	80	2.54	70	1.83
2	70	1.83	60	1.25	90	3.41	90	3.41	80	2.54	70	1.83
3	60	1.25	70	1.83	—	(3.41)	90	3.41	60	1.25	80	2.54
4	50	0.79	70	1.83	90	3.41	100	4.42	70	1.83	—	(2.54)
5	50	0.79	60	1.25	90	3.41	110	5.60	70	1.83	80	2.54
6	80	2.54	70	1.83	70	1.83	120	6.95	—	(1.83)	70	1.83
7	90	3.41	80	2.54	70	1.83	120	6.95	—	(1.83)	70	1.83
8	80	2.54	60	1.25	—	(1.83)	120	6.95	70	1.83	60	1.25
9	—	(2.54)	60	1.25	70	1.83	130	8.48	80	2.54	70	1.83
10	80	2.54	60	1.25	80	2.54	130	8.48	90	3.41	80	2.54
11	60	1.25	50	0.79	80	2.54	130	8.48	90	3.41	90	3.41
12	60	1.25	50	0.79	80	2.54	130	8.48	90	3.41	—	(3.41)
13	50	0.79	40	0.46	90	3.41	130	8.48	100	4.42	90	3.41
14	50	0.79	50	0.79	90	3.41	130	8.48	100	4.42	100	4.42
15	70	1.83	60	1.25	90	3.41	120	6.95	100	4.42	100	4.42
16	80	2.54	70	1.83	100	4.42	120	6.95	90	3.41	—	(4.42)
17	60	1.25	80	2.54	100	4.42	130	8.48	90	3.41	100	4.42
18	50	0.79	80	2.54	100	4.42	—	(8.48)	—	(3.41)	100	4.42
19	50	0.79	70	1.83	100	4.42	—	(8.48)	80	2.54	90	3.41
20	60	1.25	70	1.83	90	3.41	130	8.48	70	1.83	90	3.41
21	60	1.25	70	1.83	—	(3.41)	120	6.95	100	4.42	80	2.54
22	50	0.79	70	1.83	90	3.41	120	6.95	100	4.42	80	2.54
23	50	0.79	60	1.25	80	2.54	120	6.95	80	2.54	80	2.54
24	50	0.79	60	1.25	80	2.54	110	5.60	80	2.54	—	(3.41)
25	50	0.79	60	1.25	80	2.54	110	5.60	80	1.83	90	3.41
26	70	1.83	70	1.83	70	1.83	100	4.42	70	3.41	70	1.83
27	40	0.46	70	1.83	70	1.83	90	3.41	90	3.41	70	1.83
28	50	0.79	70	1.83	70	1.83	90	3.41	90	3.41	60	1.25
29	50	0.79	80	2.54	80	2.54	80	2.54	—	—	60	1.25
30	50	0.79	90	3.41	70	2.54	80	2.54	—	—	70	1.83
31	40	0.46	—	—	80	2.54	80	2.54	—	—	90	3.41
		41.54		49.03		91.73		197.31		82.09		85.55

Sum of Daily Instantaneous Flows (l/s)

NOTE: Monthly total = sum of daily instantaneous flows × 86.4 (m³)
       Total winter flow = sum of monthly flows
                        = 47 282 m³

Catchment area assessed at 40 ha

Appendix H Typical flood assessment

H1 EXAMPLE

H1.1 Information available

Assume the following parameters for the same catchment in eastern England, as Example G1:

- catchment area A_c = 40 ha, with no external influences on the catchment
- reservoir area A_w = 3 ha
- isolated ponds totalling 3 ha in area
- a concentrated area of hardstanding, buildings and glasshouses amounting to 2 ha (CA_R) near the upstream end of the catchment
- a flat catchment with SI085 value of 8 m/km
- mean catchment slope g = 3°
- commonly waterlogged soils within 600 mm of ground surface during winter with medium permeability soils and 600 mm depth to impermeable horizon over 40% of the catchment
- rarely waterlogged soils with medium permeability soils and depth to impermeable horizon in excess of 800 mm over remaining 60% of the catchment
- newly planted trees, with drainage ditches, over 40% of the catchment, located entirely on the rarely waterlogged soil.

H1.2 Assessment of flood producing rainfall (Figure 7.2)

RSMD = 24 mm by interpolation (Figure 7.1)
A_c = 40 ha
hence Q_m/A_c = 0.14 m³/s/ha (Figure 7.3)

H1.3 Assessment of catchment area, water surface areas and areas of rapid runoff (Figure 7.5)

A_R = 2 ha
Thus A_R/A_c = 2/40 and less than 0.1

Hence specialist advice not required.

$$A_w = 3 \text{ ha}$$
$$\text{Thus} \quad A_w/A_c = 3/40 \text{ and less than } 0.1$$

Hence ignore effects of water areas.

$$A = 40 \,(1 + (2/40)) \text{ ha}$$
$$= 40 \times 1.05 \text{ ha}$$
$$= 42 \text{ ha}$$

H1.4 Assessment of stream and catchment slopes (Figure 7.6)

$$SI085 = 8 \text{ m/km}$$
$$\text{Thus} \quad G = 0.94 \text{ (Table 7.1)}$$

H1.5 Assessment of soil and geology (Figure 7.7)

40% of catchment is Soil Type S4
60% of catchment is Soil Type S2 } Tables 7.2, 7.3 and 7.4

$$\text{Thus} \quad SOIL = 0.40 \times 0.40 + 0.30 \times 0.60 \text{ (Figure 7.7)}$$
$$= 0.34$$
$$\text{Note} \quad S_I = S_3 = S_5 = 0$$
$$\text{Hence} \quad V = 0.82 \text{ by interpolation (Table 7.5)}$$

H1.6 Assessment of vegetation (Figure 7.8)

$$A_f = 40\% \text{ of catchment on Soil Type S2}$$
$$A_f/A_c = 40\% \text{ and greater than } 0.1$$
$$F \text{ (Soil Class 2)} = 1.15 \text{ by interpolation (Table 7.6)}$$

Hence
$$W = 1 + (F-1)\frac{40}{100}$$
$$= 1 + (1.15 - 1) \times 0.4$$
$$= 1.06$$

H1.7 Assessment of maximum flood (Figure 7.9)

$$Q_{max} = (Q_m/A_c)\,AGVW$$
$$= 0.14 \times 42 \times 0.94 \times 0.79 \times 1.06 \text{ m}^3/\text{s}$$
$$= 4.804 \text{ m}^3/\text{s}$$

Assume $Q_{max} = 4.9 \text{ m}^3/\text{s}$ for design purposes.

It should be noted that the calculated value of Q_{max} must be rounded up to the next two significant figures, not down, to arrive at the Q_{max} value for design purposes.

Appendix I Effects of water surface areas

I1   EFFECTS OF THE RESERVOIR ON FLOOD FLOWS

The flood rise due to a flood entering a reservoir can be reduced or attenuated, as discussed in Section 7.2.2. The flood rise, h_u, is reduced to Rh_u, where R is the attenuation ratio, which is always less than unity. R may be determined from Figure I2. Hence a reduced, or attenuated design flood flow, Q_o, passes out of the reservoir.

R is dependant on the following:
- design flood, Q_i (m³/s)
- slope factor, X, based on the terrain. X can be estimated using:

Terrain	X
mountainous	1455
hilly	1676
undulating	1866
flat	2498

It should be noted that X is not dimensionless.
- maximum unattenuated flood rise, h_u (m)
- reservoir area at a height of $h_u/2$ above top water level (which may be considered to be equal to the reservoir area at top water level), a (m²)
- catchment area, A_c (ha)
- average annual rainfall AAR (mm)

I2   EFFECTS OF OTHER WATER AREAS ON FLOOD FLOWS

Other water areas may be sufficient to reduce flood flows in the watercourse upstream of the reservoir. Their effects can be treated individually and any reduction will only be significant if their individual areas are greater than 10% of their respective catchment areas. The reduction in flow should be calculated in the manner described in Section H1 using the appropriate values for the various factors.

I3   EXAMPLES

I3.1   Information available

Assume the same catchment parameters as Examples G1 and H1 except that the reservoir has a water area of 10 ha. The reservoir is upstream of a community. The unattenuated flood rise is 600 mm over a single overflow. The reservoir area will now form a substantial part of the catchment area at 25% of the total and will offer some attenuation to the outflow from the reservoir.

I3.2 Assessment of design flood (Section 7.2.1)

Maximum flood, Q_{max} = 4.9 (Example H1)
Flood multiplier, M = 1.0 (Table 6.7)
Hence design flood, Q_1 = $MQ_{max} = 4.9$ m³/s

I3.3 Assessment of storage ratio (Figure I1)

X = 2498 (flat catchment) (Section I1)
h_u = 600 mm
a = 10 ha = 100 000 m²
A_c = 40 ha
Hence S = $(ah_u)/(XQ_i A_c^{1/4})$
= $(0.6 \times 100,000) / (2,498 \times 4.9 \times 40^{1/4})$
= 1.95

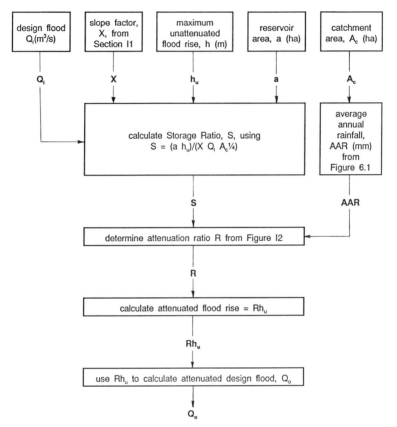

Figure I1 Flood attenuation procedure

13.4 Assessment of attenuation ratio (Figure I2)

Average annual rainfall (AAR) = 587 mm
Hence R = 0.68
Maximum alleviated flood rise = 0.68 × 600 mm
= 408 mm

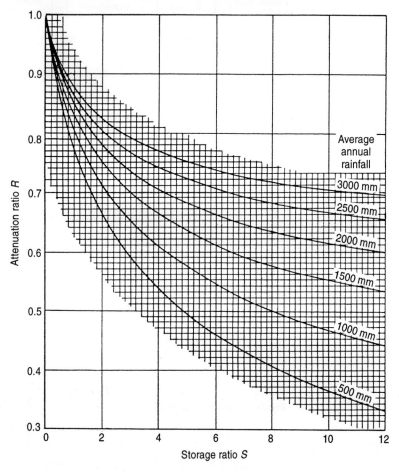

Figure I2 Determination of flood attenuation ratio, R (Ref 54)

Appendix J Sources of information

Table J1 lists information from maps and plans which should readily be available in local libraries or other sources of reference or may be purchased from HMSO and other bookshops.

Table J2 lists information that may be available detailing past mining activities and aerial photographs.

The local planning authority and the relevant licensing authority may have extensive information including records of:

- previous land use
- regional plans for land use
- planning controls
- previous planning applications
- records of ground conditions
- groundwater levels
- flood levels
- flows in watercourses
- groundwater management and protection plans.

Local utilities shall have records of their service locations but may also have information, particularly where it is considered to affect their interests. British Rail, the Department of Transport and similar public bodies may also hold relevant information.

Information may also be available following local development or construction works. This may be available not only for actual projects which have been built, but for schemes which have not proceeded to construction. The information may be of commercial value, however, and may not readily be available from private and commercial organisations.

The information referred to in the previous two paragraphs is likely to consist mainly of the results of ground investigations and construction operations. Information may be available, however, from earlier information searches, records which have now become lost or unavailable, and personal experience or information which is unobtainable elsewhere.

Information on past building or construction problems may be very useful, particularly if these are related to settlement, instability or drainage problems. Similarly, information concerning problems or difficulties relating to groundwater aspects and levels will be of importance.

Previous land filling or waste disposal operations in the area of interest warrant special consideration. Past control of such operations, and adequate records, were normally absent. If the area of interest is found, or suspected, to have been used for such purposes, a detailed search for information is necessary to try to discover the nature and extent of the infill. The potential effects of the possible hazardous ground on the embankment and elsewhere regarding settlement and watertightness must be ascertained as far as is practicable to any ground investigation work on site.

Other local societies and educational establishments may also have useful information. These could include:
- archaeological societies
- local history societies
- natural history societies
- geological societies
- schools
- universities and colleges.

Table J1 Available information from maps and plans

	Maps		Plans			
	Available scales	Typical information provided	Available scales	Typical information provided	Site details to note	Source
Topographical survey	1:63360 (1 inch to the mile) 1:50000 1:25000 (several editions available, old copies may date back to the 1700s)	provides a general overview of the site and topographic setting 1:25000 particularly useful for catchment hydrology and flood assessment	1:10560 (6 inches to the mile) 1:10000 1:2500 1:1250 (several editions available, old copies may date back to the 1700s) 1:200 site survey plan	detailed plans for site specific topographic information base plan to use for pre-construction drawings, visual observations, maintenance	• past and present land use • boundaries : walls, fences, buildings • rivers, streams, springs, ditches • roads, tracks, access points • landslipped areas • contours indicative of slope steepness • change in site conditions and land-use with time	Local Library Local Authority British Map Library Bookshops stocking Ordnance Survey County Record Office
Geological survey • Drift • Solid and Drift • Solid	1:63360 (1 inch to the mile) 1:50000	provides a general overview of geology of the site, especially if read in conjunction with published geological memoirs	1:10560 (6 inches to the mile) 1:10000	detailed plans for site specific geological information	Drift • lateral variation of deposits across site • thickness of deposits • composition of deposits • slope instability Solid • geological succession • faults • dip and dip direction of beds • slope instability • previous mineral extraction details (eg old mines)	Local Library Local Authority British Geological Survey Offices HMSO Bookshops County Record Office
Soil survey	1:63360 (1 inch to the mile) 1:5000 1:2500	provides a general overview of main soil types across the site, especially useful when drift geological maps are not available	Larger scales may be available for selected areas	detailed plans for site specific soils information	• agricultural soil types • drainage features • indications of underlying geology	Local Library County Library British Soil Survey Offices

Table J2 Information from archive sources

	Available scales	Potentially relevant information	Source
Mining records	1:625000 1:100000 1:63360 (1 inch to the mile) 1:10560 (6 inches to the mile) 1:5000 1:2500 also frequently available at obscure scales, especially the older records	• location of concealed mineshafts and adits • extent of underground workings • location of quarries, clay, sand and gravel pits • location of filled ponds and settlement lagoons • changes in topography and drainage • areas of subsidence and landslip associated with mine workings.	unpublished mine plans may be available from: • British Coal • Mines Record Office • Local Authorities • Institution of Mining and Metallurgy • Health and Safety Executive (non-coal mines) published mine plans may be available from: • British Geological Survey • British Map Library • Department of the Environment • County Record Office
Aerial photographs	1:25000 1:10000 1:5000	• topography • soil and rock boundaries • faults • landslips and shallow surface ground movements • swallow holes, solution features • drainage patterns and features • erosion features	• Local Authorities • Local Libraries • County Libraries • Central Register for England, Wales and Scotland • Ordnance Survey • County Record Offices

Appendix K Typical field sketch

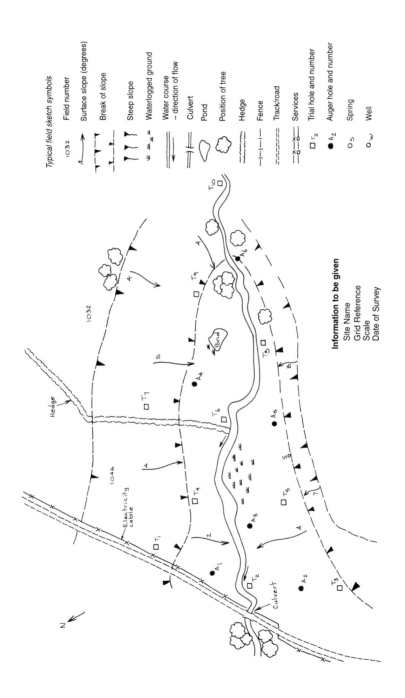

Appendix L Reservoir liners

L1 GEOMEMBRANES

L1.1 Choice of materials

A number of different synthetic materials are used as geomembrane liners. These are classified as:

- polyethylene materials
- chlorinated polymer materials
- rubber-based materials.

Plastic liners are susceptible to deterioration when exposed to ultra-violet light and protection is essential. Chlorinated polymers are not affected to the same extent, but must also be protected to ensure a reasonable life. Some chlorinated polymers can suffer from leaching of the plasticiser resulting in brittleness and premature failure. Both liners are available in thicknesses from 0.3 mm to over 3 mm . The thicker grades provide a more robust lining, particularly with respect to tearing and puncture and a minimum thickness of 1.5 mm is recommended. Both materials exhibit moderate changes in length with reduced temperature.

Rubber-based materials are generally more expensive but have a higher resistance to ultra-violet light degradation and weathering. They possess a greater elasticity and hence have a greater resistance to tearing. Reinforced materials are also available and comprise synthetic fibres, usually of polypropythene or nylon, embedded in the impermeable sheet. This enhances their resistance to tearing and, to a lesser extent, puncturing, but has no effect on their weathering or ultra-violet degradation resistance. Jointing is more difficult than for the unreinforced material.

The liners are marketed under different product names. A selection of proprietary geomembranes available in the UK has been included in Tables L1, L2, L3 and L5. The choice of a geomembrane should be based upon:

- cost
- suitability of geomembrane
- durability of geomembrane
- evidence of successful applications in the past
- experience of supplier and installation staff
- maintenance requirements.

A number of suppliers should be approached. Specialist advice is necessary to ensure that a suitable geomembrane liner is selected.

Table L1 Geomembranes – butyl rubber-based products

Product name	Thickness available (mm)	Manufacturer/supplier	Brief description
Butyl	0.75	Aquatics Ltd	butyl rubber / EPDM mix
Butyl	0.75, 1.00, 1.50	Franklin Hodge Industies Ltd	butyl rubber
Varnamo	0.75, 1.00, 1.50, 2.00	Russetts Developments Ltd	butyl rubber
Butylite	0.75, 1.00, 1.50, 2.00	Singleton Flint and Co Ltd	butyl rubber
Stapelite	0.75	Stapeley Water Gardens Ltd	butyl rubber
Butylite AA	0.75, 1.00, 1.50	White Cross Rubber Products Ltd	butyl rubber
Butylite AX	0.75, 1.00, 1.50	White Cross Rubber Products Ltd	butyl rubber
Butylite AF/AN	0.75, 1.00, 1.50	White Cross Rubber Products Ltd	butyl rubber
Bytylan	0.75	Hertel (UK) Ltd	butyl rubber
Butyl	0.75, 1.00	Butek Ltd	butyl rubber
Butyl	0.75, 1.00	Gordon Low Ltd	butyl rubber

Note: The list is not exhaustive and does not imply recommendation

Table L2 Geomembranes – polyethylene products

Product name		Thickness available (mm)	Manufacturer/supplier	Brief description
MDPE		0.50, 1.00, 1.50, 2.00, 2.50, 3.00	Saftex Ltd	medium-density polyethylene
HDPE		1.50, 2.00, 2.50		high-density polyethylene
Exceliner		0.50 – 2.00	LEC Geosystems Ltd	low-denstiy polyethylene
HDPE LDPE Recycled HDPE		all at 0.50, 1.00, 1.50, 2.00	Euro Erosion Engineering Ltd	high-density polyethylene low-density polyethylene recycled high-density polyethylene
Monarflex	500 700	0.50 0.70	Monarflex Geomembranes Ltd	low-density polyethylene
Blackline	500 750 1000 1500	0.50 0.75 1.00 1.50	Monarflex Geomembranes Ltd	low-density polyethylene
HDPE LLDPE		all at 1.00, 1.50, 2.00, 2.50	Butek Ltd	high-density polyethylene very low-density polyethylene
Gundline Gundline	HD VL	1.00, 1.50, 2.00 0.75, 1.00, 1.50, 2.00	Gundie Lining Systems Inc	high-density polyethylene very low-density polyethylene
Nicotarp	100 100-HD 100-2HD	0.40 0.80 0.85	MMG Civil Engineering Systems Ltd	high/low-density polyethylene with low-density polyethylene coating on both sides
SLT Sheet		1.00, 1.50, 2.00, 2.50, 3.00, 3.50, 4.00, 4.50, 5.00	SLT Lining Technology GMBh	high-density polyethylene
Politarp		0.375, 0.500, 0.75, 1.00, 1.50, 2.00, 2.50	Visqueen	polyethylene film

Note: The list is not exhaustive and does not inply any recommendation

Table L3 Geomembranes – polyvinyl chloride (PVC)-based products

Product name	Thickness available (mm)	Manufacturer/supplier	Brief description
PVC	0.75	Butek Ltd	polyvinyl chloride
Aquatex	0.50	Gordon Low Ltd	polyvinyl chloride
PVC	0.35	Aquatics Ltd	polyvinyl chloride
PVC	0.35	Singleton Flint and Co Ltd	polyvinyl chloride
PVC	0.35, 0.50	Stephens (Plastics) Ltd	polyvinyl chloride

Note: The list is not exhaustive and does not inply any recommendation

Table L4 Bentonite liners – bentonite mats

Product Name	Manufacturer/Supplier	Brief description
Rawmat	Rawell Marketing Ltd	granular american sodium bentonite sandwiched between polypropylene geotextile and a protective scrim
Bentomat CS50	Voclay Ltd	granular american sodium bentonite sandwiched between polypropylene geotextile
Bentofix	Netlon Ltd	granular american sodium bentonite sandwiched between Netlon geotextiles

Note: The list is not exhaustive and does not inply any recommendation

Table L5 Geomembranes – other lining products

Product name	Thickness available (mm)	Manufacturer/supplier	Brief description
Hypalon	0.75, 0.90, 1.20	Dunstable Rubber Co Ltd	chlorosulfonated polyethylene
Hertaion	1.00	Hertel (UK) Ltd	ethylene propylene diene monomer
Aqualast Rubber	1.00	Gordon Low Ltd	rubber
Hypofors	3.00 – 5.00	MMG Civil Engineering Systems Ltd	reinforced bitumen

Note: The list is not exhaustive and does not inply any recommendation

L1.2 Design aspects

Reservoir slopes flatter than 1V:3H may need to be adopted to ensure the stability of the bedding material, geomembrane liner and covering material. The appropriate slope should be determined from the findings of the site investigation and specialist advice may be required.

Adequate ground preparation is essential to prevent damage to the geomembrane liner during and after installation. A firm foundation surface for the geomembrane and bedding material should exclude stones, topsoil or other unsuitable material.

It is essential that the geomembrane liner is placed upon a suitable bedding material to prevent any damage during subsequent covering and compaction after jointing. This should be a compacted sand layer of 100 mm minimum thickness. If the foundation material is sand, then this bedding layer may not be required.

An anchor trench is required along the crest and the reservoir perimeter to secure the embankment geomembrane liner. This should ideally be a U-shaped trench (not less than 600 mm × 600 mm). A typical detail for an anchor trench is shown in Figure L1.

There are three basic methods of jointing geomembrane liners in the field. These depend on the type of geomembrane material as follows:

- solvent / adhesive   –   (rubber-based materials and chlorinated polymer materials)
- hot air / hot weld   –   (polyethylene materials)
- extrusion bonding   –   (polyethylene materials).

Whichever method a manufacturer/supplier recommends, it is important to ensure that the material to be joined is clean and that the proposed method is followed carefully. Joints are the major potential source of leakage in a lined reservoir.

Consideration should be given at the design stage to any pipework or structures that are within the area to be lined as special connection details to the geomembrane will be required. Advice on these details should be sought from geomembrane suppliers/manufacturers.

Small non-impounding reservoirs have been designed without a covering layer in the past. This practice exposes the geomembrane to sunlight, cyclical temperature change and wetting and drying. Whilst the long-term behaviour of exposed geomembrane liners is claimed to be satisfactory, it is considered prudent and good practice to provide protective cover to geomembrane lined reservoirs. It is also important to ensure that cover to the geomembrane is placed correctly so as not to overstress, rupture or otherwise damage the geomembrane liner and joints. A minimum thickness of 100 mm should be provided, as shown in Figure L1.

Where high groundwater levels are present, these could lead to large uplift pressures on the membrane when the reservoir is empty or at a low level. The granular bedding layer will facilitate water migration and more widely distributed uplift pressures. Specialist advice should be sought and appropriate means of drainage or counterweighing provided.

Similarly if it is considered that gas could collect beneath the membrane, precautionary measures, including venting, must be provided. This requires specialist advice to ensure the safe construction, operation and maintenance of the reservoir, including the safe venting of the gases, and is outside the scope of this guide.

L1.3   Construction requirements

The site should be prepared by:

- removal of all standing water
- cutting of all vegetation and removal of the topsoil and roots. Where appropriate, a weed killer should be used
- excavation and regrading of the reservoir area to the required gradients and profile

- removal of all unsuitable materials as detailed in Section 9.3.6, including stones or any other fragments which could cause damage to the geomembrane.

The geomembrane installation should proceed as follows:

- carry out pre-installation tests on geomembrane test joints to ensure that the jointing method is suitable
- place and compact the required thickness and type of bedding
- excavate an anchor trench for the geomembrane as shown in Figure L1
- install the geomembrane in the anchor trench and backfill with excavated material
- unroll the geomembrane from the anchored end down the reservoir slope
- avoid pulling out all slack in the lining as the geomembrane will undergo some deformation as the reservoir is filled
- ensure that the overlaps between adjacent sheets are at least 300 mm and run down to the slope
- visually inspect all joints
- test all joints using the most suitable method to ensure a watertight seal
- ensure that joints to concrete or other structures have been constructed adequately (many manufacturers have typical details for this purpose)
- ensure that any areas subject to erosion from an inlet are adequately protected. A double layer of geomembrane is often used
- ensure that any areas of damage are repaired and tested prior to placement of the protection.

Protection to the liner should be placed at the earliest opportunity as follows:

- place a minimum thickness of a 100 mm fine granular material (shingle or hoggin may be used)
- where erosion due to wave action is anticipated, a 300 mm minimum thick layer of crushed stone (maximum particle size 200 mm) should be placed on top of the covering layer along the reservoir slopes
- if excessive erosion from an inlet is anticipated, a concrete pad should be constructed to protect the reservoir lining and covering material in the immediate area. A nominal thickness of 100 mm should be suitable. Dimensions will vary depending on the inlet details
- fill the reservoir as soon as possible after installation of the covering layer and any additional protection, at a rate which will not cause any damage to the protection measures.

L2 BENTONITE

L2.1 Choice of materials

The raw material, bentonite, occurs naturally in two forms: calcium bentonite and sodium bentonite. Calcium bentonite may be treated with soda ash to form a calcium/sodium bentonite called 'activated' bentonite. Fullers earth is a naturally occurring calcium bentonite that is chemically treated to become an activated bentonite. The principal differences between these forms are outlined below and it is important to select the right one for the particular application.

- natural sodium bentonite (most extensively used)
    - swelling capacity 10 to 15 times original volume;
    - able to absorb up to five times its own dry weight of water;
    - will hydrate, dehydrate and rehydrate.
- natural calcium bentonite (limited use)
    - swelling capacity three to four times original volume;
    - able to absorb up to one-and-a-half times its own dry weight of water.
- activated bentonite (limited use – generally smaller applications)
    - swelling capacity eight to nine times original volume;
    - able to absorb two to three times its own dry weight of water;
    - will hydrate and dehydrate but will not rehydrate to any appreciable extent.

There are essentially three methods that are in general use for the construction of a reservoir liner using bentonite in a powder or granular form. These are:

- mixed insitu blanket – layer of bentonite and insitu soil mixed on the reservoir floor
- mix and place blanket – mixed bentonite and sand, subsequently placed on the reservoir floor
- pure blanket – layer of bentonite placed on the reservoir floor.

Bentonite mats comprise a core of sodium bentonite between geosynthetic materials. They are used in a similar way to the pure bentonite method for bentonite powders but save considerable time during construction as the liner is in a much more manageable form.

Advice on the appropriate choice of bentonite and method of use should be obtained from a specialist adviser. Selections of proprietary loose bentonite and bentonite mats available in the UK have been included in Table L4.

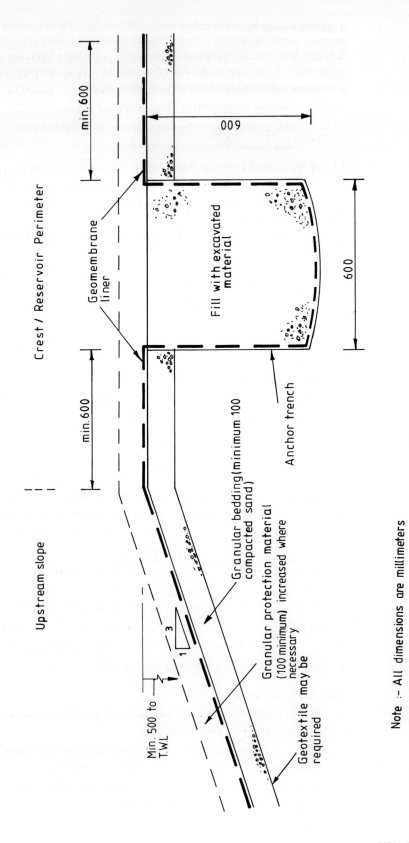

NOTE: Top water level at least 500 mm below the top of the geomembrane liner

Figure L1 Anchor trench details

L2.2 Design aspects

The design aspects described for geomembranes in Section L1.2 should generally be followed, except that a flatter maximum slope may be required. The design details and placing requirements of the manufacturer/supplier should be followed closely to ensure the most suitable choice of application.

L2.3 Construction requirements

The site should be prepared as detailed for the geomembranes in Section L1.3 and the bentonite placed in accordance with the following particular requirements.

Bentonite – mixed insitu blanket

- Select application rate after discussion with the suppliers and specialist adviser
- spread bentonite evenly across the whole of the reservoir site using hand rakes or suitably designed equipment
- rotovate the reservoir area thoroughly in two directions (a farm rotovator may be used)
- ensure mixed, uncompacted layer is not less than 150 mm thick
- thoroughly compact with a smooth roller (a sheepsfoot roller should not be used)
- the thickness of the compacted layer should be not less than 100 mm
- cover and protect as described in Section L1.3.

L2.3.3 Bentonite – mix-and-place blanket

- Use a bentonite powder/sand mix in the proportions and at the application rates selected after discussion with the suppliers and specialist adviser
- mix the materials thoroughly in a dry state (a cement mixer may be used)
- spread mixed material evenly across the whole of the reservoir site using hand rakes or suitably designed equipment
- rotovate the reservoir area thoroughly in two directions (a farm rotovator may be used)
- ensure mixed, uncompacted layer is not less than 150 mm thick
- thoroughly compact with a roller (a sheepsfoot roller should not be used)
- the thickness of the compacted layer should be not less than 100 mm
- cover and protect as described in Section L1.3.

Bentonite – pure blanket

- Select a suitable thickness for the bentonite layer after discussion with the suppliers and specialist adviser
- place a geotextile separating fabric on the prepared reservoir foundation material (overlap of adjacent sheets 300 mm minimum)
- spread granular bentonite evenly across the whole of the reservoir site using hand rakes or suitably designed equipment

- place a geotextile separating fabric over the bentonite layer (overlap of adjacent sheets 300 mm minimum)
- thoroughly compact with a smooth roller (a sheepsfoot roller should not be used)
- cover and protect as described in Section L1.3.

Bentonite mats
- Roll the foundation area after preparation with three passes of a smooth drum roller
- excavate an anchor trench for the bentonite mat along the crest of the embankment and around the perimeter of the reservoir. The ground level at the trench must be at least 500 mm vertically above the water line and the trench side not less than 500 mm back from the edge of the crest
- the dimensions of the anchor trench should be not less than 600 mm wide × 600 mm deep with a U-shaped profile
- install the bentonite mat in the base of the anchor trench and backfill the anchor trench with excavated material, as shown in Figure L1, for a geomembrane liner
- carefully unroll the bentonite mat from the anchored end down the reservoir slope
- ensure the mats are lying flat
- ensure the overlaps between adjacent mats are parallel to the slope and that the minimum overlap is 250 mm
- apply bentonite slurry between the mats and over the full width of the overlaps
- place a layer of granular material (shingle or hoggin may be used)
- compact covering layer with a smooth drummed roller – under no circumstances should a sheepsfoot or grid roller be used. A minimum covering material thickness of 300 mm between the roller and bentonite mats is essential
- only install sufficient bentonite mat that can be covered and compacted in the same day
- protect the bentonite mat at all times during installation from waterlogging
- cover and protect as described in Section L1.3.

Appendix M Geotextiles

M1 GENERAL

The more readily used geotextiles are given in Table M1. This is an area which is constantly changing and reference should be made to the manufacturers for the latest products.

Table M1 Geotextiles

Product name		Thicknesses available (mm)	Manufacturer/ supplier	Brief description
Fibertex	G-100	0.6	Tex Steel Tubes Ltd	non-woven
Fibertex	F-2B	0.95		polypropylene
Terram	500	0.4	Exxon Chemical	thermally bonded non-
	700	0.5	Geopolymers Ltd	woven polypropylene
	1000	0.7		and polyethylene
Amoco	6060	0.4	Amoco Fabrics	polypropylene woven
	6061	0.8		tape fabric
	6062	0.5		
Lotrak	16/15	0.3	Don and Low Ltd	non-woven polypropylene
Nicolon	F180	0.7	MMG Civil Engineering	woven polyethylene/
	F250	0.8	Systems Ltd	polypropylene
	F300	0.5		
Polyfelt	TS500	1.6	Chemie Linz UK Ltd	non-woven
	TS600	2.1		polypropylene
Ecofelt	021T	1.2	Chemie Linz UK Ltd	woven polypropylene
	022T	1.3		

Note: Table M1 list only some of the products available under this heading. This is not an exhaustive lists and does not inply any recommendation.

The geotextile size given in Section 9.5 may normally be used in cohesive soils to give a satisfactory drainage arrangement. In some instances, or where a **granular** material is present, a more detailed approach is necessary and the following methods may be used. Unless the undertaker is experienced in the use of geotextiles, it is recommended that the geotextile assessment is carried out by a specialist adviser using this appendix.

M2 DESIGN OF GEOTEXTILES

The design criteria may be summarised as follows:
Predominantly granular soils:

- $D_{85} \geq O_{90}$ to prevent soil loss into the drain
- $D_{15} \leq O_{90}$ to ensure adequate permeability into the drain

Predominantly cohesive soils:

- $O_{90} \leq 0.12$ mm or D_{85}, whichever is smaller
- $O_{90} \geq 0.05$ mm or D_{15}, whichever is larger

Where: D_{85} = grain size of the insitu soil where 85% of the soil particles are smaller by size

D_{15} = grain size of the insitu soil where 15% of the soil particles are smaller by size

O_{90} = pore size of the geotextile where 90% of the pores are smaller by size

The boundary between the predominantly granular and predominantly cohesive soils is shown in Figures M1 and M2.

The values of D_{85} and D_{15} can be determined from particle size distribution tests, as discussed for classification purposes in Section 8. Where the values are to be determined for a cohesive soil to design a geotextile, the test results are dependent on the method of testing. This should be carried out using only water obtained from one of the trial holes or, if this is not available, from the watercourse itself or another source of water to be used for filling the reservoir. No dispersant agent should be used as for the standard tests.

M3 EXAMPLE FOR COHESIVE SOIL

Four samples of cohesive soil were taken from a trial hole excavated within an area of poorly drained ground. This area requires drainage measures. These soil samples were taken to a laboratory and particle size distribution tests were carried out on each sample using water collected from the trial hole. The results of these tests are shown in Figure M1 and indicate the soil to be a slightly clayey silt. The range of D_{85} and D_{15} values are as follows:

D_{85} = 0.01 – 0.04 mm
D_{15} = 0.004 – 0.006 mm

The smallest D_{85} and largest D_{15} values should be adopted for design purposes. The design criteria are thus:

$O_{90} \leq 0.01$ mm
$O_{90} \geq 0.05$ mm (as D_{15} less than 0.05 mm)

A geotextile should be selected, therefore, with a O_{90} range of 0.01 to 0.05 mm

M4 EXAMPLE FOR GRANULAR SOIL

Four samples of granular soil were taken from a trial hole excavated within an area of poorly drained ground. This area requires drainage measures.

These soils samples were taken to a laboratory and particle size distribution tests were carried out on each sample using water collected from the trial hole. The results of these tests are shown in Figure M2.

These indicate the soil to be a silty, medium to fine sand. The range of D_{85} and D_{15} values are as follows:

D_{85} = 0.15 – 0.6 mm
D_{15} = 0.03 – 0.1 mm

The smallest D_{85} and largest D_{15} should be adopted for design purposes.

The design criteria are thus:

$O_{90} \leq 0.15$ mm

$O_{90} \geq 0.10$ mm

A geotextile should be selected, therefore, with a O_{90} range of 0.1 to 0.15 mm.

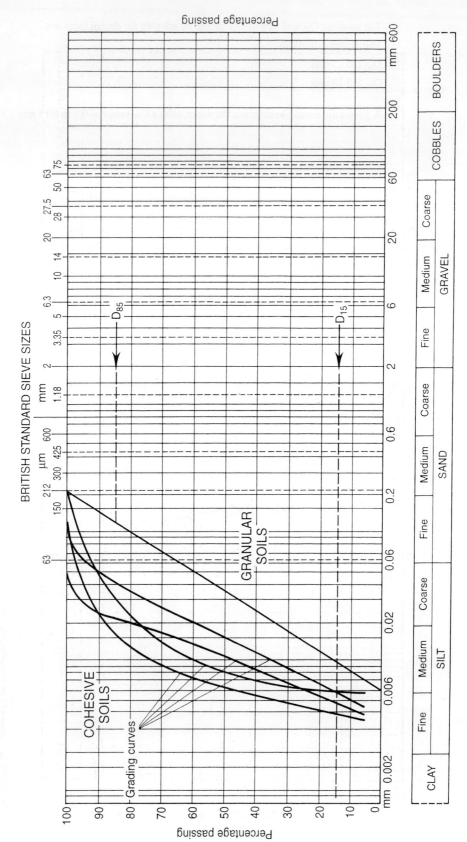

Figure M1 Soil grading curves – cohesive soil

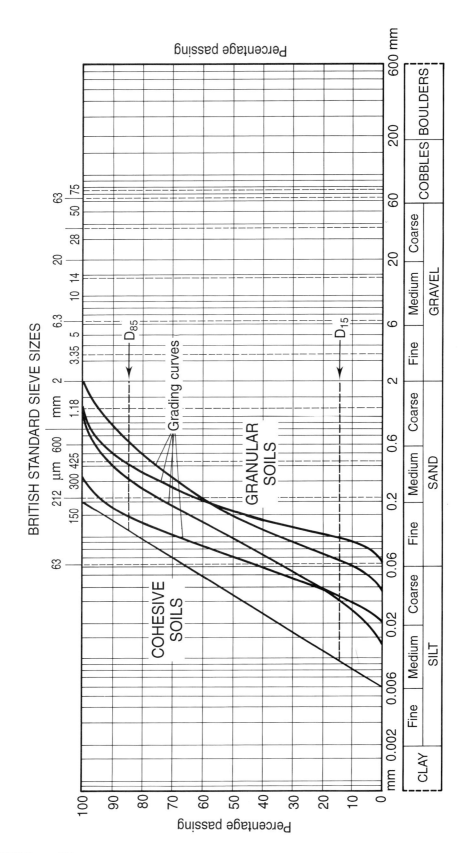

Figure M2 Soil grading curves – granular soil

Appendix N Concrete

N1 SUMMARY OF THE MAIN PROVISIONS OF BS 8007

Provision	Clause in BS 8007
Explicit assumptions about qualifications and experience of designers and builders	Foreword
Maximum design surface crack width of 0.2 mm for reinforced concrete	2.2.3.3
Only aggregates with low or medium coefficients of thermal expansion to be used. Shrinkable aggregates to be avoided	2.6.2.2
Cement content to be the minimum consistent with durability requirements	2.6.2.2
Cements with lower rates of heat evolution to be used	2.6.2.2
Concrete to be prevented from drying out	2.6.2.2
Minimum reinforcement area and spacing are prescribed	2.6.2.3
Steel to be as near to surface as is consistent with cover for durability (minimum of 40 mm)	2.6.2.3
Designer to provide maintenance and operation instructions	2.7.2
Methods of providing movement and construction joints, including their locations, are to be prescribed by the designer (this is a fundamental aspect of design to BS 8007)	Section 5
Waterstops and joint sealing compounds to be used in expansion and contraction joints	Section 5
Concrete to be grade C35A	6.3
Cement content not to exceed 400 kg/m^3 of ordinary Portland cement. Other limits specified for partial cement additions	6.3
Water/cement ratio not to exceed 0.55	6.3
Where walls or floors are founded on the ground a 75 mm (minimum) concrete screed of not less than C20 concrete to be placed over the ground	6.6
Completed structure to be tested for impermeability	9.1 and 9.2

N2 CONCRETE MIXES

N2.1 Designed mix

The mix is specified primarily by its required performance in terms of compressive strength. In addition, limits are placed on other mix parameters such as minimum cement content and maximum water/cement ratio to ensure that important performance requirements other than strength (e.g. durability) are met. Designed mixes range from Grade C7.5 (with a 7.5 N/mm^2 compressive strength) up to C60 (60 N/mm^2 compressive strength).

N2.2 Prescribed mix

The mix is specified by its component parts (rather than by strength) so as to produce a concrete with the required performance. The purchaser must specify at least the following:

permitted type(s) of cement

required nominal maximum size of aggregate

- permitted type(s) of aggregate
- mix proportions (by weight) of each constituent
- workability
- any quality assurance requirements.

N2.3 Standard mix

There is a range of five standard mixes, ST1 to ST5. The constituents and proportions of the mixes are set out in BS 5328: Part 2. Typical standard mix applications and examples of mix proportions for standard mixes are shown below. BS 5328: Part 1 does not list a typical application for an ST5 mix. Aggregates for mixes are to be batched by weight, except for mixes ST1, ST2 and ST3, which may be batched by volume as:

Constituent	Standard mix		
	ST1	ST2	ST3
Cement	2 parts	4 parts	3 parts
Damp concreting sand	5 parts	9 parts	6 parts
20 mm nominal maximum size aggregate	8 parts	15 parts	10 parts

Standard mixes are normally used as follows for foundation and other ground work. The slump is a measure of the workability of the wet concrete and a high slump is needed where concrete is to be placed with little or no compaction. Lower slumps are required where compaction can be carried out with a consequent decrease in the cement and water content and increase in the amount of aggregate.

Application	Standard mix	Slump (mm)
Mass concrete fill	ST1	75
Blinding	ST2	75
Strip footings	ST3	75
Mass concrete foundations	ST4	75
Trench fill foundations	ST4	125
Trench backfill	ST2	75

Mix proportions to produce approximately 1 m^3 of concrete require:

Standard mix	Constituent	Nominal maximum size of aggregate (mm)			
		40 Slump 75	40 Slump 125	20 Slump 75	20 Slump 125
ST1	Cement (kg)	180	200	210	230
	Total aggregate (kg)	2010	1950	1940	1880
ST2	Cement (kg)	210	230	240	260
	Total aggregate (kg)	1980	1920	1920	1860
ST3	Cement (kg)	240	260	270	300
	Total aggregate (kg)	1950	1900	1890	1820
ST4	Cement (kg)	280	300	300	330
	Total aggregate (kg)	1920	1860	1860	1800
ST5	Cement (kg)	320	340	340	370
	Total aggregate (kg)	1890	1830	1830	1770
ST1 ST2 ST3 ST4 ST5	Fine aggregate (percentage by mass of total aggregate) Fine aggregate (percentage by mass of total aggregate)	30–45	30–45	35–50	35–50
	Grading limits C	30–40		35–45	
	Grading limits M	25–35		30–40	
	Grading limits F	25–30		25–35	

N2.4 Designated mix

Where concrete is supplied by ready-mixed concrete suppliers operating an approved Quality Assurance Scheme, the recommended method of specifying is by designated mixes. The mixes are designed to be fit for their intended use and that is all that most purchasers will need to know. The QA scheme operated by the supplier, with its independent checks, means that no testing of concrete on-site is necessary.

There are four ranges of designated mixes:

- GEN1, GEN2, GEN3 and GEN4 mixes. These are approximately equivalent to the standard mixes, ST1 through to ST5, under a QA scheme
- RC30, RC35, RC40, RC45 and RC50 mixes which are equivalent to designed mixes C30 through to C50 under a QA scheme
- FND2, FND3 and FND4 mixes are for works below ground where sulphate resistance is required
- PAV1 and PAV2 mixes are for pavings where resistance to de-icing salts is an important requirement.

N2.5 Mixes for reinforced concrete

BS 8007 permits only one concrete mix specification for concrete in water-retaining structures. It is called a C35A mix and must have a cement content between 325 and 400 kg/m^3 and a maximum water/cement ratio of 0.55 (see Section 8.12.5). (Note C35A is equivalent to C40 or RC40 in BS 5328). Most concrete used in small reservoir construction will be ready mixed concrete and specified as an RC designated grade. There may be occasions when concrete is mixed on site. The British Cement Association has produced guidance on mix proportions for site-mixed concrete to produce concrete equivalent to RC designated mixes. These show that a mix equivalent to RC40 should contain:

- cement   50 kg (equivalent to 375 kg/m^3)
- concreting sand   90 kg
- aggregate (20 mm nominal maximum size)   160 kg

This will give a total yield of 0.14 m^3 and quantities can be factored accordingly.

The total amount of cement in a concrete mix is an important factor in determining the impermeability and durability. Generally, a higher cement content produces a denser, less permeable and more durable concrete. The recommendations for minimum cement content for OPC in BS 5328 range from 220 kg/m^3 for unreinforced concrete in 'mild' exposure conditions to 400 kg/m^3 for reinforced concrete subject to 'most severe' conditions. High cement content mixes stiffen more quickly, however, and without admixtures, are more difficult to handle and place than leaner mixes. The concrete is also more likely to be affected by thermal cracking and shrinkage cracks on drying. This is clearly undesirable in structures intended to be impermeable and for this reason, BS 8007 specifies a maximum cement content of 400 kg/m^3 in addition to the minimum of 325 kg/m^3.

N3   ADMIXTURES FOR CONCRETE

Admixtures	Effect and application
Accelerators	Accelerate setting and hardening times. Increase heat of hydration evolution rate. Used to permit more rapid striking of formwork and also for cold-weather concreting. Accelerators based on calcium chloride should not be used in reinforced concrete work.
Retarders	Delay setting and reduce heat of hydration evolution rate. Often used in conjunction with water-reducing admixtures. Used to increase the time available for placing and working concrete.
Water reducers (plasticisers)	Reduction in the amount of water needed for a given workability; increased workability for a given water content. Widely used to improve impermeability and durability of concrete.
Superplasticisers	As water reducers, but higher dosages may be applied to produce free-flowing or very low water/cement ratio concretes.
Air entrainers	Entrain bubbles of gas in the concrete during mixing. Allow reduction in water/cement ratio. Air-entrained concrete is resistant to effects of freezing and thawing and to de-icing chemicals.
Integral waterproofers	Lower the porosity of concrete by reducing or blocking the capillary pores. Often used in combination with water-reducing admixtures. Improved durability and chemical resistance is achieved as aggressive agents are not able to permeate the concrete. Microsilica and stearate-based admixtures are included within this group.

Appendix O Typical bill of quantities

PART: 1 **GENERAL ITEMS**

Number	Item description	Unit	Quantity	Rate	Amount £	p
1.1	Insurance of the works, constructional plant and against damage to persons and property	sum				
1.2	Maintenance of site as specification clause 211	sum				
1.3	Maintenance of bench mark as drawings and specification clause 204	sum				
1.4	Maintenance of access as specification clauses 201 and 208	sum				
1.5	Dealing with water as specification clause 405	sum				
1.6	Reinstatement as Specification clauses 217 and 407	sum				
1.7	Comply with all other obligations under Parts 1 and 2 of the specification	sum				
1.8	Setting out as drawings and specification clause 401	sum				

NOTE: Clause numbers refer to descriptions in the specification and are numbered accordingly for specific site and project requirements.

Page 1/1 carried to summary

Page: 1/1

PART: 2   TECHNICAL ITEMS

Number	Item description	Unit	Quantity	Rate	Amount £	p
2.1	Site clearance as drawings and specification clause 402	ha				
2.2	Diversion works as drawings and specification clauses 403, 404, 406, 408, 411, 416 and 417	sum				
2.3	Foundation preparation over embankment areas, trench and formation of island, as drawings and specification clauses 403, 404, 408, 410, 411 and 412	m^3				
2.4	Excavation of cutoff trench as specification clauses 403 and 412	m^3				
2.5	Works in borrow area, excluding excavation for fill materials, as drawings and specification clauses 403, 404, 408, 409 and 410	sum				
2.6	Core including fill to cutoff trench, fill to main embankment, including excavation in borrow area, selection, transportation, placing and compaction as drawings and specification clauses 403, 404, 406, 409, 410, 413, 414 and 415	m^3				
2.7	Shoulder fill. ditto	m^3				
2.8	200 millimetre topsoil. ditto	m^2				
2.9	200 millimetre thick hardcore to crest track of main embankment as drawings and specification clause 419	m^2				
2.10	300 millimetre thick stone facing to upstream slope of main embankment as drawings and specification clause 419	m^2				
2.11	Auxiliary overflow works including training bund as drawings and specification clauses 403, 404, 406, 409, 410, 413, 414 and 415	sum				
2.12	Toe drain as Drawings and Specification Clauses 403, 406, 408, 410 and 417	m				
2.13	Existing water main in trench as drawings and specification clauses 403, 406, 408, 410 and 417	m				
2.14	Main overflow works; concrete box culverts as drawings and specification clauses 403, 404, 406, 408, 416, 418 and 419	sum				
2.15	Main overflow works; side entry reinforced concrete entry spillway as drawings and specification clauses 403, 404, 406, 408 and 416	sum				
2.16	Main overflow works; protective works as drawings and specification clauses 403, 404, 406, 408, 416 and 419	m^2				

Number	Item description	Unit	Quantity	Rate	Amount £	p
2.17	Main overflow works; new channel as drawings and specification clauses 403, 406, 408, 416 and 419	m				
2.18	Fill, including 300 millimetre thick stone facing, to silt pond embankment and excavation of pond as drawings and specification clauses 403, 404, 406, 408, 409, 410, 413, 414 and 415	sum				
2.19	Reinforced concrete slab and 200 millimetre thick hardcore to silt pond embankment as drawings and specification clauses 416 and 419	sum				

Page 2/2 carried to summary

Page: 2/2

Appendix P Reference gauge post for reservoir level monitoring

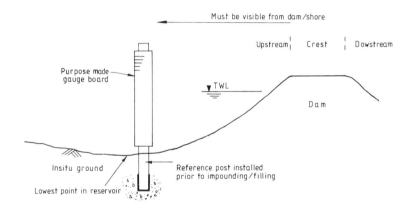

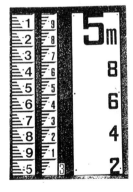

NOTE

Board set at suitable height to allow for fluctuation of water level or with zero at TWL for a non-abstraction reservoir.

Manufactured in wood, metal or glass reinforced plastic and available in a range of lengths from suppliers.

Appendix Q Surveillance indicators of possible defects

Indicators (1)	Possible causes (2)	Possible consequences (3)
EMBANKMENT UPSTREAM SLOPE		
Dislodged surface protection	Accidental damage Deformation of embankment Degradation of protection layer material Fishermen's footholds Ice action Inadequate size of protection layer material Loss of joint material Seepage erosion in underlying material Uplift pressure due to rapid drawdown of reservoir Vandalism Vegetation Wave action Wild animals	Loss of protection Embankment erosion
Damage to membrane	Accidental damage Deformation of embankment Degradation of protection layer material Insufficient joint flexibility Internal erosion in embankment Jointing materials no longer serviceable Poor installation method Uneven support from embankment Uplift pressure due to rapid drawdown of reservoir Vandalism Vegetation	Leakage from reservoir Loss of protection Embankment erosion Slope instability
Deformation and cracks over large area	Fill or foundation consolidation Erosion beneath protection layer Inadequate slope protection Instability of embankment Internal erosion Mining or natural subsidence	Reduced freeboard Overtopping Damage to structure of dam Embankment softening due to standing water
Sink holes, local depressions and cracks	Animal burrowing Inadequate localised protection layer Infilled excavations Internal erosion Growth/loss of vegetation Desiccation of surface fill materials	Embankment erosion Embankment softening due to standing water
Whirlpool in reservoir	Drawoff pipe in use	None
	Internal erosion	Severe leakage from reservoir Embankment erosion Damage to structure of dam
Sounds of flowing water	Internal erosion Ineffective anti-seepage collars on pipes through embankment	Severe leakage from reservoir Embankment erosion Damage to structure of dam

Indicators (1)	Possible causes (2)	Possible consequences (3)
EMBANKMENT CREST		
Surface erosion	Animals Flood water overtopping the dam Ice action Inadequate road drainage Public utilities Surface water Traffic (vehicular/foot) Growth/loss of vegetation Wave action	Reduced freeboard Overtopping Lack of support to top of dam Damage to structure of dam
Deformation and cracks over large areas	Consolidation Desiccation of embankment Instability of embankment Internal erosion Mining subsidence Reservoir drawdown Traffic	
Sink holes, local depressions and cracks	Animal burrowing Subsidence of infilled excavations Internal erosion Growth/loss of vegetation Desiccation of surface material	
EMBANKMENT DOWNSTREAM SLOPE AND VALLEY **(including reinforced grass overflows)**		
Surface erosion	Animals Excessive human activity Flood water overtopping the dam Inadequate drainage system Public utilities Surface water runoff Growth/loss of vegetation Vandalism	Embankment erosion Slope instability
Deformation and cracks over large areas	Fill or foundation consolidation Instability of embankment Mining or natural subsidence Desiccation of surface fill materials Internal erosion	Reduced freeboard Overtopping Embankment erosion Embankment erosion
Sink holes, local depressions and cracks	Animals Infilled excavations Internal erosion Growth/loss of vegetation	Embankment erosion
Wet areas or vegetation indicative of poor drainage (see Table 12.1) (No or minimal visible flow) Seepage flows Sound of flowing water	Defective drainage system Hillside groundwater Leakage from outlet pipe Leakage from reservoir Seepage from reservoir Surface water retained by impermeable embankment material or local depressions Internal erosion Ineffective anti-seepage collars on pipes through embankment	Slope instability Embankment erosion

Indicators (1)	Possible causes (2)	Possible consequences (3)
DRAINS		
Increased drainage flows	Leakage from reservoir	Slope instability
	Improvements in drainage system	Unlikely to have
	Surface water after rain or thaw	detrimental effects
Decreased drainage flows	Choking of drains by debris or mineral deposits	Slope instability
	Collapse of pipes	
	Blockage by vegetation/roots	
	Low reservoir level	No detrimental effects
DRAWOFF WORKS: PIPEWORK		
Low flow in low level outlet	Control valve not fully open	Inability to lower reservoir at acceptable rate
	Outlet pipework silted up	
	Outlet pipework at inlet choked	
	Outlet pipe partially collapsed	Leakage from reservoir
		Internal erosion
No flow in low level outlet	Control valve closed	Inability to lower reservoir
	Outlet pipework silted up	
	Outlet pipework at inlet choked	
	Outlet pipe collapsed	Inability to lower reservoir
		Leakage
		Internal erosion
Pipes corroded	Inadequate pipe protection	Leakage from pipe
	Inappropriate type of pipe	Collapse of pipework
		Internal erosion
		Slope instability
DRAWOFF WORKS: VALVES		
Excessive vibration	Deterioration of materials	Valve damage
Sound of flowing water when shut	Inadequate design	
	Lack of maintenance	
Will not open/close	Deterioration of materials	Valve damage
	Ice damage	Inability to control discharges
	Lack of maintenance	
	Seized operating rods	Inability to lower reservoir
	Seized or inadequate hydraulic or pneumatic operation	Damage to draw-off structure
	Silt trapped	
	Valve corrosion	
Cracked casing	Frost or ice	Leakage from valve
	Overstressed in operation	
DRAWOFF WORKS		
Deformation and distress	Foundation settlement	Damage to pipework and valves
	Ice action	
	Mining and natural subsidence	Inability to lower reservoir
	Movement of dam embankment	
	Seizing of access bridge movement joints	Loss of access to controls
	Thermal movement	
Deterioration of materials	Deterioration of concrete	Seizing of valve
	Corrosion	Structural inadequacy
	Frost damage	

Indicators (1)	Possible causes (2)	Possible consequences (3)
CONCRETE AND MASONRY OVERFLOWS AND CHANNELS		
External erosion	Floods overtopping channels Overtopping due to blockage of channels Overtopping due to chocked screens Overtopping due to vegetation growth in channels Surface water flow Uneven channel base	Erosion of embankment Slope instability
Deformation and distress	Erosion beneath structures Instability beneath structures Instability of foundations Growth/loss of Vegetation	Structural inadequacy
RESERVOIR		
Wet areas of vegetation indicative of poor drainage (see Table 12.1) No or minimal visible flow Seepage flow Sound of flowing water	Defective drainage system Hillside groundwater Surface water retained by impermeable material or local depressions	Slope instability Surface erosion
Erosion of reservoir banks	Wave action	Instability of reservoir banks Enlargement into neighbouring land
Erosion at inlets to reservoir	High inflow at low reservoir levels	Slope instability Enlargement into neighbouring land
Deformation and cracks	Groundwater in hillside Mining and natural subsidence Hillside instability	Instability of reservoir banks Overtopping of dam Flooding of ground adjacent to reservoir

NOTES: The possible causes (Column 2) may occur separately or in combination. The extent of the possible consequences (Column 3) depends on the level of supervision and maintenance. Without appropriate action, many defects can lead to a breach of the dam.